Construction
Adjudication

Construction Adjudication

Second Edition

John L. Riches

and

Christopher Dancaster

Blackwell Publishing

Blackwell Publishing,
Editorial offices:
Blackwell Publishing Ltd, 9600 Garsington Road, Oxford OX4 2DQ, UK
  Tel: +44 (0)1865 776868
Blackwell Publishing Inc., 350 Main Street, Malden, MA 02148-5020, USA
  Tel: +1 781 388 8250
Blackwell Publishing Asia Pty Ltd, 550 Swanston Street, Carlton, Victoria 3053, Australia
  Tel: +61 (0)3 8359 1011

First published 1999 by LLP Reference Publishing
Second edition published 2004 by Blackwell Publishing

Library of Congress Cataloging-in-Publication Data
Riches, John L.
  Construction adjudication/John Riches & Christopher Dancaster.–2nd ed.
    p. cm.
  Rev. ed. of: Construction adjudication/Robert Stevenson and Peter Chapman. 1999.
  Includes bibliographical references and index.
  ISBN 1-4051-0635-2 (hardback: alk.paper)
  1. Construction contracts – Great Britain.   2. Arbitration and award – Great Britain.   1. Dancaster, Christopher.   Title.
  KD1641.S75   2004
  343.41′078624–dc22

                                                                                          2004000879

ISBN 1-4051-0635-2

A catalogue record for this title is available from the British Library

Set in 9.5/11.5pt Palatino
by DP Photosetting, Aylesbury, Bucks
Printed and bound in Great Britain
by MPG Ltd, Bodmin, Cornwall

The publisher's policy is to use permanent paper from mills that operate a sustainable forestry policy, and which has been manufactured from pulp processed using acid-free and elementary chlorine-free practices. Furthermore, the publisher ensures that the text paper and cover board used have met acceptable environmental accreditation standards.

For further information on Blackwell Publishing, visit our website:
www.thatconstructionsite.com

Contents

Preface

There are those who say that, until the courts had explained to the construction industry what adjudication is and how it operates, it was a case of the blind leading the blind. When we look back on what we wrote in the first edition in 1998, however, we are pleasantly surprised that much of what we said is as valid now as it was then.

It was thus with a light heart that in the spring of 2002 we embarked on what we thought would be a simple update to take into account the various decisions of the courts that affected the practice of adjudication. At that time there were about 100 published judgments of the courts dealing with adjudication matters.

Little did we imagine what the task would develop into. During the period that we originally set for the production of this edition (we originally aimed to finish in October 2002) the courts made a further 30 or so judgments. From then to December 2003 there have been nearly 40 more. Each time we thought that we were near finishing, yet another judgment would appear that put a slightly different nuance on the way that adjudicators should conduct themselves. The end result has been a substantial review of the whole of the text.

We have managed to unearth a total of 172 judgments up to December 2003 and we set out a list of these in Appendix 18. Most of these have come from the websites produced by adjudication.co (www.adjudication.co.uk) and BAILII (the British and Irish Legal Information Institute) (www.bailii.org). Others have come from the Court Service website (www.courtservice.gov.uk) and the Scottish Courts website (www.scotcourts.gov.uk). A few have come from those involved in enforcement proceedings who have been kind enough to let us have notes of the judgment of the court. We must acknowledge our debt to all these organisations and people. We have no doubt that there are readers who will be aware of other matters that have reached the courts for enforcement or other proceedings in connection with adjudication. We should be pleased to hear from anyone who can help in this way.

Some may say that this is an inordinate number of cases needed to explain how the adjudication procedure works. We would counter that suggestion by saying that it is our understanding, gleaned in the absence of any hard information, that after more than five years there have probably been between 9,000 and 12,000 notices of adjudication. This is an extrapolation based upon the number of nominations made by the various adjudicator nominating bodies, our own experience and that of other adjudicators who receive appointments direct from the parties. In that context the number of cases where parties have had to resort to the courts is less than 2% of the total number of adjudications and probably nearer $1\frac{1}{2}$%. It must always be remembered that, as most of these judgments relate to enforcement, there are many that deal with almost exactly the same matters and the actual issues dealt with by the courts are limited in number.

The seminal judgment relating to adjudication was that of Mr Justice Dyson in *Macob* v. *Morrison*[1] which was made in February 1999. This decision gave the lie to those who suggested that adjudication would not work, and made it clear that the courts would fulfil the intention of Parliament that adjudicators' decisions would be enforced.

At one time the view was that an adjudicator's decision would be enforced almost come what may, but as the number of cases reaching the courts for enforcement has risen it has become more and more clear that there are two guiding principles that must apply to the adjudication process if the adjudicator's decision is going to be enforced. These are: the adjudicator must have jurisdiction to act, and in conducting the proceedings the adjudicator must be unbiased and fair.

The courts have laid down guidelines for adjudicators to follow in order that their actions may not transgress into areas that are seen by parties and the courts as being unfair, guidance that is particularly necessary where the process itself could well be described as intrinsically unfair. It does seem, however, that as long as the adjudicator follows the principles of the 'rules of natural justice', albeit slightly modified to allow for the extremely restricted time-scales involved, all other things being equal the courts will enforce the decision.

The other principal area that has exercised the courts has been the question of the adjudicator's jurisdiction. If the adjudicator has no jurisdiction, any decision that is produced is unenforceable. The questions that have been addressed by the courts in respect of jurisdiction have been many and varied, but in the main they boil down to two questions: 'Is there a dispute?' and 'Is there a contract to which adjudication applies?'. Many of the courts' judgments that have been handed down during 2003 relate to these questions and interestingly enough the penultimate case in our list in Appendix 16, *Galliford Try* v. *Michael Heal Associates*[2], a decision of Judge Richard Seymour QC of the Technology and Construction Court, deals with questions which included whether or not a contract was concluded and if it was, whether it was a contract in writing.

The whole edifice of the courts' support for adjudication was initially, quite understandably, built upon judgments made in the Technology and Construction Court, that is, at first instance. A particular benefit of the delay in the production of this second edition is that there have, in October and November 2003, been three cases relating to the enforcement of adjudicators' decisions that have been heard in the Court of Appeal. Two of these deal with questions relating to an adjudicator's jurisdiction. In one of them, *Tally Wiejl* v. *Pegram*[3], Lord Justice May reviews the adjudication process, considering the tensions between the intentions of Parliament as set out in the Housing Grants, Construction and Regeneration Act 1996 and challenges to jurisdiction. There are now eight Court of Appeal judgments and all are on the BAILII website; anyone interested in the process of adjudication should access them and read them carefully.

So we make no apology for the length of time that it has taken to produce this second edition. We thank all those who have regularly asked us 'When is it coming out?' for their patience, and also thank Julia Burden and all at Blackwell Publishing for bearing with us over a lengthy gestation period. We think that the end product has been well worth the wait and hope that they do too.

December 2003

---

[2] *Galliford Try Construction Ltd* v. *Michael Heal Associates Ltd* [2003] EWHC 2886 (TCC 1 December 2003).
[3] *Tally Wiejl (UK) Ltd* v. *Pegram Shopfitters Ltd* [2003] EWCA Civ 1750, CA (21 November 2003).

Acknowledgements

We are grateful to the following for permission to publish extracts from contracts, Rules and associated documents: Association of Consulting Engineers, Centre for Effective Dispute Resolution, Civil Engineering Contractors Association, Construction Confederation, Construction Industry Council, Institution of Chemical Engineers, Her Majesty's Stationery office, RIBA Enterprises, Technology and Construction Solicitors Association, Thomas Telford Publishing.

Chapter 1
What is Adjudication?

General introduction

Terminology

Before we start to look at adjudication in detail it is worth laying down a few ground rules.

Throughout this book, where the words 'the Act' are used they refer to the Housing Grants, Construction and Regeneration Act 1996. Where we refer to other legislation we name the Act concerned by its full title. In order to avoid confusion, for example where there is reference to the Act in close proximity to other legislation, we add the abbreviation HGCRA.

There is secondary legislation derived from the Act. In England and Wales this is the Scheme for Construction Contracts (England and Wales) Regulations 1998, which provides the default position where a construction contract does not comply with the requirements of the Act. We refer to this throughout as 'the Scheme'. Scotland and Northern Ireland have provisions with different titles but these are essentially the same in concept, albeit certain amendments have been made in them. There is also an Exclusion Order, the Construction Contracts (England and Wales) Exclusion Order 1998. This excludes certain types of contract from the definition of construction contracts in the Act.

The parties who wish to use adjudication can be described in a number of ways. The party who starts the process can be described as the 'referring party', 'applicant' (applying for adjudication) or by the title they are given in the contract, such as employer, contractor or sub-contractor. The other party is the 'responding party', the 'respondent', the 'other party' or again the title by which they are referred to in the contract. It would not be correct to use terms such as 'plaintiff' and 'defendant' as these belong to the field of litigation. The terms 'claimant' and 'respondent' are used in arbitration and could be used in adjudication. In practice this arbitral terminology is rarely used. We believe that the simplest and least confusing method of identifying the parties is to use their actual names or an abbreviated form of their names in the body of the decision. This avoids error when directing who shall pay what to whom, or in directing who shall do what for whom.

One aspect that often causes confusion is the nomenclature used when describing the final written product of the adjudication process. It is easy to call it an 'award' but we are of the view that this is not appropriate. The term 'award' is appropriate to arbitration, as is 'judgment' to litigation. Throughout this book we describe the final written product as 'the decision'. We note that neither the Act nor the Scheme use the word 'award'; reference is always to 'decision'.

Why is it that in the courts there is often a reference in enforcement proceedings to the adjudicator's award and not to the decision? What the legislation clearly describes is a decision. That is what the adjudicator does, simply decide. When the matter goes to enforcement, however, the element of the decision which is enforced is the amount or part of the decision which awards something, hence the use by the courts of the term 'award' in the context of enforcement.

We have to prepare the reader for an element of confusion here. It is the house style of our publishers to avoid the use of capital letters in describing things such as a document that is entitled a decision or a person such as an adjudicator. Thus we have no easy way of differentiating between the individual decisions that the adjudicator makes on the several issues contained within the decision document, and that document itself.

Before 1 May 1998

Adjudication had been an option in a number of standard contracts in the construction industry for many years. It had been generally limited in its scope. The extent to which these provisions had been used is unknown. It was the authors' experience that in the period leading up to the introduction of the statutory right to adjudication, with the higher profile of adjudication generally, those adjudication provisions that did exist were used rather more frequently. There had certainly been an increase, particularly in the civil engineering field, in the number of contracts with provisions for the setting up of dispute review boards where adjudication-type arrangements had come to the fore. This trend accelerated as a result of the publication of the Latham report *Constructing the Team*, culminating in the legislation entitled the Housing Grants, Construction and Regeneration Act 1996.

The first introduction into a standard construction contract of any form of adjudication was in 1976 when the 'Green Form' of Nominated Sub-Contract was amended to include an adjudication procedure. A similar amendment was made to the 'Blue Form' of domestic sub-contract. These forms, of course, applied to sub-contracts under the 1963 edition of the Joint Contracts Tribunal (JCT) Contract (previously known as the RIBA form). This inclusion of the adjudication process related to the introduction of elaborate provisions regulating the rights of main contractors to set off costs against monies otherwise due to their sub-contractors.

This amendment emanated from the decision by the House of Lords in *Gilbert Ash* v. *Modern Engineering*[1]. This decision reversed the Court of Appeal decision in *Dawnays* v. *Minter*[2], which decided that monies certified under the Green Form represented a special type of debt to which the normal rules of set-off did not apply. *Gilbert Ash* decided that an employer is entitled to exercise the normal rights of set-off that arise by operation of law unless the contract excludes those rights.

The sub-contractors identified that unless a main contractor's rights of set-off were controlled by changes in the sub-contract terms there was great opportunity for abuse. This led to the 1976 amendments to the Green and Blue Forms of subcontract. When the JCT Forms were revised in 1980 the Green and Blue Forms disappeared and the NSC/4 and DOM/1 replaced them with the set-off adjudication provisions included. The JCT Forms are now in their 1998 editions and are undergoing further revision. Both main and subcontract forms now reflect the requirements of the Act.

During the 1980s the policymakers, in an endeavour to avoid the delays and costs that occur in the more traditional forms of formal dispute resolution, started to look at a wider application of adjudication as a means of obtaining resolution of disputes in a summary fashion.

---

[1] *Gilbert Ash (Northern) Limited* v. *Modern Engineering (Bristol) Limited* [1974] AC 689.
[2] *Dawnays* v. *Minter* [1971] 1 WLR 1205; 1 BLR 16.

In addition to matters relating to set-off, adjudication clauses were developed in other contracts to resolve a far wider range of disputes. Three of these contracts are noted here.

The Association of Consultant Architects' form of contract was originally produced in 1982. A second edition, known as ACA2, published in 1984, included three alternative dispute resolution procedures: adjudication by a named adjudicator, arbitration or litigation. The adjudication provisions were limited to a specific list of disputes and the adjudicator was allowed five days to give his decision. The British Property Federation version of this contract included a mandatory requirement for adjudication.

The JCT 1981 With Contractor's Design Contract had supplementary provisions, including adjudication, added in 1988. These were optional inclusions and the adjudication provisions were not all encompassing, being restricted to certain categories of dispute.

The Institution of Civil Engineers (ICE) New Engineering Contract (NEC), First Edition, introduced in 1993, the precursor to the Engineering and Construction Contract, included adjudication as a front line dispute resolution system and was the first contract to require the application of adjudication to all disputes that arose. Apart from the NEC, the ICE did not include adjudication in any of its other contracts. It did have an optional system of conciliation, which was first introduced into its Minor Works Contract in 1988, followed subsequently by the ICE Sixth Edition in 1991, and then the ICE Design and Construct Conditions in 1992. Conciliation in these forms of contract was a two-stage procedure; the first was a discussion between the parties and the conciliator with the aim of reaching a settlement of the dispute. If that was successful there was no second stage. If, however, a settlement was not reached, the conciliator made a 'recommendation' as to how in his opinion the dispute should be settled. This recommendation was similar to an adjudicator's decision in that it was final and binding unless referred to arbitration or litigation. The ICE produced a formal Conciliation Procedure in 1994 and this procedure still forms a part of the dispute resolution regime in current ICE contracts, even after their amendment to comply with the Act.

None of the adjudication clauses mentioned above complied with the Act and as a result all were discarded in favour of compliant clauses in 1998.

After 1 May 1998 New adjudication – 'Sea change' or Hurricane?

The Act became operative on 1 May 1998 in England, Wales and Scotland (1 June 1998 for Northern Ireland). Any construction contract formed after that date had to have the new right to adjudication incorporated in the contract in a form which satisfied the requirements of section 108 of the Act or it would be incorporated by default with the provisions of the Scheme.

Prior to this date there was great scepticism in the trade press, particularly by many of the lawyers, coupled with the poor drafting of Part II of the Act and lack of any provision in the legislation for enforcement, that new construction adjudication would ever work. There were also those who thought that it would have only deterrent value and would rarely be used in practice.

A major last minute publication[3] sought to discredit the whole of the legislation and the system for adjudication itself, but only on the basis that it should not be a statutory right.

---

[3] John Uff, QC *et al., Construction Contract Reform: A Plea for Sanity*. Centre for Construction Law and Management, London, 1997.

The sceptics nevertheless remained supportive of voluntary contractual adjudication schemes.

There had been extensive training for new adjudicators prior to 1 May 1998 and a number of professional institutions and trade bodies had formed panels of adjudicators and set themselves up as adjudicator nominating bodies ready for the industry's uptake of new adjudication. At the eleventh hour the government's listing of approved adjudicator nominating bodies in the Scheme was abandoned. This move was disappointing at the time for those who had taken the trouble to 'qualify' as approved bodies, but the logic of not publishing an approved list was impeccable. This was because it would restrict any new organisation qualifying as an adjudicator nominating body without a change each time to the secondary legislation.

The initial uptake of adjudication was slow. This is not surprising as the legislation only operated on contracts formed after 1 May 1998. A party wishing to exercise its right to adjudication had to both get into contract and get into dispute after 1 May 1998. As many construction practitioners know, it is probably easier in the construction industry to find oneself in dispute than in contract.

However the sea change, if one was needed, came from the courts. The question of enforceability of an adjudicator's decision was put beyond doubt in the case of *Macob Civil Engineering* v. *Morrison*[4]. This case established beyond doubt that adjudicators' decisions would be enforced, thus removing one of the major scepticisms in the system of new adjudication. This confirmed the authors' views expressed in the first edition of this book.

Whether this first judgment of itself led to a rapid increase in numbers of adjudications or whether they were already on the increase is unknown. Figures from a survey by the Construction Industry Council (CIC)[5] give one of the more accurate appraisals of the growth of the numbers of adjudications.

Review and consultation

When the Act was under development, Government made an undertaking that a review of how the legislation was working would take place within two years of the operative date for adjudication. In the relatively short period since the birth of new adjudication, not only has a substantial review taken place of the legislation in particular but also of other aspects of adjudication. Other than generating a lot of comment, this review had no effect in that nothing was done about amending either the primary or secondary legislation. A further review is being undertaken under the auspices of the Construction Industry Council, which is rather surprising given the absence of any change as a result of the two-year review.

The first review was undertaken by wide-scale consultation with industry and users of adjudication. It is the authors' experience that, regrettably, such reviews rarely attract grass roots feedback and are usually fuelled by non-practitioners and those with vested interest but no practical experience of the matter in debate.

The CIC, through the newly formed Construction Umbrella Bodies (CUB) Adjudication Task Group, which analysed the response from industry, sought to redress this defect in the consultation process by including practitioners in order to provide a balanced view.

Very little has come from the consultation process. Minor amendments have been

---

[4] *Macob Civil Engineering Ltd* v. *Morrison Construction Ltd* [1999] BLR 93.
[5] *Adjudication – The First Forty Months*. A Report on Adjudication under the Construction Act. Construction Industry Council, 2002.

proposed to the Scheme, which would only affect the legislation in England[6]. Scotland and Wales can now make their own legislation following changes brought about through devolution.

The message seems clear: 'it ain't broke so don't try to fix it'. There is nothing in the proposals that is likely to involve new regulation or amendment to the Scheme for Construction Contracts (England and Wales) Regulations 1998 and it is likely, rightly in the authors' view, that the Scheme will remain unchanged.

There has been some case law concerning whether or not adjudicators should have the power to award party to party costs. The industry view on this is divided and there are strong views in both camps. Any real problem with the legislation will need amendment to primary legislation rather than secondary legislation.

In the authors' view this particular aspect of adjudicators' decisions is unimportant. We can see no reason why consenting parties should not have an adjudicator deal with costs and there is nothing to prevent such agreement.

There is, however, a less attractive scenario on costs which is of considerable concern to those representing sub-contractors and which does require the assistance of amended primary legislation. This arises in the case of *Bridgeway* v. *Tolent*[7]. This case concerned an adjudication commenced by a sub-contractor against a main contractor under an amended CIC Model Adjudication Procedure. The amendment in question incorporated provisions that the party who issued the Notice of Adjudication would be responsible for both parties' costs in the adjudication. These costs were to include the adjudicator's fees and expenses and any legal and experts' costs. The court held that this term of the contract was not void. It did not breach the Act in any way and did not provide a 'device' that prevented an aggrieved party going to adjudication at any time.

It is unfortunate that Parliament did not deal with this aspect at the time primary legislation was introduced. It is a simple matter both to maintain the parties' right to agree that an adjudicator can deal with costs, and to ban any agreement prior to the dispute occurring that one side should bear all the costs in any event. This could be achieved by adopting section 60 of the Arbitration Act 1996. This would make void all pre-agreements on costs prior to the occurrence of any dispute and would redress any imbalance in bargaining position at the time the contract is made. Unfortunately at the time of writing the Parliamentary time for such an amendment is unlikely to be made available in the immediate future.

Training and quality of adjudicators

Another issue that was raised in the consultation process was the quality of adjudicators. It was thought that the initial training may have been inadequate or that the quality of performance of some of the adjudicators was inadequate. The authors themselves had then, and still have, a concern that the three-day course originally set up as a 'conversion course' for those already trained as an arbitrator, has been adopted as the standard for training those aspiring to be adjudicators without any background as arbitrator.

A study was carried out by the CIC to establish a common theme and minimum criteria for training adjudicators. Subsequent to this study the RICS independently decided to examine the quality of its adjudicator panel and have appraised continuing professional

---

[6] The Scheme for Construction Contracts (Amendment) (England) 2001.
[7] *Bridgeway Construction Ltd* v. *Tolent Construction Ltd* (11 April 2000).

development (CPD) and provided updated training courses. This is part of an ongoing programme to enhance the quality of adjudicators available to the construction industry.

Other institutions that provide adjudicator panels and act as adjudicator nominating bodies are also reviewing their training and CPD for their panellists.

One proposal that indicates that adjudication has 'arrived' is the commencement of preliminary investigations into setting up a diploma course in adjudication. Such a course would obviously take some time to develop but once it is in place it should go a long way to address any problems in respect of training.

Advice

The final area that the consultation process examined was that of advice. The initial thinking was that there should be advisory documents for adjudicators. The RICS had already taken the initiative and published guidance for its adjudicators[8].

The CUB Adjudication Task Group has also published a similar guidance document[9]. This does not address one of the major problems identified by practising adjudicators: the difficulty that some parties have in putting together a case for adjudication without using the services of expensive advisers. This resulted in the production by the CIC of the *Users Guide to Adjudication* in April 2003.

The rest of the world

The UK legislation has stimulated interest in adjudication across the rest of the world. There is the Building and Construction Industry Security of Payment Act 1999 No. 46 in New South Wales in Australia and the Construction Contracts Act in New Zealand which came into force in 2003. There is also considerable interest elsewhere in Europe as well as in China and the West Indies.

The authorities in Australia and New Zealand took considerable interest in the HGCRA and its application to payment and disputes in the UK. Their legislation does not follow the UK pattern entirely and in some ways their approach is much tougher.

Public or private law?

There is a debate as to whether or not adjudication is governed by private or public law. An excellent paper by Anthony Speaight QC[10] explored this concept on the then case law on enforcement.

The importance of the point is simply that when it comes to enforcement, if adjudicators' decisions fall within the remit of public law they would be subject to judicial review and therefore a wider investigation than has presently been the case at enforcement. The English courts have been very supportive of adjudication from the outset. The basic premise has been that adjudicators' decisions are enforceable unless they fall into limited exceptional circumstances. These limited circumstances are considered later in this book.

---

[8] *Surveyors Acting as Adjudicators in the Construction Industry*. RICS guidance note.
[9] *Guidance for Adjudicators*. Construction Umbrella Bodies Adjudication Task Group, July 2002.
[10] *The Construction Act – Time for Review*, Chapter 8. King's College London, November 2000.

Without entering the debate on what constitutes public law, the regime for enforcement of adjudicators' decisions in England and Wales has essentially followed the law developed in expert determination where matters such as mistakes in law and fact are considered subordinate to the process itself, and therefore ignored. What remains important is whether the right question has been answered and, as now clarified by the courts, whether the procedure has been fair and the adjudicator has acted within his jurisdiction.

Mr Speaight's paper suggests that if there is a public law angle this would only apply where the construction contract is non-compliant and therefore the Scheme applies by default. Any Scheme adjudication would therefore be subject to judicial review. If adjudicators' decisions are subject to judicial review the merits would be examined more closely than is presently the case at enforcement.

The three broad grounds of judicial review are still probably those stated by Lord Diplock in the GCHQ case[11]:

'(1) Illegality: the decision maker must understand correctly the law that regulates the decision-making and must give effect to it.

(2) Irrationality: the test which is frequently referred to as "Wednesbury[12] unreasonableness", that is a decision which is so outrageous in its defiance of logic or accepted standards that no sensible person could have arrived at it.

(3) Procedural impropriety: this is a failure to observe basic rules of natural justice or a failure to act with procedural fairness.'

Under these grounds remedies would be available in public law for an aggrieved party where there would be no remedy available in private law.

There are no cases in English law where judicial review has been used to resist enforcement of an adjudicator's decision. It is doubtful from any of the current cases that a challenge will be successful by judicial review.

The position is different in Scotland:

'One difference between Scots and English law (in procedure at least), however, is (as I have already mentioned) that judicial review is not confined under Scots law to issues of public law, but extends to powers conferred by a contract upon a third party to determine the rights of the parties to the contract *inter se*. In particular, judicial review under Scots law extends to arbitration and is not uncommon in the context of arbitration under building and engineering contracts.'[13]

Adjudication defined

Dictionaries provide a number of definitions of the verb 'adjudicate'. For example, the Oxford English Dictionary includes '(Of a judge or court) decide upon (claim etc.)' and 'sit in judgment and pronounce sentence'. The word has its roots in the Latin *adjudicare*, to award to (judicially): *ad* (= to) + *judicare* (= to be a judge, from *judex*, a judge).

These definitions show that adjudication is something that judges and arbitrators do every day. It is an action that they carry out as part of their overall function as judge or arbitrator. It is in principle no more than making a decision. In these days of the court Civil Procedure Rules and case management by judges and the active management of the arbitration by arbitrators, the making of a decision, or the adjudication element of their task, is only a part

---

[11] *Council of Civil Service Unions* v. *Minister for the Civil Service* [1985] AC 374 HL.
[12] *Associated Provincial Picture Houses* v. *Wednesbury Corporation* [1949] 1 KB 223 CA.
[13] *Ballast PLC* v. *The Burrell Company (Construction Management) Ltd* (21 June 2001) Outer House, Court of Session.

of their function. The remainder of their task can be described as 'arbitrating' or, in the case of a judge, 'case managing'.

Parliament sought to define adjudication in a proposed amendment to the then Housing Grants, Construction and Regeneration Bill. The definition was: 'For this purpose "adjudication" means a summary non-judicial dispute resolution process that leads to a decision by an independent person that is, unless otherwise agreed, binding upon the parties for the duration of the contract, but which may subsequently be reviewed by means of arbitration, litigation or by agreement'.

In the authors' view it was wise that this amendment was rejected and not included in the legislation. There are certainly some elements of this definition that are attractive but others which conflict with the balance of the legislation in its final form. Our reasons for saying this are developed below.

Adjudication as set out in the Act is not just that part of the process that involves the making of a decision, albeit that this is the one thing an adjudicator must do. It encompasses the whole procedure from the initial notice of intention to refer a dispute to adjudication, through to the making of the decision by the adjudicator. When adjudicating, the adjudicator is not only deciding on the merits, he is also administering the process.

There are others with the title adjudicator, for example VAT or immigration adjudicators. They are similar to the construction industry adjudicator in that their title encompasses their whole function, not just the decision-making aspect.

The nearest definition to the authors' view of construction industry adjudication as brought in by the Act is included in the Reader's Digest Universal Dictionary: 'to hear and settle (a case) by judicial procedure'. There is only one word in that definition which, in the authors' opinion, is not applicable as a definition of construction industry adjudication and that is the word 'procedure', of which more later. There are those, no doubt, who will be surprised to hear construction industry adjudication described in any way with a word such as 'judicial'. We are firmly of the view, however, that it is a 'judicial process' and we shall develop that theme later in this chapter. This is distinct from judicial procedure.

There has been little assistance from the courts, in the 171 cases that we have found, in defining the nature of new adjudication. The majority of the judgments in the courts are first instance which recycle, by analogy, law from arbitration and expert determination. There are more indications of what adjudication is not rather than of what it is. There are a number of quotations in the current case law which make observations on what the courts think adjudication might be:

> 'An adjudicator may not be a classic judicial tribunal but in practice an adjudication will probably be closer to an arbitration rather than expert determination. Whatever similarities adjudication may have with arbitration the decision has only temporary effect and validity.'[14]

> 'Adjudication is a special creature of statute.'[15]

> 'Adjudication is intended as a summary process. There is implicit within it a risk of injustice; but Parliament has considered that risk to be acceptable because an adjudication is of limited temporal effect and only of an interim nature.'[16]

> 'In the first place, the adjudication was meant to produce a rough-and-ready answer as a stop-gap solution. Secondly, it would be absurd if an adjudicator could "decide" that he had too little

---

[14] *Glencot Development & Design Co Ltd* v. *Ben Barrett & Son (Contractors) Limited* (13 February 2001).

[15] *Herschel Engineering Ltd* v. *Breen Property Ltd* (14 April 2000).

[16] *Shepherd Construction Limited* v. *Mecright Limited* (27 July 2000).

information to make a determination ... It appears from the cases cited to me that different views have been taken as to the appropriate legal framework within which to address the issues raised by adjudicators' decisions: in particular whether the adjudicator is to be regarded as a statutory decision-maker, albeit one whose statutory powers and duties have been clothed in contractual form (the approach adopted by Lord Macfadyen in *Homer Burgess Ltd* v *Chirex (Annan) Ltd*, as I understand his Opinion), or whether adjudication should be regarded as a contractual procedure (as Dyson J appears to have regarded it in, for example, *Macob Civil Engineering Ltd* v *Morrison Construction Ltd* [1999] BLR 93).'[17]

'The question for me is whether the adjudication procedure which I have outlined is "quasi legal proceedings such as arbitration or not". On behalf of the applicant it has been argued strongly that it is not. Mr Royce argued that it is the equivalent of some decision by an expert or a valuer giving a certificate. Decisions such as this, it seems to me, are largely matters of impression but I have come to the clear conclusion that the adjudication procedure under section 108 of the Act and/or clause 41 is quasi legal proceedings such as an arbitration within the classification of Vice-Chancellor Browne-Wilkinson in *Re. Paramount Airways*. It seems to me that it is, in effect, a form of arbitration, albeit the arbitrator has a discretion as to the procedure that he uses, albeit that the full rules of natural justice do not apply. The fact that it needs to be enforced by means of a further application does not stop it from being an arbitration. It is the precursor to an enforceable award by the court. It seems to me that it is "other proceedings" within section 11(3) and in my judgment accordingly leave is required.'[18]

'This case concerns the provision of the Housing Grants, Reconstruction [sic] and Regeneration Act 1996, which sets up a system of adjudication to protect parties to construction contracts who may be caught up in the web of large contractors making allegations, pursuing large counterclaims, not being paid themselves, and therefore depriving, sometimes rightly and sometimes wrongly, the sub-contractor of their rightful monies on an immediacy basis to enable the sub-contractor to continue working. We are all familiar with the old situation where a main contractor would string along a sub-contractor until he went bust.
So the 1996 Act provided for a system of adjudication during the contract works whereby persons who were the contracting parties, either the main contractor or the sub-contractor, could go along and get decisions and statements of monies owing at any particular time within a very short period.'[19]

'The Housing Grants, Construction and Regeneration Act 1996 introduced changes of some importance to the construction industry. It gave an entitlement to stage payments, made provision for the date when payments under a construction contract became due, dealt with the need to give effective notice of intention to withhold payments, provided a right to suspend performance for non-payment, prohibited conditional payment provisions and imposed a statutory set of contractual provisions which in default took effect as implied terms of the contract concerned. It also gave the important and practical right to refer disputes to adjudication so as to provide a quick enforceable interim decision under the rubric of "pay now, argue later".'[20]

'The question is whether the European Convention on Human Rights Article 6 applies to proceedings before an adjudicator. In the first place, the proceedings before an adjudicator are not in public, whereas the procedure under Article 6 has to be in public. I can see that the problems arise over whether one refers to a decision as a final decision or whether one has to consider whether Article 6 applies to a decision that is not a final decision, but it seems to me that if Article 6 does apply to proceedings before an adjudicator it is manifest that a coach and horses is driven through the

---

[17] *Ballast Plc* v. *The Burrell Company (Construction Management) Ltd* (21 June 2001).
[18] *A Straume (UK) Ltd* v. *Bradlor Developments Ltd* (7 April 1999).
[19] *Bridgeway Construction Ltd* v. *Tolent Construction* (11 April 2000).
[20] *RJT Consulting Engineers Ltd.* v. *DM Engineering (Northern Ireland) Ltd*, CA (8 March 2002).

whole of the Housing Grants, Construction and Regeneration Act. Maybe it is, of course because the Convention is something which though not at the moment binding by way of statute soon will be and the courts have to take account of it.

In my judgment, Article 6 of the European Convention on Human Rights does not apply to an adjudicator's award or to proceedings before an adjudicator and that is because, although they are the decision or determination of a question of civil rights, they are not in any sense a final determination. When I say that, I am not talking about first instance or appeals, but merely that the determination is itself provisional in the sense that the matter can be re-opened.

In those circumstances, therefore, I think the fact that the procedure before the adjudicator is very much a rough and ready procedure cannot of itself be regarded as a reason for not ordering summary judgment. I do not think the special circumstances to which I have referred alter that position.'[21]

There are eight cases that have been subject to appeal in England and two in Scotland. In his judgment in *Ferson* v. *Levolux*[22], Lord Justice Mantell sets out in sequence various comments that had been made by the courts in explaining the nature of adjudication as included in section 108 of the Act:

'The scheme provided by section 108 was explained by Dyson J in *Macob Civil Engineering Ltd* v. *Morrision Construction Ltd* [1999] BLR 93 at para 24:

"The intention of Parliament in enacting the Act was plain. It was to introduce a speedy mechanism for settling disputes in construction contracts on a provisional interim basis, and requiring the decision of adjudicators to be inforced pending the final determination of disputes by arbitration, litigation or agreement: see section 108(3) of the Act and paragraph 23(2) of Part 1 of the Scheme. The timetable for adjudications is very tight (see section 108 of the Act). Many would say unreasonably tight, and likely to result in injustice. Parliament must be taken to have been aware of this. So far as procedure is concerned, the adjudicator is given a fairly free hand. It is true (but hardly surprising) that he is required to act impartially (section 108(2)(e) of the Act and paragraph 12(a) of Part 1 of the Scheme). He is, however, permitted to take the intitiative in ascertaining the facts and the law (section 108(2)(f) of the Act and paragraph 13 of Part 1 of the Scheme). He may therefore, conduct on entirely inquisitorial progress, or he may, as in the present case, invite representations from the parties. It is clear that Parliament intended that the adjudication should be conducted in a manner which those familiar with the grinding detail of the traditional approach to the resolution of construction disputes apparently find difficult to accept. But Parliament has not abolished arbitration and litigation of construction disputes. It has merely introduced an intervening provisional stage in the dispute resolution process. Crucially, it has made it clear that the decisions of adjudicators are binding and are to be complied with until the dispute is finally resolved."

Further explanations are to be found in *Bouygues* v. *Dahl-Jenson* 2000 BLR 522 and *CB Scene Concept Design Ltd* v. *Isobars Ltd* 2002 EWCA Civ 46. In the former, at para 2 Buxton LJ described the section as being:

"To enable a quick and interim, but enforceable, award to be made in advance of what is likely to be complex and expensive disputes."

In the same case at para 26 Chadwick LJ stated:

"The purpose of those provisions is not in doubt. They are to provide a speedy method by which disputes under construction contracts can be resolved on a provisional basis. The adjudicator's

---

[21] *Elanay Contracts Ltd* v. *The Vestry* (30 August 2000).
[22] *Ferson Contractors Ltd* v. *Levolux AT Ltd* [2003] BLR 118.

decision although not finally determinative, may give rise to an immediate payment obligation. That obligation can be enforced by the courts. But the adjudicator's determination is capable of being reopened in subsequent proceedings. It may be looked upon as a method of providing a summary procedure for the enforcement of payment provisionally due under a construction contract."

In the latter, Sir Murray Stuart-Smith lent further emphasis to the draconian character of section 108 at para 23:

"The whole purpose of section 108 of the Act, which imports into construction contracts the right to refer disputes to adjudication, is that it provides a swift and effective means of resolution of disputes which is binding during the currency of the contract and until final determination by litigation or arbitration section 108(3). The provisions of 109-111 are designed to enable the contractor to obtain payment of interim payments. Any dispute can be qucikly resolved by the adjudicator and enforced through the courts. If he is wrong, the matter can be corrected in subsequent litigation or arbitration."

The case of *Bouygues* is a good illustration of the scheme put into practice. The adjudicator had made what was acknowledged to be an obvious and fundamental error which resulted in the contractor recovering monies from the building owner whereas in truth the contractor had been overpaid. The Court of Appeal held that since the adjudicator had not exceeded his jurisdiction but had simply arrived at an erroneous conclusion, the provisional award should stand. In this context the court adopted the test formulated by Knox J in *Nikko Hotels (UK) Ltd* v. *MEPC Plc 1991* 2 BGLR 103 at 108B:

"If he answered the right question in the wrong way his decision will be binding. If he has answered the wrong question, his decision will be a nullity." '

There is thus little to assist in case law in giving a single definition of new adjudication. The cases give observations on what the courts think new adjudication might be in given circumstances in enforcement proceedings. There is no doubt that the various angles on what adjudication might be will continue to develop in subsequent court proceedings.

Construction industry adjudication is thus a unique process. It is not adjudication as judges and arbitrators know it. It is far wider than that. It is also not mediation or conciliation. It is not expert determination, although it probably remains closer to this form of dispute resolution than any other. It is not arbitration. It is certainly not litigation.

Adjudication has been described as a procedure where, by contract, a summary interim decision-making power in respect of disputes is vested in a third-party individual (the adjudicator) who is usually not involved in the day-to-day performance or administration of the contract, and is neither an arbitrator nor connected with the state. This definition was in a paper published in 1991 before the advent of the Act.[23]

Seeking to define the word adjudication using previous definitions or referring to other areas where the word is used, for example in statutory contexts in the field of public law, such as prisons, asylum and immigration, social security, tax, complaints against solicitors and car parks in London, may prove unhelpful. In this field of the law, adjudication is supported by statute or regulation. Many of the cases reported in this field concern the failure to apply the rules of natural justice which has, of course, been a major aspect of many of the headline cases on adjudication.

Adjudication in the construction industry is also provided for by statute as a default position, excepting those areas where it has already been agreed to incorporate adjudication

---

[23] Mark C McGaw, *Adjudicators, Experts and Keeping Out of Court.* Centre for Construction Law and Management, September 1991 Conference, Current Developments in Construction Law.

provisions in contracts on a consensual basis. Why then cannot adjudication in construction be described as statutory adjudication? The answer is because the statute does not actually demand that parties take their disputes to adjudication. All the statute does is give the right to a party to take a dispute to adjudication if that party so wishes and the other party has to go along with it. The statute imposes new terms in construction contracts in a similar way to terms imposed on transactions for the sale of goods by the Sale of Goods Act 1979.

Although the right to adjudication is in statute, its source is in contract. Even where a contract does not comply with the minimum requirements of section 108 of the Act and therefore the Scheme applies by default, the requirements of the Scheme are treated as implied terms of the contract. Adjudication is therefore a creature of contract. Although adjudication is not consensual, it is not a requirement of the Act; all the Act is doing is regulating certain terms that must be contained within the contract.

The standard forms of contract which were redrafted in 1998 and shortly after to comply with the requirements of the Act do, in many instances, go beyond the minimum criteria required by the Act. These new clauses will be given their full effect through the law of contract and not through any requirement in the statute. It is to be noted that the court is prepared to enforce decisions that arise out of contracts that are not governed by the Act (*Parsons Plastics* v. *Purac*[24])

Construction is a risky business. A building contract is the means by which the parties to that contract allocate that risk between themselves. Parties to a contract may well not fully understand how that risk is allocated between them. That is why in so many building contracts, reference has hitherto had to be made to an arbitrator or to the courts, with all the time and cost that these processes can entail. Arbitration and the courts decide the dispute save for limited rights of appeal. Adjudication has been introduced to provide an inter-mediate step which, whilst its primary purpose is to determine the dispute on a temporary basis, can explain to the parties the allocation of risk that they have undertaken. It does not finally decide the dispute but it does give the parties an impartial view, which has the benefit of being obtained after the expenditure of relatively little time or cost and may assist them in understanding their respective rights and obligations under the contract and thus enable them to resolve the dispute themselves.

Adjudication is often grouped into the general category of dispute resolution processes. This is not, strictly speaking, correct. A number of procedures that are described as dispute resolution processes do not actually resolve the dispute. They may facilitate the resolution of the dispute but that is all. The resolution of the dispute depends on more than the process itself. A true dispute resolution process has at its end a decision, determination, award or judgment that brings the dispute to an end.

An adjudicator's decision does not do this. All adjudication can do is to put the parties into the position of understanding their rights and obligations under the contract that they have entered into. Adjudication is not final and binding. The dispute is not settled by adjudication, although it may be if the parties choose to accept the decision. Once an adjudicator has made his decision, it is for the parties to decide whether they should agree to accept those conclusions, or have the dispute heard again by a tribunal that will give them a decision that is final and binding. They are quite at liberty not to accept the conclusions, albeit they are bound by the decision, temporarily at least. Adjudication by itself does not and cannot resolve the dispute; for the dispute to be resolved each party has to decide for itself that it is going to take the matter no further.

---

[24] *Parsons Plastics (Research & Development) Limited* v. *Purac Limited* [2002] BLR 334, CA.

The adjudicator states what his conclusions are and this decision m.
very limited information. It certainly will have been made in a limit.
factors may well mean that the answer that the adjudicator reaches leaves a
The adjudicator will certainly not have had sufficient time to make a full fc.
gation. It may be that the adjudicator considers that the dispute is of such a nat.
totally impossible for him to reach any sort of conclusion. We examine this situa.
where when considering making the decision. In the vast majority of disputes, howev.
to be hoped that the skilful adjudicator, experienced in the industry, will be able to utilise
information produced by the parties, together with his own knowledge and the results of h.
own investigations, to reach a decision that will in fact reflect the rights and obligations of
the parties under their contract; and that those parties will accept the decision as it stands or
perhaps in some instances use the decision as a basis for negotiations and thus the dispute is
resolved. To that extent only, adjudication may become a dispute resolution process.

Construction adjudication as introduced by the Act provides for a process that gives the
opportunity for improvements in cash flow and for the resolution of disputes, at least on a
temporary basis, before they are allowed to escalate into something that requires vast
amounts of time and money to resolve. The process is quick and relatively cheap, particularly in comparison to litigation and to those arbitrations that have somewhat unfortunately,
for whatever reason, been allowed to become more time consuming and costly than they
should have been.

All this does perhaps leave us with the simple statement 'it is what it is' but this will not
satisfy the enquiring mind. Because the word adjudication as adopted by the Act encompasses the whole process and as the process itself has many aspects that cannot easily be
encapsulated in a few words, the definition has to be rather complex. If this were to be set out
in a single sentence it would be of inordinate length so the constituent parts are broken down
below.

Adjudication in the construction industry is a process that:

- provides for the referral of a dispute arising under the contract at any time, to a person (an
  adjudicator) who,
- acting impartially,
- on the basis of such information as the parties to the dispute are able to provide him, or he
  is able to ascertain for himself,
- in a very limited time-scale,
- reaches conclusions as to the parties' rights and obligations under their contract on the
  basis of that information,
- those conclusions being set out in a decision that is contractually binding on the parties
  until the original dispute is finally determined by legal proceedings or by arbitration (if
  the contract so provides or the parties so agree) or by agreement between the parties.

Is adjudication a judicial process?

It may come as a shock to many readers but the authors are firmly convinced that adjudication is a judicial process. This is not to say its procedures are judicial, far from it, but it
certainly is a judicial process.

To understand how this conclusion has been reached the nature of the adjudication
process must be examined. Adjudication arises out of a contract. The parties have rights and
obligations under that contract. For an adjudication to be necessary there must be a dispute

ɔ those rights and obligations. A good way of explaining this is to look at a dispute as to quality of work.

The man in the street might think that a dispute regarding plaster relates merely to a matter of it being good plastering or bad plastering. This is, however, a misconception. What the dispute actually relates to is whether the plasterwork as executed complies with the specification. The specification is a part of the contract. The dispute is actually whether the plasterer has complied with the duties and obligations that are placed upon him by the contract.

The task of the adjudicator is to decide whether the plasterer has in fact complied with the specification. In other words, has he complied with the contract? He is deciding upon the obligations that the plasterer has entered into when he made the contract. A person who makes such decisions is undertaking a judicial process. Determining the rights of the parties by applying the law to the facts is a judicial process. This is what every adjudicator is required to do under the Act, the Scheme and the new standard form contracts. This is a judicial function and not a commercial decision.

What the adjudicator is not doing is carrying out a judicial procedure. A judicial procedure, in our view, has connotations of pleadings, discovery, further and better particulars, evidence under oath and all other such matters beloved of the legal fraternity. This is not an appropriate way of conducting construction industry adjudication. There is not the time or the necessity to adopt a judicial or formal court procedure.

Adjudication does not generally require judicial procedures such as those found in the court rules. It would be impossible to conduct adjudications within the framework of very formal procedures. The management of the process needs inventiveness to understand a case at a very early stage and tailor the procedures to suit the dispute itself and the statutory timeframe. This has been recognised by the courts which have accepted the concept of procedural fairness as a yardstick for the conduct of adjudications within the necessarily limited time-scale.

This is not to say that there are not sometimes exceptional circumstances where the parties to adjudication proceedings with many millions of pounds at stake agree to a procedure that involves a rather more formal approach, but this requires both parties to be willing and a lengthy extension of time.

Comparison of adjudication with other forms of dispute resolution

The adjudication process is similar to, but different from, a number of other processes that are utilised in the resolution of disputes in the construction industry. These are, as mentioned above, mediation, conciliation, expert determination, arbitration and litigation.

The mediator, conciliator, independent expert and the adjudicator are all creatures of the contract and their powers are specifically constrained by it. In this they are distinct from the arbitrator who has powers created by statute, albeit the parties can amend them in their arbitration agreement or subsequently. Arbitration is consensual because the parties volunteer to put it in their contract, but the arbitrator is not a creature of the contract but of the arbitration agreement and statute. The arbitration agreement under which arbitration takes place has a separate existence by virtue of section 7 of the Arbitration Act 1996 and is not necessarily dependent upon the contract in which it was originally included. While there are requirements in the Act regarding what the contract must include in respect of adjudication, it is not the statute that governs the adjudication but the contract. Even in a situation where the contract does not comply with the Act and the Scheme applies, the process is still contractual. In that event the provisions of the Scheme become terms of the construction contract.

Mediation and conciliation

For the purposes of this comparison we shall refer to mediation only. The only substantive difference that we can identify is that in the ICE procedure conciliation allows the conciliator to make a recommendation at the end of the process. That is not the generally accepted situation in mediation.

Adjudication is similar to mediation only in that it is not a dispute resolution process in itself; they are both processes which may or may not result in the dispute being resolved. The role of the mediator is, however, to seek to facilitate an agreement between the parties by negotiation. There is no place for negotiation in adjudication, albeit the adjudicator may meet with the parties in an attempt to ascertain matters that he would otherwise not be able to find out in the time allowed. That will in all likelihood, however, take the form of an inquisitorial process rather than a negotiation.

Expert determination

Adjudication is in some ways quite similar to expert determination but the principles are rather different. The basic similarity is that they are both contractual procedures and they involve an impartial third party carrying out a review of the parties' rights and obligations under the contract.

In expert determination, the expert uses the results of his own knowledge and investigations in making his determination. The whole process relies on this procedure. The expert has no obligation at all to tell the parties the extent of his own knowledge that he uses or the results of his investigations into the facts and the law in reaching his determination.

Adjudication is similar but only to the extent that the adjudicator is permitted to ascertain the facts and the law for himself. There is, however, no obligation for him to do this. Time may in any event preclude such an option. If the adjudicator does use his own knowledge or ascertain matters of fact or law he must be careful to ensure that he acquaints the parties of such things or the courts will not enforce his decision (*RSL* v. *Stansell*[25]).

The expert's determination is contractually binding on the parties with no longstop provision, whereas the adjudicator's decision, whilst binding, is only so until the original dispute is finally determined by other means.

Arbitration

Adjudication is also similar in many ways to arbitration but again it is also very different. Both arbitration and adjudication involve the production of evidence before a tribunal and the tribunal reaching a decision on the parties' respective rights and obligations under the contract. Arbitration is definitely a judicial process as, in the authors' opinion, is adjudication.

---

[25] *RSL (South West) Ltd* v. *Stansell Ltd* (16 June 2003).

The difference is in the procedures used. There is an absolute obligation in arbitration that the arbitrator shall give each party a reasonable opportunity of putting its case and dealing with that of his opponent. There is also a statutory obligation on the arbitrator to adopt procedures suitable to the circumstances of the particular case, so as to provide a fair means for the resolution of the matters falling to be determined.

This can, in the case of complex disputes, lead to the utilisation of procedures that can be as costly and time consuming as litigation. This should not occur to such a degree in future given the introduction of the Arbitration Act 1996 with its provisions for the arbitrator to adopt suitable procedures and to cap costs.

In adjudication the time factor is paramount. All the requirements of arbitration, which can be loosely defined as being the 'rules of natural justice' and which are still important in adjudication, are secondary to the requirement that the adjudicator shall reach his decision in a very limited period.

The adjudicator may be prevented by the exigencies of a process that has to be completed so very quickly from giving each party an exhaustive opportunity of putting its case and dealing with that of his opponent. The adjudicator may also, by the time constraints, be put in the position of having to do things that might, in arbitration, if done by the arbitrator, be considered to breach the requirement to provide a fair means of resolving the matters to be determined. As has already been mentioned, the idea of procedural fairness is adopted by the courts to fulfil the requirement to be as fair as can be within the time limitations.

The adjudicator is empowered to take the initiative in ascertaining the facts and the law. The arbitrator has a like power, albeit that it can be limited or removed by the parties by agreement. In neither process is there any general obligation upon the arbitrator or adjudicator to do so save for one exception. That is when an adjudicator is operating under the Scheme for Construction Contracts. Paragraph 12(a) of the Scheme requires that he 'shall reach his decision in accordance with the applicable law in relation to the contract'. This effectively means that he has an absolute obligation to ascertain the law himself unless the parties provide him with everything that he needs in this respect. There is certainly no question of the adjudicator being under an obligation to make such investigations in the standard forms of contract unless such a procedure is agreed by the parties as an amendment. The non-legally qualified adjudicator should always be aware of the possibility that he could be taking on something that he is not competent to do and should always look carefully at any enquiry from an adjudicator nominating body with this in mind.

The adjudicator's decision is binding but not final unless the parties so agree. It is, however, not appealable. It is the dispute itself that can be reviewed subsequently. The arbitrator's award is by definition final and binding and only subject to very limited rights of appeal.

Litigation

In principle there is no comparison between litigation and construction industry adjudication. They are so utterly different.

It is worth making the point that, in the end, adjudication, like arbitration, has to be subject to the court system. It is a matter of contract and of enforcement. Ultimately, the only way any decision of an adjudicator can be enforced is for the courts to do it. There may be an intermediate stage where an arbitrator makes an award enforcing the adjudicator's decision, but if this award is not honoured the courts will have to enforce it.

The obligations placed upon the adjudicator

There are only two specific obligations placed upon an adjudicator by the Act. These are, first, that he reaches his decision within 28 days or such longer period as he may have as a result of an extension of this time period and, secondly, that he must act impartially.

The Act requires that the contract between the parties shall include these provisions. There is no requirement as to the form of the contract between the adjudicator and the parties. It is extremely unlikely that those who undertake the task of adjudicator will have time to enter into a formal contract with the parties to the contract out of which the dispute arises. The adjudicator will, however, be deemed to be aware that by taking on the task of adjudicator he is bound by these obligations. They will become express or implied terms of his contract. If he fails to comply with them he will be in breach of his contract.

The courts in their role as enforcers of adjudicators' decisions apply two specific obligations: has the adjudicator acted within his jurisdiction and has he complied with the requirements of procedural fairness?

The only other obligations placed upon an adjudicator will arise out of any procedures that may be included in the contract between the parties and any rules that apply to the adjudication process as a result of, or derived from, the Scheme for Construction Contracts if that is the regime that governs the adjudication. Such matters will also govern the adjudicator's contract and it is therefore most important that any adjudicator, before agreeing to act, ensures that he obtains precise information as to any rules or other provisions with which he will have to comply. It is of particular note that the JCT and ICE contracts widen the scope of the Act.

The powers of the adjudicator

It must be implied into the agreement under which the adjudicator operates that he has power to take any steps that he needs to in order to reach his decision within 28 days or any agreed extension to that.

As with the obligations, there are only two powers that the Act requires are given to the adjudicator in the contract. These are the power to take the initiative in ascertaining the facts and the law, and the power to extend the 28-day period by up to 14 days with the consent of the applicant. These powers are reflected in the Scheme and the various rules.

The adjudicator's powers relate solely to the contract under which he is appointed.

The various published sets of adjudication rules and the Scheme deal with the adjudicator's powers in slightly differing ways but almost all of them set out to reinforce the fact that the adjudicator has complete discretion as to procedure. Most set out a list of the powers that are available to the adjudicator. These are by way of examples only and are non-exhaustive. Some forms of contract set requirements relating to the time for submissions and the adjudicator should always be aware of such a possibility.

It is just as important that the adjudicator seeks to ascertain his powers before appointment, as it is that he ascertains his obligations. His powers may be restricted in some way and if the enquiry or referral documents give any indication that the contract is likely to contain provisions with which he is not familiar, he should take steps to find out the details of that contract at the earliest possible moment.

It must be remembered that the whole adjudication process is subject to the overall supervision of the court and it will generally be the forum for enforcement. If the adjudicator has failed to comply with the express and implied terms of his contract with the parties, there may be a problem with enforcement. It is therefore vital that the adjudicator operates strictly within the contract under which he is appointed.

Possible liabilities of the adjudicator

The Act requires the contract to provide that the adjudicator is not liable for anything done or omitted in the discharge or the purported discharge of his functions as adjudicator unless the act or omission is in bad faith.

This immunity applies only to the parties themselves and does not provide protection to the adjudicator should a third party decide to sue him.

It may be that the adjudicator will have the opportunity to obtain an indemnity from the parties in respect of suit by others. This is extremely unlikely where the adjudicator is nominated by an adjudicator nominating body because of the time restrictions. It will probably only occur where the adjudicator's agreement is in a standard form or the adjudicator is named in the contract between the parties and the contract provides such an indemnity. Even if an indemnity is given, the adjudicator should always be aware that the parties giving the indemnity may not always exist and in the case of insolvency on the part of both parties, a not unknown situation, the indemnity will disappear with them. In any event, professional indemnity insurance should be a must for any adjudicator.

What personal qualities should the adjudicator have?

The adjudicator's task is not an easy one. In that respect it is no different from any similar position where someone is required to make a decision that will affect others.

There are, as we have already identified, certain aspects of the adjudication process that make it unique and it presents similarly unique problems to those who adjudicate. Anyone can adjudicate but those who make a success of the process have clearly mature personal qualities.

We are not talking here about things like technical knowledge, awareness of adjudication procedures or knowledge of contracts and contract law. These can be learnt. We are not talking about professional experience. What we are considering are those rather less definable qualities that relate to the personality of the adjudicator. These are the qualities that enable the adjudicator to make a success of the process.

Adjudication can be a tricky business. The timescale involved makes this almost inevitable. The adjudicator cannot just sit back and wait for things to happen; if he is to make a success of the process he has to be pro-active and drive forward any adjudication to which he is appointed. In anything other than the simplest of disputes he will have to be setting procedures that will enable him to be in possession of the fullest information when he sits down to write his decision. This is the only way that the adjudication process will provide the opportunity for the parties to settle their dispute.

The adjudicator has to exercise his judgement. This is not the judgement that he will need when reaching his decision. This is judgement in the way that he deals with the procedural aspects of the adjudication and the way that he deals with the parties. It is the judgement that he needs when faced with difficulties that he cannot avoid and that have to be resolved.

An adjudicator must also be decisive. It is no use the adjudicator taking so long to make up his mind over a procedural issue that the delay affects the adjudication as a whole. This decisiveness will be brought to the fore in the need to decide upon the most appropriate procedure for the adjudication at a very early stage, probably on the basis of limited information.

Decisiveness and judgement also come to the fore when the decision on the issues that have been put to him is to be made. There is also a need for an analytical ability, logic and clarity of thought at this time.

These facets must be mixed up with such things as integrity, flexibility and initiative, and much of this comes down to simple common sense.

All these qualities when put together produce an adjudicator who exudes an air of competence and has what has been described as gravitas or judicial capacity. The parties will respect such an adjudicator who will, when combining these attributes with his knowledge and experience, provide the parties with a decision that they can use as a basis for resolving the dispute.

Summary

Adjudication cannot be categorised readily with other forms of dispute resolution. It is closest to expert determination in that both are creatures of contract. They differ in that expert determination is consensual; the parties choose to provide for expert determination in their contract and expert determination is binding and finally resolves the dispute. Construction adjudication is completely new. It is not consensual; it is forced upon the parties by the requirements of the Act. It is not binding; the decision is only temporary until the dispute is resolved by another means. Construction adjudication is probably *sui generis*, the only one of its kind.

Chapter 2
The Act: the Overarching Provisions
Sections 104–107 and 114–117, 146, 148–151

This chapter examines the provisions of Part II of the Act other than the specific sections that apply to adjudication, payment and suspension of performance which are considered in Chapters 3 and 4. The majority of this chapter relates to sections 104 to 107 which define the ambit of application of the adjudication and payment provisions, but we also consider various other sections mainly of an administrative nature. It is not intended to provide a legal treatise on the Act. It is however necessary to provide the reader with both legal and lay interpretation of the Act so that adjudicators and those using adjudication understand the full effect of the legislation.

There have been nearly 200 judgments of the court relating to adjudication at the time of writing. These primarily deal with the enforcement of adjudicators' decisions and it is really only when sections 104 to 107 inclusive are examined by the courts that any law is made. This is because judicial interpretation is required to decide whether or not the contract in question is a qualifying construction contract within the meaning of the Act. Those court judgment decisions relating to section 108 onwards are only in essence an interpretation of what the Act says.

Section 104: Construction contracts

Section 104 gives the basic definition of construction contracts. The points to note are that the definition is complex in that it relies on a further set of definitions for construction operations. Its coverage is also much wider than was anticipated in the Latham Report. The types of contract to which the adjudication and payment provisions apply are defined. These include professional service contracts in relation to construction contracts.

The secretary of state had the option to further define types of construction contract that would fall within or outside the provisions of the Act. This option has been exercised in the form of the The Construction Contracts (England and Wales) Exclusion Order 1998. This is discussed fully in Chapter 5.

The territorial operation of the Act originally applied to England, Wales or Scotland. This has been extended to Northern Ireland under the powers in section 149.

Section 104(1)

(1) In this Part a 'construction contract' means an agreement with a person for any of the following–
    (a) the carrying out of construction operations;
    (b) arranging for the carrying out of construction operations by others, whether under sub-contract to him or otherwise;

(c)   providing his own labour, or the labour of others, for the carrying out of construction operations.

This sub-section gives a wide definition of a construction contract. The word 'person' includes corporations, limited companies and individuals by reference to the Interpretation Act 1978.[1] 'Person' also includes the Crown. This is dealt with in detail in section 117.

Construction operations, which are an essential part of the definition of a construction contract, are defined in detail in section 105. There is a two part test. The basic requirements of section 104 must be satisfied before examining the definitions of construction operations in section 105. If the basic test of the section 104 parameters cannot be satisfied there is no need to examine section 105. In the broad sense a construction contract will include the traditional main contracts, sub-contracts and sub-sub-contract arrangements that the industry is familiar with. Many of the collateral warranties used in the construction industry also fall within the scope of section 104(1)(a).

Section 104(1)(b) which uses the term 'arranging for the carrying out of construction operations' places management contracts and other similar methods of procurement within the scope of the Act.

Section 104(1)(c) covers labour only, self-employed and gang master-type arrangements.

Agreements which vary a construction contract[2] are not construction contracts in themselves. The reference point is the original contract.

A dispute under a compromise agreement is not a construction contract albeit that the contract it settled was a construction contract.[3]

Section 104(2)

(2)   References in this Part to a construction contract include an agreement–
(a)   to do architectural, design, or surveying work, or
(b)   to provide advice on building, engineering, interior or exterior decoration or on the laying-out of landscape,
in relation to construction operations.

The services of construction professionals in relation to construction operations also constitute construction contracts. As a result the professional relationship between the employer and the design team is within the provisions of the Act. Equally design work carried out by professionals for contractors under design and build arrangements is also within the provisions of the Act. Providing advice in connection with construction operations is also within the provisions of the Act, but giving advice on the merits of a claim or a dispute concerning construction is not; neither is acting for one of the parties to a construction dispute or the giving of evidence as an expert. All the normal post contract services that involve, amongst other matters, claims and disputes would fall within surveying work and are thus within the definition in the Act.

This facility is a 'double-edged sword' for professionals. The right to pursue disputes over outstanding fees by using adjudication is counterbalanced by the opposing right of the employer to pursue damages for professional negligence. Professional negligence arises

---

[1] Interpretation Act 1978 Chapter 30 Schedule 1, 'Person' includes a body of persons corporate or unincorporate.
[2] *Earls Terrace Properties Limited* v. *Waterloo Investments Limited* (14 February 2002).
[3] *Shepherd Construction Limited* v. *Mecright Limited* (27 July 2000); *Quality Street Properties (Trading) Limited* v. *Elmwood (Glasgow) Limited* (8 February 2002).

predominately under contract[4]. The position was thought to be different in Scotland until the *Gillies Ramsay* v. *PJW Enterprises*[5]. This case held that the adjudicator has the power to award damages for breach of contract and professional negligence.

In the Gillies Ramsay case it was argued that the role of contract administrator under the terms of that particular contract did not fall within the provisions of section 104(1) and 104(2) of the Act in the following terms:

> '...a contract administrator ... was not "carrying out construction operations" within section 104(1)(a) of the 1996 Act, nor was he "providing labour" within section 104(1)(c). Further, in relation to section 104(1)(b), counsel submitted that the contract administrator was not "arranging" for others to carry out work: he was taking decisions about parties' rights under the contract, for example, by granting extensions of time, or by issuing instructions. So far as section 104(2)(a) was concerned, counsel submitted that contract administration was not "surveying work". A contract administrator was the man in the middle. He had to determine parties' rights. It was wrong that such a man should have claims made against him. It was one thing for an adjudicator to order that a contractual payment should be made; it was quite another matter to have claims for damages for professional negligence being decided by an adjudicator.'

The Scottish Courts rightly rejected these arguments and upheld that the 'man in the middle nature' of contract administration did not exclude this type of work from the provisions in section 104.

When, however, the services provided relate to arbitration or litigation by way of expert witness work or advice or acting as a witness of fact, they do not fall within section 104.

> 'The result is that the adjudicator did not in my judgment upon the true construction of S.104 (2) of the 1996 Act have jurisdiction to rule upon the entitlement of the defendant to payment for the services rendered by it as a witness of fact or by way of assistance at the arbitration..'[6]

A novation agreement may fall within section 104 even where this would give the legislation retrospective effect over the original contract[7].

Section 104(3)

(3)   References in this Part to a construction contract do not include a contract of employment (within the meaning of the Employment Rights Act 1996).

This is straightforward. The employer/employee relationship is not subject to the Act.

Section 104(4)

(4)   The Secretary of State may by order add to, amend or repeal any of the provisions of subsection (1), (2) or (3) as to the agreements which are construction contracts for the purposes of this Part or are to be taken or not to be taken as included in references to such contracts.
No such order shall be made unless a draft of it has been laid before and approved by a resolution of each of *(sic)* House of Parliament.

---

[4] *Heyman and Another* v. *Darwins Limited* AC [1942] 1 All ER 337, HL; [1942] AC 356.
[5] *Gillies Ramsay Diamond* v. *PJW Enterprises Ltd* (27 June 2002).
[6] *Fence Gate* v. *James R Knowles Limited* (31 May 2001).
[7] *Yarm Road Limited* v. *Costain Limited* (30 July 2001).

There is a drafting error in the last sentence of this subsection. It is clear however that any amendment to the scope of the preceding subsections would require approval by resolution of both the House of Commons and the House of Lords.

Section 104(5)

(5) Where an agreement relates to construction operations and other matters, this Part applies to it only so far as it relates to construction operations.
An agreement relates to construction operations so far as it makes provision of any kind within subsection (1) or (2).

It would have been much simpler if, where an agreement is a mixture of construction operations and non-construction operations, the whole of the contract had been subject to the Act. This sub-section provides severability between parts of the agreement that contain construction operations and those parts that do not. The parts which do contain construction operations will be subject to the Act; the remainder of the contract will not. Where an agreement is made for design and off-site fabrication of goods or components but the designer/fabricator is not to fix those goods or components, the design work will be subject to the Act but the fabrication work will not[8].

There is nothing however to prevent the parties from agreeing that the whole of the agreement is subject to adjudication and/or payment provisions. These may comply with the Act or they may be wider in scope. Such an agreement would then be a contractual matter and enforceable through the contract itself and not through the provisions of the Act. The standard form building contracts have been amended to take account of the provisions of the Act. These are discussed in later chapters. They do not attempt to make the distinction between construction and non-construction contracts. Therefore any contracts that are entered into on the basis of these standard forms will be subject to a contractual right to adjudication and the payment provisions of the contract. The Act becomes redundant for these purposes and the contract prevails. Parties who contract on this basis will not be able to use the Act to seek to limit contractual provisions that provide more than required by the Act.

Equally there is nothing to prevent the parties from identifying the parts of the contract which fall within the Act and those which do not. However, where the parties identify the parts of their agreement to which the Act applies incorrectly, the Act would still apply by virtue of this sub-section insofar as the work relates to construction operations.

This is one of the areas of difficulty for adjudicators and the parties. If the work over which the dispute exists does not fall within the provisions of the Act, the adjudicator will not have jurisdiction to deal with it[9]. It is likely that there will be some contracts where the work which constitutes construction operations and that which does not is either difficult to distinguish or indistinguishable.

This difficulty has been examined in a number of cases[10]. The simple rule from these cases is to distinguish between those parts of works which are construction operations and those parts which are not. It is only on those parts which are construction operations that an

---

[8] See section 105(2)(d).
[9] See Chapter 9.
[10] See *Homer Burgess Limited* v. *Chirex (Annan) Limited* (10 November 1999); *ABB Power Construction Ltd* v. *Norwest Holst Engineering Ltd* (1 August 2000); *Gibson Lea Retail Interiors Limited* v. *Makro Self Service Wholesalers Limited* (24 July 2001); *Palmers Limited* v. *ABB Power Construction Limited* (6 August 1999); In the Petition of Mitsui Babcock Energy Services Limited (13 June 2001).

adjudicator will have jurisdiction. This does of course only apply where the contract is not one where the parties have agreed that the adjudication provisions apply to the whole of the work involved and thus to both those parts of the works that come within the statutory definition of a construction contract and those that do not.

The final part of this sub-section refers back to sub-sections (1) and (2) above. So far as the agreement makes provision of any kind, those parts of the agreement are within the Act. It is not sufficient to seek to argue that the primary purpose of the agreement is not construction operations and that the construction operations are subsidiary to the overall agreement. Those parts that are construction operations are caught within the Act.

Section 104(6)(a)

> (6) This Part applies only to construction contracts which–
>   (a) are entered into after the commencement of this Part, and

The commencement order (Statutory Instrument 1998 No. 650 (C. 13) The Housing Grants, Construction and Regeneration Act (England and Wales) (Commencement No. 4) Order 1998) made Part II sections 104 to 117 operative from 1 May 1998. This gave rise initially to a number of difficulties where a main contract entered into before the operative date which was not subject to the Act could have sub-contracts which were. The main contractor was not in a position to start his own adjudication with the employer to offset the effect of a sub-contractor's adjudication that results from the employer's actions. This was a temporary problem and apart from the possibility of an adjudication on a very long outstanding final account is now unlikely to occur.

Section 104(6)(b) and (7)

> (6) (b) Relate to the carrying out of construction operations in England, in Wales or Scotland.
> (7) This Part applies whether or not the law of England and Wales or Scotland is otherwise the applicable law in relation to the contract.

Section 104(6)(b) gives the locality of the carrying out of the construction operations. Any work involving construction operations carried out in England, Wales or Scotland is covered by the provisions of the Act.[11]

Section 104(7) deals with the applicable law of the contract. The parties may choose to make a contract governed by any regime of law that they wish. One or both of the parties may be an overseas concern. The Act applies to any construction contract carried out in England, Wales or Scotland whatever the law stated in the construction contract.

The Act applies to Northern Ireland. This was brought in under the Act through the powers in section 149. The instrument made was the Construction Contracts (Northern Ireland) Order 1997 No. 274 (NI 1). The NI Order follows the Act repeating the provisions of sections 104 to 117. The Order, the Scheme and the Exclusion Order became operative for Northern Ireland on 1 June 1999.

---

[11] For Northern Ireland see section 149 later in this chapter.

Section 105: Meaning of 'construction operations'

Those matters included as construction operations and those matters excluded are listed in two sub-sections. The 'in' or 'out' lobbies had their extensive interests aired when the Bill which formed this part of the Act was debated in Parliament. The definition of construction operations is an essential part of what constitutes a construction contract and therefore that work which is within the Act and that which is not covered. The definition is taken from section 567 of the Income and Corporation Taxes Act 1988.

It is arguable that the more liberal definition of construction found in Regulation 2 of the Construction (Design and Management) Regulations 1994 (SI 1994 No. 3140 would have been more appropriate and consistent with the 'mischief' the Act is seeking to correct.

The definition of construction operations read in conjunction with section 104 gives the scope of the types of work and agreements that are covered by the Act and are therefore construction contracts.

Whether or not a contract is in fact a construction contract as defined by the Act has proved to be an area rich in arguments concerning jurisdiction (see Chapter 9). Disputes concerning definitions often result in litigation. Some of those that have been subject to adjudication are discussed below.

The definition of construction operations is in two parts: those types of work which are included in sub-section 105(1) and those types of work which are excluded in sub-section 105(2).

Section 105(1)(a)

(1) In this Part 'construction operations' means, subject as follows, operations of any of the following descriptions–

 (a) Construction, alteration, repair, maintenance, extension, demolition or dismantling of buildings, or structures forming, or to form, part of the land (whether permanent or not);

Sub-section 105(1)(a) gives a broad definition of works which are construction operations. The governing factors are 'forming, or to form, part of the land (whether permanent or not)'. Temporary works such as falsework, formwork and scaffolding and any form of enabling works that do not form part of the final product are within this definition of construction operations.

Part of the land is an important point. Land covered by tidal land is within the confines of section 567 of the Income and Corporation Taxes Act 1988. This matter was visited in *Staveley*[12] which concerned works for the supply and installation of fittings into steel modules which were being constructed in England. The modules were intended for use as living quarters for operatives of an oil platform in the Gulf of Mexico. They were to be towed to location and welded onto platforms which were supported by legs founded in the bed of the sea. The court ruled 'structures which were, or were to be, founded in the sea bed below low water mark were not structures forming, or to form, part of the land for the purposes of section 105(1) of the 1996 Act'.

A shop-fitting contract was also subject to a dispute as to whether the shop-fittings formed part of the land[13]:

---

[12] *Staveley Industries Plc (t/a El.WHS)* v. *Odebrecht Oil & Gas Services Ltd* (28 February 2001).
[13] *Gibson Lea Retail Interiors Limited* v. *Makro Self Service Wholesalers Limited* (24 July 2001).

'What might be involved in a structure or fittings "forming part of the land" is not something which is addressed in the Act. However, in the context of the law of real property the concept of a fixture is well-established, and it seems to me that that to which the part of the definition of "construction operations" in section 105(1) of the Act which I have just set out is directed is whether the particular structure or fittings will, when completed, amount to a fixture or fixtures. In the law of real property one of the factors which is relevant to a determination of whether a chattel attached to a building is a fixture or not is whether the attachment is intended to be permanent – see, for example, *Billing* v. *Pill* [1954] 1 QB 70.'

Judge Seymour QC thought it clear that shop-fitting did not amount to construction operations unless it consisted of the construction of 'structures forming, or to form, part of the land (whether permanent or not)' or 'installation in any building or structure of fittings forming part of the land', as per sections (105)(1)(a) and (c) of the HGCRA. None of the items supplied by Gibson Lea were fixtures. The Act did not apply.

The maintenance of heating systems in housing has been held to be operations of repair or maintenance forming or to form part of the land[14].

Section 105(1)(b)

(1) (b) Construction, alteration, repair, maintenance, extension, demolition or dismantling of any works forming, or to form, part of the land, including (without prejudice to the foregoing) walls, roadworks, power-lines, telecommunication apparatus, aircraft runways, docks and harbours, railways, inland waterways, pipe-lines, reservoirs, water-mains, wells, sewers, industrial plant and installations for purposes of land drainage, coast protection or defence;

Sub-section 105(1)(b) lists in further detail the types of work which are included. The wide field of works covered by the construction industry, collectively in building and civil engineering, are included. In both sub-sections 105(1)(a) and (b) repair and maintenance are included. The word maintenance was a late addition at the draft bill stage. It is difficult to distinguish repair and maintenance. The construction industry has an industry within itself simply devoted to repair and maintenance. To leave maintenance work outside of the definition would have left a major part of construction free to carry on its disputes in a way the Act was seeking to prevent. The inclusion of maintenance work will bring facilities management contracts, which involve a maintenance function, within the Act.

Schedule 17 to the Communications Act 2003 makes the following amendment to sub-section 105(b):

137 In section 105(1)(b) of the Housing Grants, Construction and Regeneration Act 1996 (c. 53) (meaning of 'construction operations'), for 'telecommunication apparatus' there shall be substituted 'electronic communications apparatus'.

Section 105(1)(c)

(1) (c) installation in any building or structure of fittings forming part of the land, including (without prejudice to the foregoing) systems of heating, lighting, air-conditioning, ventilation, power supply, drainage, sanitation, water supply or fire protection, or security or communications systems;

---

[14] *Nottingham Community Housing Association Limited* v. *Powerminster Limited* (30 June 2000).

Sub-section 105(1)(c) includes the whole of the mechanical, electrical and engineering services industries. In the CDM Regulations, Regulation 2(1)(e) includes commissioning of the services installations listed. It is not thought that the omission of the word commissioning here is sufficient to exclude such an integral part of the services installation from the Act. The list of services is not exhaustive and will include any service that constitutes fittings forming part of the land. There is a distinction between fittings forming part of the land and chattels. Chattels in this sense are movable, tangible articles of property. The movement of loose furniture in an existing building as part of the preparatory work for refurbishment would not be caught by the Act.

> 'materials worked by one into the property of another becomes part of that property. This is equally true whether it be fixed or moveable property. Bricks built into a wall become part of the house, thread stitched into a coat which is under repair, or planks and nails and pitch worked into a ship under repair, become part of the coat or the ship.'[15]

Section 105(1)(d)

(1) (d) external or internal cleaning of buildings and structures, so far as carried out in the course of their construction, alteration, repair, extension or restoration;

Sub-section 105(1)(d) would exclude separate cleaning contracts in connection with the building after its completion but would include the cleaning operations executed before handover.

Section 105(1)(e)

(1) (e) operations which form an integral part of, or are preparatory to, or are for rendering complete, such operations as are previously described in this subsection, including site clearance, earthmoving, excavation, tunnelling and boring, laying of foundations, erection, maintenance or dismantling of scaffolding, site restoration, landscaping and the provision of roadways and other access works;

Sub-section 105(1)(e) is a 'catch all' provision to cover any eventualities not covered in the previous sub-sections. It will cover such operations as geotechnical surveys, dredging, exploration work and the provision and relaying of services by utility companies.

A boiler plant and the supporting steelwork have been held to be construction operations under this sub-section:

> 'The nature, size and method of fixing into position of the steel structure and the boiler itself clearly have the consequence that the boiler forms part of the land once assembled and fixed into position. Indeed, it would be hard to conceive a more rigid and permanent structure than the steelwork in question. The fact that much of the boiler is assembled on the site but away from its permanent resting place and then lifted into position cannot affect the conclusion that a construction activity is involved. Since much industrial plant will be assembled and erected in this way and since such plant is expressly included in the definition of a construction operation, the only reasonable conclusion is that ABB's work is a construction operation.'[16]

It followed that scaffolding work in connection with these works was also a construction operation.

---

[15] See *Appleby* v. *Myers* [1867] LR 2CD 651, Blackburn J.
[16] *Palmers Limited* v. *ABB Power Construction Limited* (6 August 1999).

The hire of plant and labour has been held to fall under this sub-section where the plant and labour is preparatory to or for the purposes of rendering work complete:

'It is common ground that a contract for mere plant hire is not a construction contract within HGCRA. Baldwins' case is that the labour element in the contract is crucial. The crane plus the labour provided by Baldwins was an integral part of the building works being carried out by Barr in the construction of the football stadium. The supply of a mobile crane plus labour is clearly an operation within the scope of section 105(1)(e) because it is one which forms an integral part of, or is pre-paratory to, or is for rendering complete, works of "…construction, alteration, repair, main-tenance…" etc., being Barr's works in building the stadium. This was a contract for the hire of a crane with operator for use by Barr in construction operations, i.e. construction of a new football stadium and thus a construction contract.[17]

Section 105(1)(f)

(1) (f) painting or decorating the internal or external surfaces of any building or structure.

Sub-section 105(1)(f) covers the painting and decorating processes to the whole of the work covered in the previous sub-sections. This covers existing buildings as well as new build-ings.

The following sub-section 105(2) defines operations that are not construction operations. The industry would contend that some of the exclusions would normally be accepted as construction operations in practice. They include the trades and skills that are used throughout the industry. They are generically part of the industry even if not for the pur-poses of this Act.

Section 105(2)(a)

(2) The following operations are not construction operations within the meaning of this Part–
      (a) drilling for, or extraction of, oil or natural gas;

Sub-section 105(2)(a) excludes the activities of the gas and oil industries only for the pur-poses of extraction of their products. This does not exclude the pipelines or installations for distribution but see sub-section 105(c) below.

Section 105(2)(b)

(2) (b) Extraction (whether by underground or surface working) of minerals; tunnelling or bor-ing, or construction of underground works, for this purpose;

Sub-section 105(2)(b) excludes the activities of the mining industry. This only excludes the extraction process itself. Any works which are ancillary to that process will fall within the Act, e.g. buildings, roads, services, etc.

Section 105(2)(c)

(2) (c) assembly, installation or demolition of plant or machinery, or erection or demolition of steelwork for the purposes of supporting or providing access to plant or machinery, on a site where the primary activity is–

---

[17] *Baldwins Industrial Services PLC* v. *Barr Limited* (6 December 2002).

    (i)   nuclear processing, power generation, or water or effluent treatment, or

    (ii) the production, transmission, processing or bulk storage (other than warehousing) of
         chemicals, pharmaceuticals, oil, gas, steel or food and drink;

At first glance this seems to be a wide ranging exclusion but it applies only to work on a site where the primary activity is those listed in (i) and (ii) and then the only operations that are exempt are those to do with the plant and machinery itself and the steelwork for the purposes of giving support or access. This does mean that precisely the same works can be outside or caught by the Act simply on the distinction of the primary activity of the site. A small sewage treatment plant would be outside the Act if sewage treatment were the primary purpose of the site. The provisions of the Act would catch exactly the same installation if it were merely part of a larger development of which the primary purpose was not sewage treatment. It is only the work connected with the process that is exempt from the provisions of the Act. The buildings that enclose the process plant and equipment are within the Act. The exemption only applies to those industries listed in (c)(i) or those processes listed in (c)(ii).

    This is the type of work often covered by the Institution of Mechanical Engineers/ Institution of Electrical Engineers Model Forms (MF/1 and MF/2) and Institution of Chemical Engineers standard forms of contract (the Red, Green, Yellow, Brown, and Orange Books). This exemption was secured by the persistent lobbying of the Process Industries Latham Group (PILG).

> 'Divergent views of the process industry and the building and civil engineering industries and their history regarding dispute resolution.'[18]

The process industry exemption was reduced during the passage of the Bill:

> 'I want to make it clear that we do not intend all the work on a process engineering site to be excluded from the fair contracts provision. We want to exclude only work on the machinery and plant that is highly specific to the process industry, together with work on steelwork that is so intimately associated with that plant and machinery that it could not possibly be reasonably considered apart. To that end, we have made it clear that the steelwork mentioned in the exclusion is only that which relates to support and access. . . I repeat that all normal construction activities on a process engineering site will be subject to the provisions of the Bill. That includes building roads, erecting fences, laying foundations, and building offices or factories even if they are made of steel.'[19]

Somewhat to the authors' surprise, the IChemE did include adjudication provisions within its contracts but it is interesting that the introduction to its adjudication rules almost seems to be an apology for its inclusion.

    The distinction between plant and construction may be difficult in some instances, e.g. sewerage and power generation plant. It is not unusual that plant and the buildings that house the plant will be procured under a single contract. In the eventuality of a dispute the work which constitutes construction will have to be separated from that which is plant and connected with the plant.

    In practice the definition of plant and machinery ought to be straightforward. The distinction of what constitutes plant and machinery as opposed to construction may be assisted by those cases which have been before the courts in relation to the legislation on capital allowances (now the Capital Allowances Act 1990). The distinction often proves difficult.

---

[18] Viscount Ullswater, *Hansard*, Vol. 570, No. 70, col. 1845.
[19] Robert Jones, MP, Minister for Construction at Committee Stage in the House of Commons. Official Report – Standing Committee F: 13 June 1996, cols 301–302.

Expenditure on an underground sub-station for transforming electricity was held not to be expenditure on plant[20]. A grain silo has been held to be plant[21]. The test is whether the structure in question is forming a plant-like function.

For a distinction between plant and building services see *Comsite* v. *Andritz*[22]. Although the primary activity on this site was water treatment, the ordinary building services for the buildings themselves were held not to be part of the plant or machinery.

It does not matter that the qualifying activity will not come into existence until after the work on site is complete:

> 'This in itself may show that nobody thought that the exemption in section 105(2)(c) did not extend to work on site where the qualifying activity would only come into existence on completion of the work.
>
> In any event even if I am wrong in concluding that section 105(2)(c) applies as much to the future as to the present the evidence of Mr Merton shows that the installation work is taking place on a site where the primary activity is now power generation. The facts that a fence has been erected and that for operational reasons one side is designated a construction site are in my view irrelevant. Even if the fence is not required for reasons of health and safety or under the CDM Regulations, it denotes no more than the customary separation of the "live" side. For the purposes of section 105(2)(c) there is here only one site.'[23]

Pipework linking pieces of equipment, plant, have been held to be plant for the purposes of this sub-section[24].

Section 105(2)(d)

> (2) (d) manufacture or delivery to site of–
>   (i)   building or engineering component
>   (ii) materials, plant or machinery, or
>   (iii) components for systems of heating, lighting, air conditioning, ventilation, power supply, drainage, sanitation, water supply or fire protection, or for security or communications systems,
>   except under a contract which also provides for their installation;

This sub-section provides a surprising exemption from the Act. Over a number of years the industry has been encouraged to mechanise and industrialise its operations so that more is carried out off-site. This is currently one of the key thrusts in the Egan Report.[25] This sub-section exempts all supply only arrangements even where extensive work is carried out by way of pre-fabrication off-site. If a manufacturer of components is carrying out the whole of his work on a supply only basis, with no installation and the components are delivered late, a dispute with the manufacturer which arises from a delay to the main contract is not within the Act and would not be subject to the adjudication provisions. The contractor might however find himself in receipt of an adjudication referral from the employer in respect of the same delay, as the main contract includes for installation as well as manufacture and delivery and is thus not exempt.

---

[20] *Bradley (Inspector of Taxes)* v. *London Electricity*, The Times, 1 August 1996.
[21] *Schofield* v. *R & H Hall* (1974) 49 TC 538.
[22] *Comsite Projects Limited* v. *Andritz AG* – TCC Birmingham (30 April 2003).
[23] *ABB Power Construction Ltd* v. *Norwest Holst Engineering Ltd* (1 August 2000).
[24] *Homer Burgess Limited* v. *Chirex (Annan) Limited* (10 November 1999).
[25] Department of the Environment, Transport and the Regions, *Rethinking Construction*, 16 July 1998.

Windows manufactured off-site and delivered for installation by the main contractor or one of his sub-contractors would be an exempt contract. If the supplier carries out the installation the contract would then not be exempt. Similarly, where the manufacture of a boiler includes for the testing and commissioning this would be a non-exempt contract. The testing and commissioning element would be 'for rendering complete' under sub-section 105(1)(e).

Section 105(2)(e)

(2) (e) the making, installation and repair of artistic works, being sculptures, murals and other works which are wholly artistic in nature.

Wholly artistic would suggest if there were some functional benefit from the work that it would not be exempt.

Section 105(3) and (4)

(3) The Secretary of State may by order add to, amend or repeal any of the provisions of subsection (1) or (2) as to the operations and work to be treated as construction operations for the purposes of this Part.

(4) No such order shall be made unless a draft of it has been laid before and approved by a resolution of each House of Parliament.

The promised 'watching brief' has borne fruit in terms of the consultation process that resulted in the proposed amendments to the Scheme. This has been dealt with in Chapter 1. It is unknown at the time of writing whether or not any amendment will be made to primary legislation.

Section 106: Provisions not applicable to contract with residential occupier

The provisions of the Act do not apply to residential occupiers as defined in section 106. There is a strong argument that says that a householder in dispute with a builder would greatly benefit from the adjudication provision in the Act. The short and effective means of resolving a dispute, at least in the immediacy, must be more beneficial than the alternatives. There is nothing that prevents the parties making an arrangement for adjudication in their contract, as they will if the JCT minor works contract is used, but note should be taken of the position with regard to consumers which we discuss below.

Section 106(1)

(1) This Part does not apply —
 (a) to a construction contract with a residential occupier (see below), or
 (b) to any other description of construction contract excluded from the operation of this Part by order of the Secretary of State.

Sub-section 106(1)(a) exempts construction contracts with a residential occupier from the Act. There is nothing to prevent the parties to such a contract from making their own terms which include adjudication as one of the means of resolving their disputes. This would then provide for contractual adjudication and would not be under the statutory rights to adjudication. For an example of this see *Jamil Mohammed* v. *Dr Michael Bowles*[26]. Interestingly

---

[26] *Jamil Mohammed* v. *Dr Michael Bowles*, High Court of Justice Bankruptcy proceedings (11 March 2003).

enough in this case, Mr Mohammed included an adjudication provision in his contract with Dr Bowles. Having lost the adjudication he tried to avoid enforcement on the grounds that Dr Bowles was a residential occupier. He failed.

This is of itself an interesting point in view of the degree of protection afforded to consumers under European and UK consumer protection legislation. There may be instances that even where the consumer has entered into a contract which contains an adjudication clause there is still an exemption. This does not arise under section 106 but under the Unfair Terms in Consumer Contracts Regulations 1999. See *Picardi* v. *Cunibert*[27]. This is a complex case. One of the bases on which an adjudicator's decision was not enforced was the exemption as a consumer and more precisely that the inclusion of adjudication had not been explained by Mr Picardi who was the Cunibertis' architect and who sought to recover his fees by adjudication:

> 'I conclude that a procedure which the consumer is required to follow, and which will cause irrecoverable expenditure in either prosecuting or defending it, is something which may hinder the consumer's right to take legal action. The fact that the consumer was deliberately excluded by Parliament from the statutory regime of the HGCRA reinforces this view. Costs in an adjudication can be very significant. Unless it is properly explained to the consumer, the fact that the adjudicator is to be a neutral, even if nominated by the architect's own professional body, also may give the appearance of unfairness.'

There was a third similar case in 2003 where the residential occupiers were unable to resist enforcement on the grounds that they had put forward the contract which included adjudication provisions having been professionally advised when doing so[28].

The Act does not exempt sub-contracts made under a main contract. The main contract will be made between the main contractor and the residential occupier. The sub-contracts will be made between the main contractor and the various sub-contractors and will not be with the residential occupier. They are therefore caught by the Act and are subject to the adjudication and payment provisions. The main contractor may find himself in a dispute which is subject to adjudication and he would not be able to issue a reciprocal notice to seek to have the same matter adjudicated on with the employer.

Sub-section 106(1)(b) deals with the prospect of other descriptions of construction contract being excluded from the operation of the Act. The first of these are covered by The Construction Contracts (England and Wales) Exclusion Order 1998. This is discussed in Chapter 5.

Section 106(2)

(2) A construction contract with a residential occupier means a construction contract which principally relates to operations on a dwelling which one of the parties to the contract occupies, or intends to occupy, as his residence.

In this subsection "dwelling" means a dwelling-house or a flat; and for this purpose — "dwelling-house" does not include a building containing a flat; and "flat" means separate and self-contained premises constructed or adapted for use for residential purposes and forming part of a building from some other part of which the premises are divided horizontally.

---

[27] *Picardi* v. *Mr & Mrs Cuniberti* (19 December 2002).
[28] *Lovell Projects* v. *Legg and Carver* (July 2003).

Sub-section 106(2) gives the definition of a construction contract with a residential occupier referred to in sub-section 106(1)(a). The contract must principally relate to operations on a dwelling which one of the parties occupies or intends to occupy, as his residence. An individual having work done on his dwelling house or flat would be exempt from the provisions of the Act. However, a Residents Association commissioning work on a block of flats, as a collective, would not be exempt from the Act. The residents themselves are individual occupiers but the Association is not.

Normal housing contracts built on a speculative basis or housing built for Local Authorities or Housing Associations are within the Act. A penthouse flat built on top of an office block or a hotel is within the scope of the Act being a part of the primary construction contract.

A contract where residential occupiers were to occupy one of the units as part of a development was held not to come within the provisions of section 106 even though the part that they were to occupy was about two-thirds of the whole contract[29].

A limited company cannot be a residential occupier.

'The defendants' original contention that the contract was for work done on a residential dwelling, and therefore fell within the exception contained in section 106 of The Housing Grants, Construction and Regeneration Act of 1996 (the Act) was not pursued before me. A limited company cannot be the residential occupier of a dwelling house.'[30]

Section 106(3) and (4)

(3) The Secretary of State may by order amend subsection (2).
(4) No order under this section shall be made unless a draft of it has been laid before and approved by a resolution of each House of Parliament.

As with other sections of part two of the Act this is a provision for the secretary of state by order to amend the definition in sub-section (2). The importance of such an order to amend is emphasised in sub-section (4) by the need for approval by each House of Parliament.

Section 107: Provisions applicable only to agreements in writing

The provisions of Part II of the Act only apply to construction contracts or other agreements where they are in writing. The definition of writing is much wider than would be accepted as the common form of writing.

Section 107(1)

(1) The provisions of this Part apply only where the construction contract is in writing, and any other agreement between the parties as to any matter is effective for the purposes of this Part only if in writing.
     The expressions 'agreement', 'agree' and 'agreed' shall be construed accordingly.

This section is almost identical to section 5 of the Arbitration Act 1996. The more liberal definition of what constitutes a written agreement in the Arbitration Act 1996 is, among

---

[29] *Samuel Thomas Construction* v. *Mr & Mrs Bick* (t/a J&B Developments) (28 January 2000).
[30] *Absolute Rentals Limited* v. *Gencor Enterprises Limited* (16 July 2000).

other matters, there to avoid those arguments which have persisted over recent years concerning whether or not the arbitration agreement is incorporated in the contract[31]. The more liberal definition was intended to widen scope in the Arbitration Act 1996 of what constituted the essential criteria for writing and written agreements.

Sub-section 107(1) deals with the need for the construction contract to be in writing. There are two parts to this sub-section.

The first refers to the construction contract being in writing and the second to 'any other agreement between the parties as to any matter is effective for the purposes of this Part only if in writing'. A construction contract that is not in writing will not be subject to the provisions of the Act. Such a contract may be actionable in law but the parties will not receive the benefits and rights that the Act provides.

The second part of this sub-section probably gives wider scope to the Act than was ever intended. 'Any other agreement as to any matter' will include amendments the parties make to their contract. If such an amendment were made it would have to be in writing. The agreement may lack some of the common law characteristics to make it a contract but its scope would nevertheless be within the Act. The parties may have been silent as to their terms of payment within the original contract and rather than rely on the default provisions of the Act they may agree terms at a later date. This new agreement recorded in writing would then be governed by the Act.

It is arguable that agreements that rely on a quantum meruit for payment would be within the Act, if recorded in writing. The use of letters of intent is common in the construction industry. Often the letter of intent will cover the whole of the relationship, as it is never replaced by a contract. Depending on the terms, they may be sufficient in a letter of intent to bring the relationship within the Act.

Section 107(2)

(2)   There is an agreement in writing–
  (a)   if the agreement is made in writing (whether or not it is signed by the parties),
  (b)   if the agreement is made by exchange of communications in writing, or
  (c)   if the agreement is evidenced in writing.

Sub-section 107(2) gives the definition of what constitutes an agreement in writing. Sub-section (a) is straightforward where the parties have both signed the construction contract. It is not unusual in the construction industry for contract documents to be prepared and then to remain unsigned by one or both of the parties. In this case it will be for the party seeking to rely on the written agreement to prove that this in fact was the agreement made.

This was examined in Oakley[32] where although there was standard documentation which the parties had made some attempt to complete, there was insufficient evidence of their agreement to allow recovery of the amount in an adjudicator's decision to be pursued by a statutory demand.

---

[31] See *Aughton Ltd v. M.F. Kent Services Ltd* [1991] 57 BLR 1; *Ben Barrett & Son (Brickwork) Ltd v. Henry Boot Management* [1995] CILL 1026; *Lexair Ltd (in administrative receivership) v. Edgar Taylor Ltd* 65 BLR 87; *Smith & Gordon Ltd v. John Lewis Building Ltd* [1994] CILL 934, CA; *Giffen (Electrical Contractors) v. Drake & Scull Engineering Ltd* (1993) 37 Con LR 84; *Alfred McAlpine Construction Ltd v. R.M.G. Electrical* [1994] 29 Bliss 2; *Extrudakerb (Maltby Engineering) Ltd v. White Mountain Quarries Ltd.*, QBD Northern Ireland (18 April 1996) TLR 10 July 1996 p23.
[32] *(1) William Oakley (2) David Oakley v. (1) Airclear Envirionmental Limited (2) Airclear TS Limited* (4 October 2001).

Sub-section (b) deals with a commonplace occurrence. It is not unusual for the agreement to be reached by an exchange of documents in writing, letters, facsimile, orders, etc. There may, in the final analysis, be some difficulty as to final terms as in 'battle of the forms'[33] cases but this is no different to other non-construction contractual relationships.

Sub-section (c) only requires that the agreement is evidenced in writing. This may be something as sketchy as an invoice that refers back to an agreement[34]. This was our original view and was supported at first instance in *RJT Consulting Engineers* v. *DM Engineering*[35]:

'What the defendants to this application say is that the terms of section 107 are a widening process. In their skeleton argument DM say: ''Evidence in writing of a contract dealt with by sub-section 107(2)(c) leaves the door open for such evidence to come into existence after the commencement, or even the completion, of the contract's performance. Thus, an invoice submitted by one party to the other may be sufficient evidence, as might a confirmation of verbal instructions set out in the letter.'' . . .

It seems to me that if I were to find that it is necessary to have a recitation of the terms of an agreement when the existence of the agreement, the parties to the agreement and the nature of the work and the price of the agreement are plainly to be found in documentary form, but nonetheless in a contract worth more than three-quarters of a million pounds because the initial agreement was oral, it is not caught by the Act, then it seems to me such an attempt would run contrary not only to the terms of the Act but contrary to my duty to carry out what I believe to be the law at any particular time. And therefore, adopting that methodology, I hold that it is not necessary to have the terms identified and the extensive documentary evidence in this case is well sufficient to bring it within the adjudication proceedings and therefore I refuse this declaration.'

At the time of this judgment this appeared to support the interpretation of section 107 in the first edition of this book in that it gave wide scope as to what might constitute an agreement in writing. The Court of Appeal, however, re-examined this case and took a different view on what was necessary to be recorded in writing[36]:

'LJ WARD — On the point of construction of section 107, what has to be evidenced in writing is, literally, the agreement, which means all of it, not part of it. A record of the agreement also suggests a complete agreement, not a partial one. The only exception to the generality of that construction is the instance falling within sub-section 5 where the material or relevant parts alleged and not denied in the written submissions in the adjudication proceedings are sufficient. Unfortunately, I do not think sub-section 5 can so dominate the interpretation of the section as a whole so as to limit what needs to be evidenced in writing simply to the material terms raised in the arbitration. It must be remembered that by virtue of section 107(1) the need for an agreement in writing is the precondition for the application of the other provisions of Part II of the Act, not just the jurisdictional threshold for a reference to adjudication. I say ''unfortunately'' because, like Auld LJ whose judgment I have now read in draft, I would regard it as a pity if too much ''jurisdictional wrangling'' were to limit the opportunities for expeditious adjudication having an interim effect only. No doubt adjudicators will be robust in excluding the trivial from the ambit of the agreement and the matter must be entrusted to their common sense. Here we have a comparatively simple oral agreement about the terms of which there may be very little, if any, dispute. For the consulting engineers to take a point objecting to adjudication in those circumstances may be open to the criticism that they were taking a technical point but as it was one open to them and it is good, they cannot be faulted. In my judgment they were

---

[33] See *Petredeck Ltd* v. *Takumuru Kaiun Co. 'The Sargasso'* [1994] 1 Lloyd's Rep 162, for agreement evidenced in writing.
[34] *Butler Machine Tool Co Ltd* v. *Ex-Cell-O Corporation (England) Ltd* [1979] 1 WLR 401.
[35] *RJT Consulting Engineers Ltd* v. *DM Engineering (NI) Ltd* Liverpool TCC (9 May 2001).
[36] *RJT Consulting Engineers* v. *DM Engineering (NI) Ltd*, CA (8 March 2002).

entitled to the declaration which they sought and I would accordingly allow the appeal and grant them that relief.

LJ ROBERT WALKER – I agree that this appeal should be allowed for the reasons set out in the judgment of Ward LJ. It is the terms, and not merely the existence, of a construction contract which must be evidenced in writing. The judge aimed at a purposive approach but he did not in my view correctly identify the purpose of section 107...

LJ AULD – Although clarity of agreement is a necessary adjunct of a statutory scheme for speedy interim adjudication, comprehensiveness for its own sake may not be. What is important is that the terms of the agreement material to the issue or issues giving rise to the reference should be clearly recorded in writing, not that every term, however trivial or unrelated to those issues, should be expressly recorded or incorporated by reference. For example, it would be absurd if a prolongation issue arising out of a written contract were to be denied a reference to adjudication for want of sufficient written specification or scheduling of matters wholly unrelated to the stage or nature of the work giving rise to the reference. There may be cases in which there could be dispute as to whether all the terms of the agreement material to the issues in the sought reference are in writing as required by section 107 and it could defeat the purpose of the Act to clog the adjudicative process with jurisdictional wrangling on that account. However, there will be many cases where there can be no sensible challenge to the adequacy of the documentation of the contractual terms bearing on the issue for adjudication, or as to the ready implication of terms common in construction contracts. Section 107(5) is an illustration of the draftsman's intention not to shut out a reference simply because the written record of an agreement is in some immaterial way incomplete. It provides that an exchange of written submissions in proceedings in which the existence of an agreement otherwise than in writing is alleged by one party and not denied by the other constitutes an agreement in writing ''to the effect alleged''. If the effect of the agreement so alleged contains all the terms material to the issue for adjudication, that procedure is available notwithstanding that the agreement contains other terms not in writing which are immaterial to the issue. As Ward LJ has observed, the exchange constitutes an agreement in such terms as it may be material to allege for the purpose of the particular adjudication. In my view, it would make no sense to confine that sensible outcome to the written form of agreement provided by section 107(5) whilst excluding it in the other forms for which the section provides.'

It seems to us that this judgment has not brought clarity to what constitutes the agreement in writing for purposes of the Act. There is a conflict here between 'what has to be evidenced in writing is, literally, the agreement, which means all of it, not part of it', against the comments of Robert Walker LJ which we prefer: 'Although clarity of agreement is a necessary adjunct of a statutory scheme for speedy interim adjudication, comprehensiveness for its own sake may not be. What is important is that the terms of the agreement material to the issue or issues giving rise to the reference should be clearly recorded in writing, not that every term, however trivial or unrelated to those issues, should be expressly recorded or incorporated by reference.'

To read these potentially conflicting statements in context it must be remembered that the terms of an agreement can be minimal. It is not essential, for example, that time for performance be recorded in the terms of a construction contract and the payment terms in the Scheme may well be an adequate provision on an implied basis where they are otherwise absent.

In *Debeck* v. *T&E Engineering*[37] Judge Kirkham applied *RJT* on the basis that all the terms relevant to the claimant's claim were to be recorded in writing.

---

[37] *Debeck Ductwork Installation Ltd* v. *T&E Engineering Ltd* (14 October 2002).

In *Ballast* v. *Burrell*[38] the parts of an agreement that were in writing and those parts which were not were examined:

> 'Counsel for the respondents adopted a broader approach. He submitted that the claim as originally focused in the notice of adjudication and the subsequent referral notice was a claim for valuation of works performed within the terms of the contract. It was necessary, he submitted, to recognise the broad terms of section 107 with regard to the importance or requirement that contractual terms relied upon had to be in writing. It was not necessarily restricted, he said, to the original terms of the contract, having regard particularly to subsections 2 and 3 of section 107 of the Act. It could, he submitted, be extended to instructions subsequently reduced to writing. He accepted that there might be some parts of the claim that were encompassed by the referral notice which were not in writing or at least not supported by section 107. However, that was a matter for the adjudicator to determine as part of the exercise of his function.'

An agreement formed partly in writing and partly orally was examined again in *Cowlin Construction* v. *CFW Architects*[39]:

> '74. It appears that the contract was made partially in writing and partly orally on 21 June 2000. It is clearly evidenced in writing. Pursuant to section 107(2) of HGCRA, this was a construction contract.
> 75. It would not normally be relevant to consider documents which were created after the contract is said to have been formed, but Mr Brannigan submits that I should have regard to the correspondence after June 2000 and to the documents in the first adjudication which, he submits, throw light on whether the parties considered there was a contract. These all reinforce my conclusion as to the formation of the contract. The invoices all indicate an acceptance by CFW that there was a contract in place, as does the letter from CFW dated 29 August 2000 which makes a number of specific references to the contract.'

Section 107(3)

> (3) Where parties agree otherwise than in writing by reference to terms which are in writing, they make an agreement in writing.

Sub-section 107(3) would include oral agreements (or other non-written means) as writing insofar as they refer to written terms. There would need to be evidence that an agreement had been reached between the parties.

For example, if parties agree orally to adopt DOM/1 as the terms of their contract this would be within the provisions of the Act. It is however for the party averring that the oral contract exists, to prove it. The terms agreed can be partly written and partly oral. There is no requirement that the agreement refer to one document only. The oral agreement can refer to several documents.

The case of *Total* v. *ABB Building*[40] gives an interesting view on the scope of what amounts to oral agreements under this sub-section. The contract contained no clause to deal with variations or extra work.

The judge dealt with the matter in a pragmatic way:

> '34. What has to be considered here is not the enforceability of the contract but whether the statutory adjudication scheme can be invoked in relation to a particular construction contract. That is governed by section 107 of the Act. (supra) There is reference in sub-section 3 to an agreement

---

[38] *Ballast PLC* v. *The Burrell Company (Construction Management) Limited* (21 June 2001).
[39] *Cowlin Construction Limited* v. *CFW Architects (a firm)* (15 November 2001).
[40] *Total M & E Services Limited* v. *ABB Building Technologies Limited (formally ABB Steward Limited)* (26 February 2002).

otherwise than in writing, such an agreement, provided it refers to terms which are in writing, is an agreement in writing. In my judgment; the adjudicator made his decision on the basis of dispute arising out of the single written construction contract as varied orally by the parties. The contract as varied is clearly within the provisions of section 107. Notwithstanding that it is a contract evidenced partly in writing and partly oral. The adjudicator therefore had jurisdiction to make determinations as to the additional works.'

An appeal from this judgment was commenced but abandoned when the parties finally settled their differences.

For a different view see *Carillion* v. *Devonport*,[41] which was decided after *RJT Consulting* v. *DM Engineering* where the key to the argument was whether the project was cost reim-bursable. As this was a material term, following the Court of Appeal decision in *RJT Con-sulting*, that term must have been evidenced in writing for the dispute to be referable to adjudication. What was in issue was an alleged oral agreement that radically changed the written agreement. The change was far greater than a typical variation made pursuant to the terms of a construction contract. Thus the judge held that the adjudicator did not have jurisdiction.

Section 107(4)

(4)   An agreement is evidenced in writing if an agreement made otherwise than in writing is recorded by one of the parties, or by a third party, with the authority of the parties to the agreement.

Sub-section 107(4) deals with what evidences a written agreement. The important qualifi-cation here is that any means of recording the agreement has to be with the authority of the parties. Without such authority the recorded evidence cannot be used. There is nothing to say that the authority of the parties to the agreement has to be recorded in writing. The authority could be given orally. A simple example of a compliant record is the minutes of a meeting prepared by a third party with the consent of the contracting parties.

The concept of one of the parties obtaining permission to make a record of the agreement was examined in *Millers* v. *Nobles*[42]:

'Miss Pennifer submitted that there was no evidence that Mr Dalton had ever been authorised by Mr Dunbar to record in writing what had been agreed and that accordingly section 107(4) had not been satisfied. That submission is I consider correct. There is nothing to indicate that Mr Dunbar or anyone else from the claimant ever authorised Mr Dalton to write the letter as a record of what had been agreed and I do not accept that such an inference should be drawn from the use of the word "confirm". On its face the letter is simply a letter written by one party recording what he understood had been agreed. The fact that Mr Dunbar also wrote a similar letter to Mr Dalton also indicates to me that he had not authorised Mr Dalton to record the terms on his behalf. The inference is that he intended to provide Mr Dalton with his own record of what had been agreed. This would have been entirely unnecessary if it had been understood expressly or impliedly that Mr Dalton had been authorised to produce the record of what had been agreed. Paragraph 7 of Mr Scarisbrick's witness statement cannot be treated as a ratification of any lack of authority because Mr Dalton did not purport to write the letter as agent.'

---

[41] *Carillion Construction Ltd* v. *Devonport Royal Dockyard Ltd* TCC (27 November 2002).
[42] *Millers Specialist Joinery Company Limited* v. *Nobles Construction Limited* (3 August 2001).

Section 107(5)

(5) An exchange of written submissions in adjudication proceedings, or in arbitral or legal proceedings in which the existence of an agreement otherwise than in writing is alleged by one
party against another party and not denied by the other party in his response constitutes as
between those parties an agreement in writing to the effect alleged.

Sub-section 107(5) does not create an agreement where there was no previous agreement,
save if a party fails to deny the existence of that agreement in proceedings. This sub-section
supports agreements made otherwise than in writing. Where in adjudication, arbitral or
legal proceedings one party in its response fails to deny the existence of the agreement, this
admits that the agreement exists and is binding for the purposes of the Act. It is not binding
where the party fails to respond at all. There must be a response that fails to deny the
existence of the agreement. This may have the effect of creating an *ex post facto* (by
subsequent act) agreement in writing. A mere exchange of statements therefore may be
sufficient to create an agreement. English authority under the Arbitration Act 1950 was in
favour of such solutions in construing arbitration agreements. The mere fact that an
exchange of submissions exists may also be sufficient to create an ad hoc agreement to
arbitrate[43].

At first instance in *Grovedeck*[44] v. *Capital Demolition* Judge Bowsher QC gave an
interpretation of this subsection that we would not have anticipated:

'29. I think this is a case where it is permissible, following the decision of the House of Lords in *Pepper*
v. *Hart* [1993] AC 593, to look at Hansard. It appears from the Hansard Report of the proceedings in
the House of Lords for 23 July 1996 that section 107(5) originally contained no reference to adjudication proceedings. The House of Lords accepted a Commons amendment that after the word
"submissions" there should be inserted the words "in adjudication proceedings or". If one reads
section 107(5) without the words "in adjudication proceedings or" it is clear that the intention of
Parliament was that a contract should be treated as a contract in writing if in arbitral or litigation
proceedings before the adjudication proceedings in question an oral contract had been alleged and
admitted. I also would read the words "and not denied" as meaning that the alleged terms of the
contract were not denied. By adding the words "in adjudication proceedings or", Parliament
intended to add a reference to other preceding adjudication proceedings. There was no intention by
Parliament to provide that submissions made by a party to an unauthorised adjudication should give
to the supposed adjudicator a jurisdiction which he did not have when he was appointed.'

For subsection 107(5) to apply, the allegation that an agreement existed would have to have
been made in previous proceedings and not in a current adjudication.

This is not the interpretation placed on this subsection in *A&D Maintenance* v. *Pagehurst*[45]:

'15. In the course of the lengthy submissions, both parties made reference to the sub-contract.
Whilst the parties do not agree upon the precise terms evidenced by the sub-contract confirmation
form, nonetheless there is sufficient, in my judgement to warrant finding that those exchanges in the
reply and response complied with section 107(5) of the Act. In other words there was a further and
alternative basis for holding that there was an agreement in writing under the Act to which the
Scheme applied.'

---

[43] See *Altco Ltd* v. *Sutherland* [1971] 2 Lloyd's Rep 515 where Donaldson J had been of the view that the agreement
remained oral, but the contrary appears to have been decided in *Jones Engineering Services* v. *Balfour Beatty Building Ltd*
[1994] ADRLJ 133.
[44] *Grovedeck Limited* v. *Capital Demolition Limited* (24 February 2000).
[45] *A & D Maintenance and Construction Limited* v. *Pagehurst Construction Services Limited* (23 June 1999).

This seems to conflict with *Grovedeck*. It also seems that Judge Bowsher's views may not be supported by the Court of Appeal's decision in *RJT*[46]:

> 'The only exception to the generality of that construction is the instance falling within sub-section 5 where the material or relevant parts alleged and not denied in the written submissions in the adjudication proceedings are sufficient.'

Section 107(6)

> (6) References in this Part to anything being written or in writing include its being recorded by any means.

Recorded by any means gives immense scope in terms of present day technology. Any imaginable means of recording will apply here. The only exception is in sub-section 107(4) above where permission of the parties is required to make the record. Save for that sub-section the possibility of recording without a party's permission is permissible, the only proviso being compliance with 'the rules of evidence' to establish the validity of the recording. Strict rules of evidence do not apply in adjudication but there has to be an evidential worth and basis for anything which is sought to be proved.

The supplementary provisions

The supplementary provisions provide the necessary definitions and powers to make Part II of the Act operative.

Section 114: The Scheme for Construction Contracts

Section 114(1)

> (1) The Minister shall by regulations make a scheme ('the Scheme for Construction Contracts') containing provision about the matters referred to in the preceding provisions of this Part.

At the time the Act was enacted there was no Scheme in place. The Scheme for Construction Contracts was to be made at a later date by regulation.

Section 114(2)

> (2) Before making any regulations under this section the Minister shall consult such persons as he thinks fit.

There was a wide consultation process throughout the industry and professions on the content and drafting of the Scheme. The first draft was rejected on the grounds that, among other matters, it resembled too closely arbitration rather than the adjudication the industry was seeking. The final consultation document, entitled *Making the Scheme for Construction Contracts* was issued by the Department of the Environment in November 1996. Commencement Order 1998 No.649 made the Act and the Scheme became operative from 1 May 1998.

---

[46] *RJT Consulting Engineers Ltd* v. *DM Engineering (NI) Ltd*, CA (8 March 2002).

Section 114(3) and (4)

    (3)   In this section 'the Minister' means–
        (a) for England and Wales, the Secretary of State, and
        (b) for Scotland, the Lord Advocate.
    (4)   Where any provisions of the Scheme for Construction Contracts apply by virtue of this Part in default of contractual provision agreed by the parties, they have effect as implied terms of the contract concerned.

Sub-section (4) deals with the status of the Scheme. The Scheme is of effect by default where contracts fail to comply with some provision of the Act. The regulations in the Scheme have the effect of implied terms under the contract in question. They are not required to satisfy the legal tests for implying terms. There seems to be no doubt by this terminology that the parties cannot contract out of the effects of the Act.

Section 114(5)

    (5)   Regulations under this section shall not be made unless a draft of them has been approved by resolution of each House of Parliament.

Section 115: Service of notices, &c.

Section 115(1)

    (1)   The parties are free to agree on the manner of service of any notice or other document required or authorised to be served in pursuance of the construction contract or for any of the purposes of this Part.

This deals with the service of notices referred to in all of the preceding sections. The service of notices should form part of the contract although there is nothing to prevent the parties agreeing the manner of service subsequently.

Section 115(2)–(4)

    (2)   If or to the extent that there is no such agreement the following provisions apply.
    (3)   A notice or other document may be served on a person by any effective means.
    (4)   If a notice or other document is addressed, pre-paid and delivered by post–
        (a) to the addressee's last known principal residence or, if he is or has been carrying on a trade, profession or business, his last known principal business address, or
        (b) where the addressee is a body corporate, to the body's registered or principal office, it shall be treated as effectively served.

If there is no manner of service agreed, these provisions apply in default. Although a notice or document can be served by any effective means, sub-section (4) gives the position on valid service by post. Save for any provision in the contract there is no reason that service should not be effected by facsimile or e-mail. What was envisaged in the *Strathmore* v. *Hestia Fireside*[47] case was written form.

---

[47] *Strathmore Building Services Limited* v. *Colin Scott Greig t/a Hestia Fireside Design* (18 May 2000).

Section 115(5)

> (5) This section does not apply to the service of documents for the purposes of legal proceedings, for which provision is made by rules of court.

The service of documents in respect of legal proceedings is excluded here, the appropriate rules of court being applicable.

Section 115(6)

> (6) References in this Part to a notice or other document include any form of communication in writing and references to service shall be construed accordingly.

A notice or document includes any form of communication in writing. This is a minimum requirement and may be modified by the contract.

Section 116: Reckoning periods of time

Section 116(1)–(3)

> (1) For the purposes of this Part periods of time shall be reckoned as follows.
> (2) Where an act is required to be done within a specified period after or from a specified date, the period begins immediately after that date.
> (3) Where the period would include Christmas Day, Good Friday or a day which under the Banking and Financial Dealings Act 1971 is a bank holiday in England and Wales or, as the case may be, in Scotland, that day shall be excluded.

This section deals with the reckoning of periods of time. The periods of time are reckoned after any specified date. Therefore if a notice is issued on a given date, time begins to run on the following day. Weekends are not excluded from the periods for reckoning time. If a notice is served on a Saturday, time begins to run on the Sunday. Only bank holidays, Christmas Day and Good Friday are excluded from the days to be counted in reckoning time. If a notice is issued prior to the industry close-down at Christmas, the periods of time will run over that break period.

Section 117: Crown application

Section 117(1)–(4)

> (1) This Part applies to a construction contract entered into by or on behalf of the Crown otherwise than by or on behalf of Her Majesty in her private capacity.
> (2) This Part applies to a construction contract entered into on behalf of the Duchy of Cornwall notwithstanding any Crown interest.
> (3) Where a construction contract is entered into by or on behalf of Her Majesty in right of the Duchy of Lancaster, Her Majesty shall be represented, for the purposes of any adjudication or other proceedings arising out of the contract by virtue of this Part, by the Chancellor of the Duchy or such person as he may appoint.
> (4) Where a construction contract is entered into on behalf of the Duchy of Cornwall, the Duke of Cornwall or the possessor for the time being of the Duchy shall be represented, for the purposes of any adjudication or other proceedings arising out of the contract by virtue of this Part, by such person as he may appoint.

The Crown as an entity is included within the provisions of the Act. Contracts entered into in the private capacity of the Queen are exempt from the provisions of the Act. Contracts entered into by the Duchy of Lancaster, Duchy of Cornwall or the Duke of Cornwall are included within the provisions but appointees will deal with any matter under the Act should it arise.

Part V of the Act

There are general provisions in Part V of the Act that affect Part II.

Section 146: Orders, regulations and directions

(1) Orders, regulations and directions under this Act may make different provision for different cases or descriptions of case, including different provision for different areas.

(2) Orders and regulations under this Act may contain such incidental, supplementary or transitional provisions and savings as the Secretary of State considers appropriate.

(3) Orders and regulations under this Act shall be made by statutory instrument which, except for–
   (a) orders and regulations subject to affirmative resolution procedure (see sections 104(4), 105(4), 106(4) and 114(5)),
   (b) orders under section 150(3), or
   (c) regulations which only prescribe forms or particulars to be contained in forms,
  shall be subject to annulment in pursuance of a resolution of either House of Parliament.

There is potential for further orders to be made except under sections 104, 105, 106, 114 and 150(3). The orders will be made by statutory instrument. The orders made under Part II cannot be annulled pursuant to a resolution of either House of Parliament.

Section 148: Extent

(1) The provisions of this Act extend to England and Wales.

(2) The following provisions of this Act extend to Scotland – Part II (construction contracts), Part III (architects), sections 126 to 128 (financial assistance for regeneration and development), and Part V (miscellaneous and general provisions), except–
   (i) sections 141, 144 and 145 (which amend provisions which do not extend to Scotland), and
   (ii) Part I of Schedule 3 (repeals consequential on provisions not extending to Scotland).

(3) The following provisions of this Act extend to Northern Ireland – Part III (architects), and Part V (miscellaneous and general provisions), except–
   (i) sections 142 to 145 (home energy efficiency schemes and residuary bodies), and
   (ii) Parts I and III of Schedule 3 (repeals consequential on provisions not extending to Northern Ireland).

(4) Except as otherwise provided, any amendment or repeal by this Act of an enactment has the same extent as the enactment amended or repealed.

The Act applies to England, Wales and Scotland with the stated exceptions.

Section 149: Corresponding provision for Northern Ireland

An Order in Council under paragraph l(l)(b) of Schedule I to the Northern Ireland Act 1974 (legislation for Northern Ireland in the interim period) which states that it is made only for purposes corresponding to those of Part II (construction contracts) or section 142 (home energy efficiency schemes)–
(a) shall not be subject to paragraph 1(4) and (5) of that Schedule (affirmative resolution of both Houses of Parliament), but
(b) shall be subject to annulment in pursuance of a resolution of either House of Parliament.

Corresponding provisions were made to bring Part II into operation in Northern Ireland. The Construction Contracts (Northern Ireland) Order 1997 No. 274 (NI 1) is the instrument which commences the process of providing legislation for Northern Ireland. The commencement date for Northern Ireland is 1 June 1999.

Section 150: Commencement

(1) The following provisions of this Act come into force on Royal Assent–
     section 146 (orders, regulations and directions),
     sections 148 to 151 (extent, commencement and other general provisions).
(2) The following provisions of this Act come into force at the end of the period of two months beginning with the date on which this Act is passed–
     sections 126 to 130 (financial assistance for regeneration and development), section 141 (existing housing grants: meaning of exempt disposal),
     section 142 (home energy efficiency schemes), sections 143 to 145 (residuary bodies),
     Part III of Schedule 3 (repeals consequential on Part IV) and section 147 so far as relating to that Part.
(3) The other provisions of this Act come into force on a day appointed by order of the Secretary of State, and different days may be appointed for different areas and different purposes.
(4) The Secretary of State may by order under subsection (3) make such transitional provision and savings as appear to him to be appropriate in connection with the coming into force of any provision of this Act.

None of Part II of the Act came into effect on Royal Assent or in the two months after Assent mentioned in sub-section (2). Part II became operative on 1 May 1998.

Section 151: Short title

This Act may be cited as the Housing Grants, Construction and Regeneration Act 1996.

The parts of the Act that concern construction contracts, adjudication and payment provisions are often referred to as the Construction Act in the trade and professional press.

Chapter 3
The Act: The Adjudication Provisions Section 108

The statutory right to adjudication applies only to construction contracts as defined in sections 104 to 107 inclusive. Section 108 sets the parameters for the provisions which must be included in construction contracts to allow resolution of disputes by adjudication. These are the minimum compliance points to meet requirements of the Act. There is nothing to prevent the parties contracting on the basis of more than the Act requires, providing this does not lead to conflict with the compliance points.

Section 108: Right to refer disputes to adjudication

Section 108(1)

(1) A party to a construction contract has the right to refer a dispute arising under the contract for adjudication under a procedure complying with this section.
For this purpose 'dispute' includes any difference.

Section 108 provides that any party to a construction contract has the right to refer a dispute arising under the contract to adjudication. It should be noted that this is a statutory right but it is not mandatory that disputes be referred to adjudication. The contract may contain clauses which offer other means of dispute resolution. The parties may therefore have a number of methods and forums available in which to resolve their disputes.

The terms dispute or difference have become extremely important as they go to the basis of the adjudicator's jurisdiction. If there is no dispute there can be no adjudication. This point has been considered in many of the recent cases relating to the enforcement of adjudicators' decisions that have reached the court. For a wider discussion see Chapter 9.

It is common practice in the industry that parties contract on amended standard forms or on hybrid forms of contract and it is not unusual for the party offering the contract to make its own terms that might, among other matters, intend to exclude the right to adjudication. Can a party do this?

The commentary in *Chitty on Contracts*[1] on this point is as follows:

'Although there is a general principle that a person may waive any right conferred on him by statute (*quilibet protest renunciare juri pro se introducto*) difficulties arise in determining whether the right is exclusively personal or is designed to serve other more broad public purposes. In the latter situation, public policy would require that the right be treated as mandatory and not be waivable by the party

---

[1] See *Chitty on Contracts*, 28th edn at p.845.

for whose benefit it operates. Whether a statutory right is waivable depends on the overall purpose of the statute and whether this purpose would be frustrated by permitting waiver.

Thus in *Johnson* v. *Moreton* [1980] AC 37 the House of Lords held that a tenant could not contract out of the protection afforded by s.24 of the Agricultural Holdings Act 1948 (c. 63) as this would undermine the overall purpose of the Act in prompting efficient farming in the national interest:

"The principle which, in my view, emerges . . . is as follows. Where it appears that the mischief which Parliament is seeking to remedy is that a situation exists in which the relations of parties cannot properly be left to private contractual regulation, a party cannot contract out of such statutory regulation (albeit exclusively in its own favour) because so to permit would be to reinstate the mischief which the statute was designed to remedy and to render the statutory provision a dead letter".'[2]

It is almost inconceivable therefore that any attempt to contract out of the statutory right to adjudication would be binding. See also the comments in sub-section 108(5) which would override any such provision in a contract.

The right to refer a dispute to adjudication only applies to disputes arising under the contract. There are a number of cases in the law relating to arbitration which deal with disputes arising under the contract as opposed to disputes which arise under more widely drafted arbitration clauses.

'There is, I suggest, a broad distinction which may be drawn between those clauses which refer to arbitration only, those disputes which may arise regarding the rights and obligations which are created by the contract itself, and other clauses which show an intention to refer some wider class or classes of disputes. This distinction is obviously clear and justined as a matter of law. It may also be one which would be recognised by the parties whose contract it is, for at the very least, by making the contract, they demonstrate their agreement to create a new category of legal rights and obligations, legally enforceable between themselves. Disputes regarding this category may well be described, as a matter of language, as ones arising "under" the contract, and this meaning of that phrase has been authoritatively recognised and established, e.g. by the House of Lords in *Heyman* v. *Darwins* ([1942] 1 All ER 337, [1942] AC 356) and by the Court of Appeal in Ashville. Conversely, if the parties agree to refer disputes arising "in relation to" or "in connection with" their contract, a fortiori if the clause covers disputes arising "during the execution of this contract" (*Astro Vencedor Cia Naviera SA* v. *Mabanaft GmbH, The Damianos* [1971] 2 All ER 1301, [1971] 2 QB 588) or in relation to "the work to be carried out hereunder" a common form in construction contracts, then both as a matter of language and of authority some wider category may be intended'[3]

'Such a dispute (about mistake leading to rectification) is not as to any matter or thing ... arising under the contract ... Similarly, a dispute between the parties as to whether an innocent mis-representation, or negligent mis-statement, which led Ashville to enter into the contract . . . [is not] a dispute as to any matter arising under the contract.'[4]

'In my judgment, on the ordinary and natural meaning of words, the phrase "disputes arising under a contract" is not wide enough to include disputes which do not concern obligations created by or incorporated in that contract.'[5]

'Taking into account previous authorities, the court considered that the meaning of "arising out of", and "arising in connection with" were synonymous. It must be presumed that the parties had intended to refer all disputes arising from the transaction to arbitration and this must include rec-

---

[2] *Johnson* v. *Moreton* [1980] AC 37, Lord Simon of Glaisdale at p.69.
[3] *Overseas Union Insurance* v. *AA Mutual International Insurance* [1988] 2 Lloyd's Rep 63, Evans J.
[4] *Ashville Investments* v. *Elmer Contractors* (1987) 10 Con LR 72, Balcombe LJ.
[5] *Fillite (Runcorn)* v. *Aqua-Lift* (1989) 26 Con LR 66.

tification. This was emphasised by the provisions of the sub-clause in the GAFTA agreement. Following the decision in *Fillite (Runcorn) Ltd* v. *Aqua-Lift* (1989) 45 BLR 27, ''arising under'' alone would probably not cover rectification. However, ''arising out of'' in this context should be given a wide interpretation so as to cover all disputes. Rectification therefore being within the scope of the clause, the stay of proceedings would be granted.'[6]

The debate on the various phrases used in such clauses is almost interminable. The cases point to the phrase 'arising under the contract' being of narrower application than phrases such as 'arising out of the contract'.[7] 'In connection with the contract' has a much wider meaning. This could include matters such as misrepresentation and tort.

This is a matter that the adjudicator must determine in deciding whether or not he has jurisdiction. The only real guidance is the current case law on arbitration clauses.

In the case of *Christiani & Neilsen* v. *The Lowry Centre*[8] a question relating to claims that are wider than 'under' the contract was touched on briefly.

Under a heading of 'The Fifth Issue – Is the decision unenforceable because it decides a question concerning the rectification of the deed which the adjudicator had no power to decide?' Judge Anthony Thornton QC made the following comments:

'It follows that the adjudicator had full jurisdiction to decide the dispute in the way he did. It also follows that I do not have to decide whether an adjudicator appointed under the provisions of the HGCRA has the power to decide a claim for rectification. Such a claim would have arisen, for example, had The Lowry sought the appointment of an adjudicator to decide its entitlement to rectification. It was contended by The Lowry that since the adjudicator was appointed, by virtue of section 108(1) of the HGCRA, to decide disputes ''under the contract'', he could not have decided disputes that were merely connected with the contract. This argument is dependent on the line of authorities which are concerned with the limited jurisdiction of arbitrators whose jurisdiction is derived from an arbitration clause which refers only to disputes arising ''under the contract''. Such a clause has been held not to extend to disputes involving misrepresentation or rectification claims or claims as to the ambit and content of the contract. Interesting and important as this question is, it does not arise in this case and I express no view about it.

The ambit of this doctrine is explored in such cases as *Heyman* v. *Darwins* [1942] AC 356, HL(B); *Ashville Investments Ltd* v. *Elmer Contractors* (1987) 37 BLR 55, CA; *Overseas Union Insurance* v. *AA Mutual International Insurance Co Ltd* [1998] 2 Lloyd's Rep 62, Evans J; *Fillite (Runcorn) Ltd* v. *Aqua – Lift (a firm)* (1989) 45 BLR 32, CA.'

The important point is the distinction between 'under the contract' and 'in connection with the contract'. It is clear that an adjudicator cannot deal with rectification of the contract. Here correctly he did not do so.

It is unlikely that the phrase 'arising under the contract' is sufficiently wide to permit an adjudicator to open up, review and revise certificates of architects or engineers although there may be a specific provision in the contract that this may be done. Adjudication provisions, if they follow the minimum requirements of the Act, will simply not give the wide authority that the arbitration clause gives. An adjudicator would not therefore be in a position to open up, review and revise an architect's certificate and to substitute figures of his own.

Adjudicators can be faced with main contract disputes that concern dissatisfaction with either the amount certified or the extensions of time granted. This is less of a problem under

---

[6] *Ethiopian Oil Seed & Pulses Export* v. *Rio Del Mar Foods Inc* [1990] 1 Lloyd's Rep 86.
[7] *Arbitration*, May 1994, S K Chatterjee, 'Do Disputes Arise ''out of'' or ''under'' or ''out of and under'' a Contract?.'
[8] *Christiani & Neilsen Limited* v. *The Lowry Centre Development Company Limited* (16 June 2000).

domestic sub-contract arrangements because there is no certifier as such. It would remain a problem under a nominated sub-contract where certification is involved in the payment process and to an extent in the extension of time process.

It was thought to have been clear since the *Crouch* case[9] that the powers of the courts are not the same as those of an arbitrator.

> SIR JOHN DONALDSON MR: 'Despite the fact that the architect is subject to a duty to act fairly, these powers might be regarded as Draconian and unacceptable if they were not subject to review and revision by a more independent individual. That process is provided for by the arbitration clause. It is however, a rather special clause. Arbitration is usually no more and no less than litigation in the private sector. The arbitrator is called upon to find the facts, apply the law and grant relief to one or other or both parties. Under a JCT arbitration clause (clause 35), the arbitrator has these powers but he also has the power to "open up, review and revise any certificate, opinion, decision, requirement or notice". This goes far further than merely entitling him to treat the architect's certificates, opinions, decisions, requirements and notices as inclusive in determining the rights of the parties. It enables, and in appropriate cases requires, him to vary them and so create new rights, obligations and liabilities in the parties. This is not a power normally possessed by any court and again it has a strong element of personal judgement by an individual nominated in accordance with the agreement of the parties.'[10]

It is this judgment which gave both the courts and arbitrators difficulty without express powers such as those found in the JCT arbitration clause. A certificate could not be opened up, reviewed or revised without express terms to do so. Arbitrators found this power in the arbitration clauses in construction contracts. The courts have had some relief in this situation. Section 43A of the Supreme Court Act 1981 (inserted by section 100 of the Courts and Legal Services Act 1990) allows the courts to take on the powers of an arbitrator under the arbitration agreement provided both parties consent. The *Crouch* case has now been overruled by the House of Lords, in *Beaufort Developments* v. *Gilbert-Ash*[11]. The courts do have an inherent jurisdiction to open up, review and revise certificates under construction contracts. Without express terms in the contract an arbitrator or adjudicator would not have such powers. Most of the standard forms of contract and the Scheme do however give adjudicators such powers.

There are no cases since *Crouch* that would provide such powers without an express provision. There is some relief in this situation. There is no doubt that without the express power under the contract, there is no basis on which an adjudicator can open up and review certificates. The effect of this decision was examined in detail by Mr Recorder Roger Toulson QC in *John Barker Construction Limited* v. *London Portman Hotels Limited*[12]. This particular contract did not have the usual arbitration clause that would have given the right to open up, review and revise. The judge defined the essential points of the *Crouch* decision, reading the judgments as a whole as follows:

> '1. The contractual machinery established by the parties provided in the first instance for determination of what was a fair and reasonable extension of time by the architect.
> 2. That agreed allocation of responsibility to the architect was subject to two safeguards:

---

[9] *Northern Regional Health Authority* v. *Derek Crouch Construction Co* [1984] QB 644.
[10] Sir John Donaldson MR at p.670 in *Crouch*.
[11] *Beaufort Developments (NI) Ltd* v. *Gilbert-Ash NI Ltd and Another* [1998] 2 All ER 778.
[12] *John Barker Construction Ltd* v. *London Portman Hotels Ltd* 50 Con LR.

(a) implicitly, an obligation on the architect to act lawfully and fairly, and

(b) explicitly, the power of review by an arbitrator or, who was entitled to substitute his opinion for that of the architect.

3. If safeguard (a) failed, the court could declare the architect's decision invalid, but it could not substitute its decision for that of the architect solely because it would have reached a different decision, for that would be to usurp the role of the arbitrator.

4. If safeguard (b) failed because the arbitration machinery broke down, the court could substitute its own machinery to ensure enforcement of the parties' substantive rights and obligations that a fair and reasonable extension should be given.'

He concluded:

'It seems to me that this is a case in which the contractual machinery established by the parties has become frustrated or, put in other words, has broken down to such an extent that it would not now be practicable or just for the matter to be remitted to the architect for re-determination; and that in those circumstances the court must determine on the present evidence what was a fair and reasonable extension of time.'

The decision in *Balfour Beatty* v. *Docklands Light Railway*[13] is also of assistance, although the judgement is more limited in that it only gave jurisdiction where there was a breach on the part of the employer. In that particular case it was the employer's representative who took the place of the engineer in making engineer's decisions.

'We would be greatly concerned at the implication of accepting Mr Ramsey's argument [for DLR] if to do so would leave the contractor without any effective means of challenging partial, self-interested or unreasonable decisions (if such were shown to have been made) by the employer. We would then have wished to consider whether an employer invested (albeit by contract) with the power to rule on his own and a contractor's rights and obligations, was not subject to a duty of good faith substantially more demanding than that customarily recognised in English contract law. Mr Ramsey has, however, accepted without reservation that the employer was not only bound to act honestly but also bound by contract to act fairly and reasonably, even where no such obligation was expressed in the contract (as in some clauses, for example clause 31.5, it was). Even on a more expansive approach to good faith, it may be that no more is required in the performance of the contract. . . . If the contractor cannot prove a breach of duty, he will not be entitled to a remedy. If it cannot, and cannot establish any other breach of contract, it will be under this contract entitled to none.'

The court concluded that the Official Referee had been correct in finding that the court did not possess the power to open up, review and revise the employer's decisions, opinions, instructions, directions, certificates or valuations, but that the court could grant appropriate relief where the contractor could prove breaches of contract on the employer's part.

In *Tarmac Construction* v. *Esso*[14] a further consideration was given to the jurisdiction of the courts where a dispute was referred to litigation. Again there was no arbitration clause in the contract.

'8. Esso's formulation would inevitably lead to a trial within a trial since the unreasonableness of the engineer's decision, or any other grounds upon which the decision was to be revised would have to be demonstrated by showing the reasonableness of the case rejected. There is no obviously useful purpose to be served in such a course.

9. The clause also contemplates that the engineer may also fail to give a decision. In such an event, what is to happen? If the dispute were about some earlier expression of opinion by the engineer,

---

[13] *Balfour Beatty Civil Engineering Ltd* v. *Docklands Light Railway* [1996] 78 BLR 42, CA.

[14] *Tarmac Construction Ltd* v. *Esso Petroleum Co Ltd* [1996] 83 BLR 65.

then, even though the engineer might under clause 66 have considered it de novo..., a failure by the engineer to decide could hardly be a breakdown of the machinery, since the clause contemplates such an eventuality. On Esso's formulation, the opinion, even though arrived at as a preliminary view, would stand. This is not a reasonable interpretation of the contract.

If it is to be assumed that there was a breakdown of the machinery whereby the court may act in default to enforce the contractual rights, then the position of the party who has referred the dispute for decision might depend entirely upon whether the engineer decided to give a decision. This too is not a reasonable result compared with the certainty (and fairness) of Tarmac's submissions, which are to be preferred.

10. Lastly, as Mr Blackburn forcefully submitted, if clause 66 were to be construed as meaning that decisions can be impeached only in the limited circumstances identified in *Balfour Beatty [Civil Engineering Ltd* v. *Docklands Light Railway Ltd* [1996] 49 Con LR 1] the parts relating to the reference to litigation become otiose, for the same right existed without any such clause in *Balfour Beatty*. The parties would not gain anything, apart from the time limit on the reference to litigation, which could as well be achieved in simpler form.

For all these reasons, clause 66 clearly expresses an intention that the engineer's decision is not final and may be revised by the court. This would ensure that, if the engineer's decision were influenced (albeit unconsciously) by self-interest, the party affected would have the opportunity of an impartial decision and the mechanism chosen should enable any imbalance in the preceding provisions to be redressed. The court will be enforcing the parties' rights and will not be doing anything for which it does not have jurisdiction, or which it does not regularly do. Accordingly, Issue 1 will be answered ''yes''.'

The importance of these cases is that they deal with the situation where there is no right to open up, review or revise certificates. Where a decision is in breach of contract it is not insurmountable simply because there are no powers to open up, review or revise. The breach itself can be remedied. The power to open up, review and revise the certificates must be viewed as an automatic right of appeal. The power is there to change the content of the certificate by virtue of an express term in the contract. Where no such power exists the fact that the certificate may be viewed to be wrong is not enough. The decisions that form the contents of the certificate should have been procured as a result of a breach if there are to be any grounds to modify that certificate. The burden of proof is on the claiming party to establish that the certificate was procured in breach of the contract. This is obviously more difficult than an automatic power that permits a review of the certificate under the contract. It can be seen from the two cases above that the courts give effect to contracts. This obviously includes providing remedies for any breach of the contract. Adjudicators must remember that unless the contract so provides they have no absolute power to open up, review and revise certificates. Where no such power exists they would have authority to remedy any breaches which procured the content of the certificates, if those breaches are brought before them.

Sub-section (2) gives the minimum criteria that all construction contracts should contain in respect of adjudication provisions. The emphasis here is that the Act is importing terms into contracts. The Act is not providing statutory rights per se; it is providing contractual rights. These contractual rights must be contained in the construction contract because the Act requires so.

Section 108(2)(a)

(2)   The contract shall–
   (a)   enable a party to give notice at any time of his intention to refer a dispute to adjudication;

A party has the right to give notice at any time of its intention to refer a dispute to adjudication. It was often the case in standard forms of contract that arbitration could not be commenced until after practical completion of the works or that arbitration could only be commenced in respect of limited issues. (In most instances this has now been amended in the latest editions of the standard forms.) This is not to be the case in respect of adjudication. The Act gives the party wishing to serve a notice an unrestricted right to do so at any time. There is nothing to prevent such a notice being issued after litigation or arbitration has commenced. This was dealt with in *Herschel* v. *Breen*[15]:

'14. Mr Brannigan's submissions on behalf of the claimant are short and simple. He emphasises the fact that section 108(2) provides that the contract provides that a party can give notice at any time of his intention to refer a dispute to adjudication. Adjudication is a special creature of statute, and the jurisprudence relied on by Mr Davies in support of his first argument has no application. The 1996 Act clearly contemplates that there may be two sets of proceedings in respect of the same cause of action, and there is nothing in the Act which indicates that they may not proceed concurrently. As for Mr Davies' second argument, Mr Brannigan submits that the commencement of proceedings in court does not amount to a waiver or repudiation of the right to refer the subject of those proceedings to adjudication. Here too he relies on the fact that the 1996 Act permits a party to refer a dispute to adjudication at any time. He also raises a doubt as to whether it is correct to regard the right to refer a dispute to adjudication as a contractual right which is capable of being waived or repudiated. This is because the right to refer a dispute to adjudication is imposed on the parties by the 1996 Act.

. . .

20. In my view, there is no obvious reason why Parliament should have intended to draw a distinction between cases where litigation or arbitration proceedings have been started before a dispute is referred to an adjudicator, and those where the proceedings have been started only after an adjudication has been completed. The mischief at which the Act is aimed is the delays in achieving finality in arbitration or litigation. Why should a claimant have to wait until the adjudication process has been completed before he embarks on litigation or arbitration? If he is in a position to start proceedings, it is difficult to see why he should have to wait until a provisional decision has been made by an adjudicator. The normal rule that concurrent proceedings in respect of the same issue or cause of action will not be countenanced is justified on the grounds that (a) it is oppressive to require a party to defend the same claim before different tribunals, and (b) it is necessary to avoid the risk of inconsistent findings of fact. But it is inherent in the adjudication scheme that a defendant will or may have to defend the same claim first in an adjudication, and later in court or in an arbitration. It is not self-evident that it is more oppressive for a party to be faced with both proceedings at the same time, rather than sequentially. As for the risk of inconsistent findings of fact, on any view this is inherent in the adjudication scheme. The answer to Mr Davies' first submission has been provided clearly and unequivocally by section 108(2)(a). Parliament has decided that a reference to adjudication may be made "at any time". I see no reason not to give those words their plain and natural meaning.

21. Mr Davies points out that, if his first submission is wrong, it is possible to conceive of absurd situations arising. For example, he suggests that the hearing in the county court may be adjourned part heard for several weeks. The judge may have made adverse comment on the claimant's case. The claimant might decide to use the period of the adjournment to refer the dispute to adjudication in the hope of obtaining a favourable provisional decision from the adjudicator. As I said in the course of argument, if an extreme case of this kind were to occur and the claimant were to succeed before the adjudicator, the most likely outcome would be that the defendant would not comply with the adjudicator's decision. If the claimant then issued proceedings and sought summary judgment,

---

[15] *Herschel Engineering Ltd* v. *Breen Property Ltd* (14 April 2000), Dyson J.

the court would almost certainly exercise its discretion to stay execution of the judgment until a final decision was given in the county court proceedings. In any event, the fact that it is possible to conceive of far-fetched examples like this does not deflect me from the view that I have already expressed.'

The adjudicator's decision was enforced in *Herschel* v. *Breen*. The comments of Dyson J regarding those circumstances where a decision might not be enforced, particularly where a judgment from the court was imminent, are worth attention but they are likely to be considered 'obiter' and thus each case needs to be reviewed on its merits.

What if a party issues the adjudication notice after the issue of a final certificate and outside the period for commencement of any arbitration? There does not seem to be any time bar, save for the limitation period.

Where the contract is brought to an end by a breach, the contract is not at an end; it is only performance which ceases. Does adjudication survive repudiation or termination? This point was dealt with in *A & D Maintenance* v. *Pagehurst*[16]:

'18. Even if the contract had been terminated, the matters referred to the Adjudicator remain disputes under the contract. Where there is a contract to which the Act applies, as in this case, and there are disputes arising out of the contract to be adjudicated, the adjudication provisions clearly remain operative just as much as an arbitration clause would remain operative.'

Most of the standard forms make provision for the determination of employment rather than a termination of the contract itself. The contract itself clearly remains live and has provisions to deal with the effects of the termination of employment. The adjudicator would therefore have jurisdiction to deal with any dispute, which arose over the determination of the employment itself, as it arises under the contract.

Section 108(2)(b)

(2) (b) provide a timetable with the object of securing the appointment of the adjudicator and referral of the dispute to him within 7 days of such notice;

The timetable of seven days is for both the appointment of the adjudicator and the referral of the dispute to him. The contract must provide a timetable with the object of securing the appointment within seven days. This is an objective; the period is not mandatory. If the contract provides that objective, the responding party could not object to the validity of the appointment and therefore claim that there was no jurisdiction on the grounds that the appointment and the referral took more than the seven days. This is not a mandatory provision of any kind; the Act requires the inclusion of a 'timetable with the object of securing the appointment and referral to the adjudicator within seven days' and any attempt to resist enforcement on the grounds of a referral outside this limit is unlikely to be received very favourably by the courts. The authors have examples of referrals outside the seven-day period in their own experience in respect of which jurisdictional objections have been made vociferously by the responding party. When the adjudication is completed in these cases there has been no attempt to resist enforcement on the grounds of referral out of time. If the adjudicator is named in the contract the selection of the adjudicator will already be in place. The act of making the referral could be sufficient to perfect the appointment for that particular dispute. Where the adjudicator is not named the selection process cannot consume all of the seven days of this period. The institutions who are likely to be named in contracts as

---

[16] *A & D Maintenance & Construction Limited* v. *Pagehurst Construction Services Limited* (23 June 1999).

appointees and the adjudicator nominating bodies have already considered the timetable as part of their structure to meet the needs of the Act and most of them operate on the basis that they must nominate within the five days set out in paragraph 5(1) of the Scheme (see Chapter 5). The Act refers to the adjudicator as 'him' but this expression also includes 'her' by virtue of the Interpretation Act 1978.

Section 108(2)(c)

(2)   (c)   require the adjudicator to reach a decision within 28 days of referral or such longer period as is agreed by the parties after the dispute has been referred;

The intention has always been that adjudication will be a fast track process. The requirement that the adjudicator reach a decision within 28 days is a contractual matter and not a statutory one. The Act only requires that the contract make this provision. The reckoning of periods of time is dealt with in section 116. There has been much argument that the period may lead to a situation of the unadjudicatable dispute. This is where the responding party and the adjudicator consider that the dispute cannot be adjudicated within 28 days but the referring party insists on its right to having a decision within that time. The parties may agree a longer period for the adjudicator to reach his decision. The emphasis here is that mutual agreement of the parties is required for the period to be extended. If one party objects, the period cannot be extended. It is likely that the referring party will wish the process to be concluded as quickly as possible and there will therefore be a reluctance to permit the extension of the period. It is here that the adjudicator's managerial and persuasive skills will need to be applied.

The Act does not stipulate what form the decision should take or whether or not reasons should be given.

There is no distinction between the reaching of the decision and its publication. Until very recently the court had not considered the question of a decision which was made but not delivered to the parties while the adjudicator awaited payment of fees, and we expressed the opinion in the first edition that there was no reason why an adjudicator should not announce that his decision had been made and would be sent on the payment of his fees.

The Scottish court considered this in *St Andrews Bay* v. *HBG Management*[17] in which St Andrews Bay claimed that the adjudicator had no power to reach her decision after 5 March 2003 and the decision sent to the parties on 7 March 2003 was thus not a valid decision. Lord Wheatley concluded that a decision cannot be said to be made until it has been actually provided to the parties. Further, in the circumstances of this case, the adjudicator was not entitled to delay communication or intimation of the decision until the fees were paid. There was nothing in the Scheme or contract to allow this. No alternative arrangement had been made. However, the judge held that the failure of the adjudicator to produce the decision within the time limits, while serious, was not of sufficient significance to render the decision a nullity. It was not such a fundamental error or impropriety to render the entire decision invalid.

Section 108(2)(d)

(2)   (d)   allow the adjudicator to extend the period of 28 days by up to 14 days, with the consent of the party by whom the dispute was referred;

---

[17] *St Andrews Bay Development Limited* v. *HBG Management Limited* (20 March 2003).

This gives the power for the adjudicator to extend time for up to 14 days. The power is granted by consent of the party who made the referral (the referring party or applicant). There is nothing to require the applicant to give such consent. This could place the adjudicator and the responding party in difficulty. There is however no requirement of reasonableness on the part of the applicant, but if the adjudicator as a result is unable to meet the time period for making his decision it appears likely that it will be unenforceable unless the delay is of a very short period.

Section 108(2)(e)

(2)   (e)   impose a duty on the adjudicator to act impartially; and

This imposes a duty on the adjudicator, which is similar to the duty imposed on an arbitrator under section 33 of the Arbitration Act 1996. This duty has to be set out in the contract and should be in the express terms and conditions of appointment or any adjudication rules. Although there is a duty of impartiality there is no requirement in the Act that the adjudicator be independent. It is conceivable therefore that an adjudicator could be appointed who in some way is connected with the parties or the work in question.

The duty to act impartially does require the adjudicator to apply the rules of natural justice. These however will be interpreted having regard to the time restraint of the adjudication process. It is arguable that simply to apply the rules of natural justice is too narrow a concept. It is more appropriate to describe the duty as to not act unfairly or in bad faith. The courts will not readily imply a procedure which would lead to an unfair result[18]. The term 'procedural fairness' has been coined by Forbes J, the judge in charge of the Technology and Construction Court when describing an adjudicator's obligations.

The adjudicator should declare any interest that could mean that he has any bias towards one of the parties. He should ensure that each party has the opportunity to present its case and to answer the case submitted by the other within the time restraints of the Act.

If there is a breach of the duty to act impartially the decision will be void. It will therefore be unenforceable. For a more detailed discussion on natural justice, procedural fairness and bias see Chapter 9.

Section 108(2)(f)

(2)   (f)   enable the adjudicator to take the initiative in ascertaining the facts and the law.

This gives the adjudicator similar powers to those of an arbitrator in section 34(2)(g) of the Arbitration Act 1996.

> 'It is of course unlikely that many arbitrators will opt for the inquisitorial approach, as it remains an unfamiliar procedure in England.'[19]

That may be so in arbitration, although perhaps less so now than when Professor Merkin wrote his comment a few years ago, but it is essential in our view that adjudicators do not fear the inquisitorial approach if adjudication is to work under this Act.

This sub-section gives the adjudicator wide powers. He may visit the site, talk to the appropriate personnel on site, make his own enquiries by telephone. There is nothing to

---

[18] See *London Export Corporation Ltd* v. *Jubilee Coffee Roasting Company Ltd* [1958] 1 All ER 494; *Wiseman and Another* v. *Borneman and Others* [1971] AC 297; *Norwest Holst Ltd* v. *Secretary of State for Trade* [1978] Ch 201.
[19] Robert Merkin commentary on section 34(2)(g) Arbitration Act 1996, An Annotated Guide.

prevent the adjudicator seeking legal advice or technical advice in pursuit of his inquiries. In fact in every respect the adjudicator is 'master in his own house' provided he does not ignore the duty to act impartially. This would necessarily require the adjudicator to tell the parties what he has discovered and allow the parties to respond. This situation has been examined by Judge Richard Seymour in *RSL (South West)* v. *Stansell*[20]:

'It is elementary that the rules of natural justice require that a party to a dispute resolution procedure should know what is the case against him and should have an opportunity to meet it. In *paragraph 17 of The Scheme for Construction Contracts* it is provided in terms that *"The adjudicator ... shall make available to* [the parties] *any information to be taken into account in reaching his decision".* At one point in her oral submissions Miss Hannaford seemed to come close to relying upon the absence of an equivalent provision in clause 38A of the sub-contract as an indication that there was no similar requirement in relation to an adjudication governed by terms similar to those of the sub-contract. If and insofar as Mr Hinchcliffe, or anyone else, may have thought that the effect of clause 38A.5.7 was that an adjudicator could, subject only to giving the parties to the relevant adjudication advance notice that he was going to seek technical or legal advice, obtain that advice and keep it to himself, not sharing the substance of it with the parties and affording them an opportunity to address it, it seems to me that he or she has fallen into fundamental error. It is absolutely essential, in my judgment, for an adjudicator, if he is to observe the rules of natural justice, to give the parties to the adjudication the chance to comment upon any material, from whatever source, including the knowledge or experience of the adjudicator himself, to which the adjudicator is minded to attribute significance in reaching his decision.'

If the rules for adjudication or the provisions in the contract seek to be too prescriptive or to detail the procedures to be followed too strictly, the flexibility of this sub-section will be lost and it is encouraging that this appears not to have been a route generally followed.

Section 108(3)

(3)   The contract shall provide that the decision of the adjudicator is binding until the dispute is finally determined by legal proceedings, by arbitration (if the contract provides for arbitration or the parties otherwise agree to arbitration) or by agreement. The parties may agree to accept the decision of the adjudicator as finally determining the dispute.

The concept of adjudication under this Act was to provide 'the short ... the dirty fix that the industry asked for'[21]. In the short term, at least, an adjudicator's decision is binding on the parties. It remains binding unless the parties do something that will replace that decision with some form of permanent decision.

The options are provided in the sub-section. The dispute can be finally determined by legal proceedings. Alternatively the dispute can be finally determined by arbitration where the contract provides for arbitration or the parties otherwise agree to arbitration. The particular clauses in contracts need to be examined for the timing to commence arbitration or litigation. In many instances there are specific issues where formal action is not permitted until after the practical completion of the works. The parties may agree to accept the adjudicator's decision as final and binding at any time. It is conceivable that it may be implied that the adjudicator's decision is final and binding where the parties have conducted their relationship on the basis that the adjudicator's decision is final. The parties may

---

[20] *RSL (South West) Ltd* v. *Stansell Ltd* (16 June 2003).
[21] R. Jones, *Hansard*, 8 July 1996, col. 84.

make it a term of their contract that the adjudicator's decision is final or they may agree at any time that the decision is final.

The parties may also substitute a new agreement which is final and binding for the decision made by the adjudicator and it has been noted that parties often use an adjudicator's decision as a basis of negotiations to finally resolve the dispute.

Section 108(4)

(4)   The contract shall also provide that the adjudicator is not liable for anything done or omitted in the discharge or purported discharge of his functions as adjudicator unless the act or omission is in bad faith, and that any employee or agent of the adjudicator is similarly protected from liability.

This sub-section has similar wording to section 29 of the Arbitration Act 1996. The Arbitration Act 1996 provides the arbitrator with statutory immunity. However this Act does not provide such immunity for adjudicators. The Act only requires that the immunity is provided in the contract and it is therefore a contractual immunity and not a statutory one. The immunity extends to the adjudicator's employees or agents. This does not of itself imply that the adjudicator will be a firm rather than an individual. It simply covers those who may be connected with the adjudication through the adjudicator.

The immunity would not extend to a body who appoints an adjudicator. This is dealt with in section 74 of the Arbitration Act 1996 but is not dealt with at all by this Act. It is conceivable that a body appointing an adjudicator, which has acted negligently in the appointment or has otherwise failed to exercise its supervisory or other powers with reasonable care, is liable in contract or tort to the parties.

'There is a question as to whether or not the adjudicator could face actions in tort from third parties affected by the decision. If the decision is regular in terms of the Act this is unlikely. There is nothing to prevent the Adjudicator's terms of appointment requiring that the parties indemnify him for actions in tort as well as in contract.

Where the adjudicator acts in bad faith there is no immunity in any event. Bad faith is difficult to define. It has been held to be "malice or knowledge of absence of power to make the decision in question".'[22]

'There have been few cases in which actual bad faith has even been alleged, but in the numerous cases where misuses of power have been alleged judges have been careful to point out that no question of bad faith was involved and that bad faith stands in a class of its own.'[23]

With these provisions in the Arbitration Act 1996 and this Act, it is likely that this will be an area of the law which will develop.

Section 108(5)

(5)   If the contract does not comply with the requirements of subsections (1) to (4), the adjudication provisions of the Scheme for Construction Contracts apply.

Sub-sections 108(1) to (4) inclusive give the 'bare bones' requirements for adjudication under the Act. If the contract does not wholly comply with the requirements of these sub-sections

---

[22] *Melton Medes Ltd and Another* v. *Securities & Investment Board* [1995] 3 All ER 881.
[23] *Smith* v. *East Elloe District Council* [1956] 2 WLR 888.

the Scheme will apply (see Chapter 5). This gives the default position under the Act. It should be noted that if the contract refers to adjudication rules and these do not comply with the Act, the Scheme would still apply.

Section 108(6)

(6) For England and Wales, the Scheme may apply the provisions of the Arbitration Act 1996 with such adaptations and modifications as appear to the Minister making the scheme to be appropriate.

For Scotland, the Scheme may include provision conferring powers on courts in relation to adjudication and provision relating to the enforcement of the adjudicator's decision.

The references to the Arbitration Act 1996 caused an outcry in the industry press at the time that the first draft of the Scheme was published. The criticisms, among others, were that it followed arbitration too closely. It is worth noting that the extracts from the Arbitration Act 1996 that are incorporated into the Scheme have never been utilised as the problem seems to have been avoided totally by the positive attitude of the court to the enforcement of adjudicators' decisions from *Macob* v. *Morrison* onwards.

Chapter 4

The Act: The Payment Provisions Sections 109–113

There have been no surprises in that the majority of disputes referred to adjudication have revolved around money and payment.

In the first year or so of the operation of the Act a considerable number of adjudications related to the payment provisions of the Act. This was a reaction to the introduction of the various notice requirements in sections 110 and 111. Paying parties were failing to recognise the need to issue the required notices, and adjudications were started by those seeking payment on the basis of the absence of these notices creating an entitlement to payment. Paying parties generally have now recognised the need to issue these notices and there seems to be a reduction in the number of adjudications which relate solely to these notices. The payment provisions are however still a very relevant factor in respect of adjudication.

The provisions of the Act as to payment are approached in a different way from those relating to adjudication. Section 108 deals with all the matters that a contract is required to include in respect of adjudication. Failure to include any one of these in a contract means that any adjudication provisions that may be included in the contract are superseded by the provisions of the Scheme which then govern any adjudication that may be brought under that contract. The provisions on payment are different in that they require compliance on a section by section basis. Where the payment provisions in a contract comply only with parts of the Act, those that comply remain binding on the parties but those parts that do not comply with the Act are replaced by the equivalent provisions in the Scheme.

Section 109: Entitlement to stage payments

Section 109(1)

(1) A party to a construction contract is entitled to payment by instalments, stage payments or other periodic payments for any work under the contract unless–
   (a) it is specified in the contract that the duration of the work is to be less than 45 days, or
   (b) it is agreed between the parties that the duration of the work is estimated to be less than 45 days.

This section gives a statutory right to some form of interim payment with only the two exceptions noted. Before the Act applied, unless there were express provisions in the contract for interim payments, the assumption was that the contract was entire. This meant that completion of the whole of the works was condition precedent to payment[1]. There is

---

[1] See *Sumpter v. Hedges* [1898] 1 QB 673 and *Ibmac v. Marshall (Homes)* (1968) 208 EG 851.

however some authority at common law to suggest that interim payments can be implied into construction contracts even when the Act does not apply.

> 'Having heard their evidence as to the custom of the industry, I concluded that it was an implied term of a contract that application could be made for payment not more frequently than monthly. Payment would then be due within a reasonable period of such application, for the work done and any unused material on site, valued in each case in accordance with the contract between the parties. I announced in the course of the hearing my intention to hold that I construed a reasonable time in all the circumstances as 30 days with a period of grace to allow 42 days in all, and I now hold that 42 days was a reasonable time for payment.'[2]

This case follows evidence of custom and practice in the industry. Such terms would also be implied where there is a course of dealing between the parties that shows that they habitually adopted a monthly payment process.

Most of the forms of contract (standard and non-standard) in use in the construction industry make provision for periodic or stage payments. The usual basis on which periodic payments are made is monthly and relates to the value of work done. Traditionally the value of work done is established from measurement or remeasurement of the works. Where however a stage payment regime applies, payment depends on the completion of a set stage before any payment is due.

There is no definition in the Act as to what the terms 'instalments, stage payments or other periodic payments' mean. Instalments could mean a series of equal payments such as those found in a hire purchase agreement. Stage payments are a much clearer matter through their existing use in construction industry payment terms. A system of monthly payments reflecting work completed remains probably the most common method by which payment is made in construction industry contracts.

The statutory entitlement to a payment system envisaged by the Act is subject to the two qualifications in sub-sections 109(1)(a) and (b). If it is specified in the contract that the duration of the work is to be less than 45 days there is no entitlement to payment under the interim payments structure envisaged in this section.

It is conceivable that works may be broken into packages, each with a declaration stated in the contract that it is of less than 45 days duration so that interim payments need not be made. Alternatively the parties may agree that the duration of the work is estimated to be less than 45 days even if it is actually longer. Again there is scope for an overoptimistic estimate of the duration simply to avoid interim payments. These are matters that will be the subject of the parties' respective bargaining positions at the time that they make their contract.

Section 109(2)

(2)   The parties are free to agree the amounts of the payments and the intervals at which, or circumstances in which, they become due.

This sub-section confirms a freedom that already exists in practice in the construction industry. There may be hybrid situations in non-standard forms that seek to impose some payment mechanism that is unattractive or onerous, but market forces apply in such cases and the imposition of onerous terms will only serve to increase costs of construction projects.

---

[2] *D R Bradley (Cable Jointing)* v. *Jefco Mechanical Services* (1989) 6 CLD 7-21, Mr Recorder Rich QC sitting as an Official Referee.

(3)   In the absence of such agreement, the relevant provisions of the Scheme for Construction Contracts apply.

The scheme will provide the default position where the parties have failed to reach an agreement on payment systems. This is dealt with in paragraphs 2–4 of the Scheme.

(4)   References in the following sections to a payment under the contract include a payment by virtue of this section.

Section 109 provides the basic concept concerning payment and the subsequent sections provide more detail. On any matter concerning payment, sections 109 to113 inclusive must be read as a whole.

Section 110: Dates for payment

Section 110(1)

(1)   Every construction contract shall–
    (a)   provide an adequate mechanism for determining what payments become due under the contract, and when, and
    (b)   provide for a final date for payment in relation to any sum which becomes due.
    The parties are free to agree how long the period is to be between the date on which a sum becomes due and the final date for payment.

There is no definition in the Act of what constitutes an 'adequate mechanism' for determining payment. Adequate means 'sufficient for a particular purpose or need', 'able to satisfy a requirement', also 'standard', 'suitable'. Whatever the mechanism is, it must determine what payment is due and when it becomes due.

The concept of dates for payment in construction contracts is not new. There is a date on which payment becomes due. This is a date on which the legal liability to pay the sum occurs. Sub-section 110(1)(b) makes the distinction between this, the date when payment becomes due, and the final date on which actual payment must be made. This distinction is not unusual and already exists in the DSC/C (formerly DOM/1) Standard Form of Domestic sub-contracts and related sub-contracts. The final date for payment is not to be confused with the familiar practice in connection with final certificates. What is required here is a final date for payment which relates to each and every payment. As a result every interim payment will have a final date on which payment is to be made. This is a further feature that the adequate payment mechanism must provide. The period between a due date and the actual final payment date is a matter for agreement between the parties. The final date for payment has two effects. Where this date is exceeded the money is then overdue and action can be taken to pursue the sum. This is also the date which triggers a right to suspend performance of the works given by section 112.

Section 110(2)

(2)   Every construction contract shall provide for the giving of notice by a party not later than five days after the date on which a payment becomes due from him under the contract, or would have become due if–
    (a)   the other party had carried out his obligations under the contract, and
    (b)   no set-off or abatement was permitted by reference to any sum claimed to be due under one or more other contracts,

specifying the amount (if any) of the payment made or proposed to be made, and the basis on which that amount was calculated.

This sub-section imposes a requirement that every construction contract must provide for giving a notice of the amount due. The notice must state the amount due and the basis on which it is calculated. The obligation is on the paying party to give the notice. This goes beyond any practice previously prevalent in the industry and has led to a far greater openness as to the basis of payments made. The general practice of those making payments has not changed; it is still a notice in the form of a certificate or a notification of payment but details of calculation must now be given and payees must at least know why they are not being paid the sum that they have applied for. The notice is required even if the amount of the payment is nil due to set-off or abatement.

There is however no sanction in section 110(2) for not issuing a notice of payment. Some commentators have suggested that this makes the amount of money applied for the amount due. This is a step too far in the authors' view. The position was clarified in *SL Timber* v. *Carillion*[3]:

'19. In my opinion the adjudicator fell into error in the first place by conflating his consideration of sections 110 and 111 of the 1996 Act. In my opinion Mr Howie was correct in his submission that these sections have different effects and the notices which they contemplate have different purposes. Section 110(2) prescribes a provision which every construction contract must contain. Section 110(3) deals with the case of a construction contract that does not contain the provision required by section 110(2) by making applicable in that case the relevant provision of the Scheme, namely paragraph 9 of Part II. By one or other of these routes every construction contract will require the giving of the sort of notice contemplated in section 110(2). But there the matter stops. Section 110 makes no provision as to the consequence of failure to give the notice it contemplates. For the purposes of the present case, the important point is that there is no provision that failure to give a section 110(2) notice has any effect on the right of the party who has so failed to dispute the claims of the other party. A section 110(2) notice may, if it complies with the requirements of section 111, serve as a section 111 notice (section 111(1)). But that does not alter the fact that failure to give a section 110(2) notice does not, in any way or to any extent, preclude dispute about the sum claimed. In so far, therefore, as the adjudicator lumped together the defenders' failure to give a section 110(2) notice with their failure to give a timeous section 111 notice, I am of opinion that he fell into error. He ought properly to have held that their failure to give a section 110(2) notice was irrelevant to the question of the scope for dispute about the pursuers' claims.'

There may be unique circumstances in the contract which result in payment of the amount applied for becoming the amount that is due and payable. This is a peculiarity of the particular contract and not the general rule from the Act[4]. There is specific provision in the JCT Standard Form of Building Contract With Contractors Design 1998 edition that where there is no equivalent of the section 110 notice the amount stated in the application for payment is the amount that the employer must pay (clause 30.3.5).

Section 110(2) requires the payer to give a notice not later than 5 days after the due date and states specifically that it must be given even where nothing is due because of the circumstances identified in sub-sections (a) and (b). Where there is a failure to perform the obligations under the contract, for example failure to construct in accordance with the specification, the value of the work executed may be abated by the amount that the work is

---

[3] *SL Timber Systems Limited* v. *Carillion Construction Limited* (27 June 2001), Opinion of Lord Macfadyen.
[4] *VHE Construction PLC* v. *RBSTB Trust Co Limited* (13 January 2000).

worth less than had it been properly performed. The contract price for that work has not been earned. The contract price for that work can therefore be extinguished or reduced. This is abatement of a sum that would otherwise have been due.

> RALPH GIBSON: 'The words of [the contract]... do not affect the right of a contractor to defend a claim for an interim payment by showing that the sum claimed includes sums to which the subcontractor is not entitled under the terms of the contract or to defend by showing that, by reason of the sub-contractor's breaches of contract, the value of the work is less than the sum claimed under the ordinary right of defence established in *Mondel* v. *Steel* (1976) 1 BLR 106.'[5]

It is suggested here that there ought to be a distinction between work which wholly fails to comply with the contract and work which complies but has some defect which would be remedied in the normal course of events.

The distinction is between, on the one hand, the failure to earn the contract price under the contract and, on the other, having earned the contract price but some matter is not entirely complete or entirely in accord with the contract. The failure to earn the contract price would mean that nothing is due for that work. Work that is defective would create a situation where something is due even if it were not the whole sum in the contract. Both are reduced by abatement.

The JCT Standard Form Contracts provide for work properly executed to be included in interim certificates. If the work is not properly executed it does not warrant inclusion in the interim certificate and abatement does not apply; it is simply a matter of account under the contract.

Sub-section 110(2)(b) permits set-off or abatement from one or more other contracts. It is the authors' view that this provision is complex and unhelpful to the overall scheme of payments and adjudication. In the first instance abatement or set-off under another contract will serve to extinguish the sums due under that other contract[6]. All counterclaims are set-offs but not all set-offs are counterclaims. The questions are always whether a counterclaim is in truth a full defence, a counterclaim only or a counterclaim that is also a set-off which can be utilised as a defence.

> 'A counterclaim is a different thing entirely from a set-off. It is any claim which the defendant is entitled to bring as a separate action against the plaintiff. It might be something entirely different in nature from the claim.
>
>   Until the Supreme Court of Judicature Act 1873, counterclaims in fact always had to be a separate action, i.e., the defendant in the first case between the parties became the plaintiff in a second case.'[7]

It is not a simple matter where claims arise under different contracts between the same parties. It is exceptional for a party to be allowed to use a claim under one contract as a set-off in response to a claim made against it on the other. It may amount to a counterclaim but it will not amount to a set-off by way of a defence. In serial contracts there may be an argument that the fact that there is more than one contract is incidental and that the transactions are so closely connected that it would be unfair to consider the claim without the counterclaim.

This was considered by Lord Denning in *Anglian Building Products* v. *French*:

> 'Anglian, in respect of the M3, have brought an action claiming the sum due to them for the beams (pre-stressed concrete units) which they supplied under that contract. The claim of Anglian on the

---

[5] *Acsim (Southern) Ltd* v. *Dancon, Danish Construction & Development Company Ltd* (1989) CILL 538; (1989) 7-CLD-08-01.
[6] See Neil F Jones *Set-off in the Construction Industry* for the history and distinction of types of set-off, Blackwell, 1999.
[7] *Hanak* v. *Green* [1958] 1 BLR 1, Morris LJ.

M3 beams is £136,519.98. They are met in that action by a cross action by French for £600,000 also on the M3. Those matters in the M3 motorway dispute have been consolidated. They will have to be sent to an Official Referee or some one to try the rights and wrongs of it; and that can be tried out there.

The question is whether the counterclaim in respect of the M3 can also be used as a ground for staying the action in respect of the M4 and M6 goods. The judge has said No; he says it is a matter which should be fought out in the M3 litigation. It should not be used to stay the judgment or the execution of the judgment in the M4 and M6 action. And now French appeal to this court.

This matter of a stay is primarily for the discretion of the judge. I must say that I see nothing wrong in the way he has exercised his discretion. There is no doubt as to the solvency of Anglian Building Products Ltd. They are a subsidiary of Ready Mixed Concrete. If there are any damages payable on any cross claim, Ready Mixed Concrete will see that they are paid.

So that there is no question that French will get their money if they are right in their counterclaim. In those circumstances I do not see why this counterclaim on the M3 should be used to hold up payment for the work on the M4 and M6, for which, as I have said, French have actually had the money from the employers; they have actually been paid for these very units which have been delivered. I see no reason for interfering with the judge's discretion and I would dismiss the appeal.'[8]

There is also a distinction between common law set-off and abatement:

'It must however be considered, that in all these cases of goods sold and delivered with a warranty . . . the rule which has been found so convenient is established; and that it is competent for the defendant . . . not to set-off by a proceeding in the nature of a cross-action, the amount of damages which he has sustained by breach of the contract, but simply to defend himself by showing how much less the subject matter of the action is worth, by reason of the breach of contract; and to the extent that he obtains, or is capable of obtaining, an abatement of price on that account, he must be considered as having received satisfaction for the breach of contract, and is precluded from recovering in another action to that extent; but no more.'[9]

Abatement only applies to contracts for the sale of goods and contracts for work and materials. It has the effect, not of discharging payment, but of preventing the obligation to make payment arising in the first place. Since the work is defective (or has not been completed according to the contract) payment has not been earned.

The set-off or abatement from other contracts also needs to be permitted under the extant contract. In some of the standard forms the common law right of set-off is restricted:

'I now come to what I think is Mr Palmer's main point in relation to the claim by the defendants to a set-off. He submits that, even if the letter of 24 July 1987 from the defendants did not raise a valid claim to a set-off under condition 21 of the contract, the defendants have a valid claim to a set-off at common law which is not excluded by the provisions of condition 21.4. I will repeat the provisions of condition 21.4. It reads:
   "The rights of the parties to the sub-contract in respect of set-off are fully set out in these sub-contract conditions and no other rights whatsoever should be implied as terms of the sub-contract relating to set-off".
Mr Palmer has pointed out that the authorities show that clear words are necessary in a contract to exclude a common law right which a party would otherwise have. In my view, those clear words are to be found in condition 21.4 which says that the rights of the parties in respect of set-off "are fully set out in these sub-contract conditions". It seems to me that those words clearly indicate that the right to set-off is to be governed exclusively by the contract itself, and, more particularly, by condition 21. I consider that the words "no other rights whatsoever should be implied as terms of the subcontract

---

[8] *Anglian Building Products Ltd* v. *W. & C. French (Construction) Ltd* (1976) 16 BLR 1, CA.
[9] *Mondel* v. *Steel Court of Exchequer* (1976) 1 BLR 107, Park B.

relating to set-off" reinforce the conclusion that condition 21 was intended to limit the right to set-off to the express provisions contained in condition 21.

In my judgment, therefore, the defendants cannot in these proceedings set off their cross claims against the plaintiffs' claim. However, there is nothing to stop them from making a counter-claim.'[10]

Abatement has been examined in cases on enforcement of adjudicator's decisions. The Whiteways case established some first principles:

'There may be some difficulty about the concept of an abatement claimed to be due under other contracts under section 110(2)(b), but we are not concerned here with other contracts. Section 110 requires the giving of a notice stating the amount of any payment proposed to be made, and that notice is to be given within 5 days after the date when payment would have been due if the party had performed its contract, i.e. inter alia if there had been no ground for abatement. Section 111 provides for notice of intention to withhold payment. It is common for a party to a building contract to make deductions from sums claimed on the Final Account (or on earlier interim applications) on account of overpayments on previous applications and it makes no difference whether those deductions are by way of set-off or abatement. The scheme of the HGCRA is to provide that, for the temporary purposes of the Act, notice of such deductions is to be made in a manner complying with the requirements of the Act. In making that requirement, the Act makes no distinction between set-offs and abatements. I see no reason why it should have done so, and I am not tempted to try to strain the language of the Act to find some fine distinction between its applicability to abatements as opposed to set-offs. Of course, in considering a dispute, an adjudicator will make his own valuation of the claim before him and in doing so, he may abate the claim in respects not mentioned in the notice of intention to withhold payment. But he ought not to look into abatements outside the four corners of the claim unless they have been mentioned in a notice of intention to withhold payment. So, to take a hypothetical example, if there is a dispute about Valuation 10, the adjudicator may make his own valuation of the matters referred to in Valuation 10 whether or not they are referred to specifically in a notice of intention to withhold payment. But it would be wrong for him to enquire into an alleged overvaluation on Valuation 6, whether the paying party alleges abatement or set-off, unless the notice of intention to withhold payment identified that as a matter of dispute.'[11]

Judge Bowsher QC identified the difficulty with abatement in connection with other contracts. An adjudicator can apply abatement where it comes within the four corners of the claim. The example used was for an adjudicator to arrive at his own value of the claim where the claim concerns value of the work. We doubt whether this is abatement alone. It may be simply a matter of account or the combination of account and abatement.

Even where the payment is to be nil, the notice should show how the sum has been calculated showing which work is not in accordance with the contract and what amounts arise through set-off or abatement under other contracts.

The effect of the notice showing the amount to be paid and how it is to be calculated is to bring disputes to a head at a much earlier date than would otherwise be the case.

Section 110(3)

(3) If or to the extent that a contract does not contain such provision as is mentioned in subsection (1) or (2), the relevant provisions of the Scheme for Construction Contracts apply.

---

[10] *BWP (Architectural) Ltd* v. *Beaver Building Systems Ltd* (1988) 42 BLR 86.
[11] *Whiteways Contractors (Sussex) Limited* v. *Impresa Castelli Construction UK Limited* (9 August 2000).

The Scheme thus provides the default position where the contract does not wholly reflect the provisions of sub-sections (1) and (2).

Section 111: Notice of intention to withhold payment

Section 111 includes additional notice requirements to Section 110. These deal with the intention to withhold payment, which is one of the principal issues in many of the cases that have been brought to adjudication.

Section 111(1)

(1) A party to a construction contract may not withhold payment after the final date for payment of a sum due under the contract unless he has given an effective notice of intention to withhold payment.
The notice mentioned in section 110(2) may suffice as a notice of intention to withhold payment if it complies with the requirements of this section.

There is no right to withhold payment of a sum that is due under the contract unless an effective notice has been given. This is a strict requirement of the Act. If no notice is given at all there is no right to withhold payment. If a notice is given but it is defective in some way, for example being given out of time, again there is no right to withhold payment.

The notice given under section 110(2) may suffice as a section 111 notice as well, but to do so it must not only comply with section 110(2) but also with section 111.

There are a number of examples of the view of the courts on where a withholding notice is required.

'18. To the extent to which the sums sought to be retained are overpayments which have already been made under the contract on previous invoices, the clear effect of section 111(1) in my judgment is to prevent the defendant from exercising any right it may have had under the general law to recover that overpayment by way of deduction or retention from the 10 invoices unless the requisite notice has been given. No such notice was given. The circumstance that a previous overpayment may operate under the general law by way of an equitable set-off and thus technically be a matter of defence or that it may perhaps be able to be characterised as an abatement which technically in law prevents the amount claimed from ever becoming due does not in my judgment obviate the need for the paying party if he wishes to rely upon a right to deduct previous overpayments to give the requisite notice under section 111(1). If this were not the case, it is difficult to see what practical effect section 111 would have. The effect of section 111 is to prevent the paying party if he does not give the appropriate notice from exercising his right to retain or withhold payment of monies which would otherwise be due and payable but for the existence of some right to withhold payment. Section 111 refers to "withholding" payment generally. It must have been intended to include situations where the paying party was legitimately entitled under the general law or under the terms of the contract to withhold monies which were otherwise payable. The section clearly cannot be read simply as a provision which is restricted to requiring a notice of intention to withhold a payment to be given when there is no right to withhold any payment.

19. So far as set-off is concerned the question was considered by Judge Hicks QC in *VHE Construction Plc* v. *RBSTB Trust Co Limited* [2000] BLR 187, 192 where, at paragraph 36 of his judgment he said:

"The first subject of dispute as to the effect of section 111 is whether section 111(1) excludes the right to deduct money in exercise of a claim to set-off in the absence of an effective notice of intention to withhold payment. Mr Thomas for RBSTB submits that it does not. I am quite clear, not only that it does, but that that is one of its principal purposes... The words 'may not withhold

payment' are in my view ample in width to have the effect of excluding set-offs and there is no
reason why they should not mean what they say."

I respectfully agree. It is clear from this decision that "the final date for payment of a sum due under
the contract" may exist although technically if a valid notice had been given the paying party would
have been entitled to exercise a right of retention or to withhold the sum in question and thus not
have been obliged to make the payment.'[12]

In *Woods Hardwick* v. *Chiltern*[13] Judge Thornton QC characterised abatement as being outside
the provisions of section 111:

'10. Any abatement, properly relied on by Chiltern, would not of course be caught by section 111 of
the HGCRA, so Chiltern's abatement defence could, in principle, defeat or reduce Woods Hardwick's claims.

Chiltern's defence was not put forward in terms as an abatement. However, the nature of Chiltern's
defence was to the effect that such fees as might otherwise have been due were eliminated or
reduced because the value of Woods Hardwick's work was greatly reduced by the alleged breaches
of contract. It is for this reason that I have characterised Chiltern's principal defence as being one of
abatement.'

In *Re A Company*[14] the seriousness of failing to follow the withholding notice was examined
in winding up proceedings:

'12. Emphasising the words "payment of a sum due…" in section 111(1), Counsel for CCL argued
that the section only applies where the monies are in fact due, and that this requires the court to
consider whether the sum demanded by the contractor is irrecoverable (in whole or in part) because
of its defective performance of the contract. But the clear intent of sections 110(2) and section 111 is to
preclude the employer (in the absence of a withholding notice with specified content) from con-
tending that all or part of the sum demanded by the contractor is not in fact due. If the work is
defective, the employer retains the right to recover damages for breach of contract in subsequent
litigation or arbitration, and can obtain a provisional order to the same effect by adjudication under
section 108 of the Act. That does not however alter the position that by virtue of section 111 the
employer is obliged to pay forthwith without deduction absent a withholding notice, the rule is "pay
now, litigate later". Indeed, any other construction of sections 110 and 111 would rob them of all
practical significance.

13. Given the absence of a withholding notice in the present case, section 111 therefore requires CCL
to pay the £9,702.47 without deduction regardless of any defence which might otherwise have
existed by reason of the alleged defects in the works. The cross-claim and/or abatement thus provides no basis on which it can properly be disputed that the £9,702.47 is due and payable, GAL must
therefore be regarded as a creditor of the company with *locus standi* to present a winding-up petition.'

Judge Bowsher in *Northern Developments (Cumbria)* v. *J & J Nichol*[15] said:

'29. The [1996] Act by section 111 imposes on the parties a direct requirement that the paying party
may not withhold a payment after the due date for payment unless he has given an effective Notice
of Intention to Withhold Payment. That seems to me to have a direct bearing on the ambit of any
dispute to be heard by an adjudicator. Section 110 requires that the contract must require that within
5 days of any sum falling due under the contract, the paying party must give a statement of the
amount due or of what would be due if the payee had performed the contract. Section 111 provides
that no deduction can be made after the final date for payment unless the paying party has given

---

[12] *Millers Specialist Joinery Company Limited* v. *Nobles Construction Limited* (3 August 2001).
[13] *Woods Hardwick Ltd* v. *Chiltern Air Conditioning* (2 October 2000).
[14] *Re A Company* (Number 1299 of 2001) (15 May 2001).
[15] *Northern Developments (Cumbria) Limited* v. *J & J Nichol* (24 January 2000).

notice of intention to withhold payment. The intention of the statute is clearly that if there is to be a dispute about the amount of the payment required by section 111, that dispute is to be mentioned in a notice of intention to withhold payment not later than 5 days after the due date for payment. Equally it is clear from the general scheme of the Act that this is a temporary arrangement which does not prevent the presentation of set-offs, abatements, or indeed counterclaims at a later date by litigation, arbitration or adjudication. For the temporary striking of balances which are contemplated by the Act, there is to be no dispute about any matter not raised in a notice of intention to withhold payment.'

In *SL Timber* v. *Carillion*[16], a Scottish first instance case, Lord McFadyan said this:

'22. In my opinion, the absence of a timeous notice of intention to withhold payment does not relieve the party making the claim of the ordinary burden of showing that he is entitled under the contract to receive the payment he claims. It remains incumbent on the claimant to demonstrate, if the point is disputed, that the sum claimed is contractually due. If he can do that, he is protected, by the absence of a section 111 notice, from any attempt on the part of the other party to withhold all or part of the sum which is due on the basis that some separate ground justifying that course exists. It is no doubt right, as the adjudicator pointed out, that, if the section did require a notice of intention to withhold payment as the foundation for a dispute as to whether the sum claimed was due under the contract, it would be relatively straightforward for the party disputing the claim to give such a notice. But that consideration does not, in my view, justify ignoring the fact that the section is expressed as applying to the case where an attempt is made to withhold a sum due under the contract, and not as applying to an attempt to dispute that the sum claimed is due under the contract. Nor, in my view, is there merit in the adjudicator's concern that acceptance of the defenders' construction of section 111 would render the 1996 Act largely ineffective. I see no difficulty for an adjudicator in reaching a provisional determination of a dispute as to whether the sum claimed is due under the contract. That is what, on the adjudicator's own view of the section, the adjudicator would require to do if the party disputing the claim on the basis that the sum claimed was not due under the contract gave a notice of intention to withhold payment on that ground. In my opinion, therefore, the adjudicator erred in holding that the pursuers were relieved, by the defenders' failure to give a timeous notice of intention to withhold payment, of the need to show that the sums claimed were due under the contract.'

The Court of Appeal considered the question of the requirement for withholding notices in *Morgan* v. *Jervis*[17]. This case refers specifically to a contract where monies had been certified by an architect. It is worth repeating a considerable amount of this judgment:

'1. This appeal involves a point of some importance in the world of building contracts. Mr Iain Wallace QC correctly forecast (in an article entitled *The HGCRA: A Critical Lacuna?* (2002) 18 Const LJ 117) that a Scottish case dealing with it, *SL Timber Systems* v. *Carillion Construction* [2001] BLR 516, "seems certain to be reviewed at some future date by an appellate court in England". And here it is. The point concerns the meaning of section 111(1) of the Housing, Grants (Construction and Regeneration) Act 1996. . . .
3. . . . Under the contract the architect is to issue an interim certificate, in practice based on his scrutiny of a bill presented by the builder. In this case there was a 7th interim certificate in the sum of £44,000 odd plus VAT. The clients accept that part of that is payable but dispute the balance amounting to some £27,000. The builders seek summary judgment for the balance.
4. The clients did not give "a notice of intention to withhold payment" before "the prescribed period before the final date for payment". The builders contend that it follows, by virtue of section 111(1) that the clients "may not withhold payment". So they seek summary judgment. The clients say they

---

[16] *S L Timber Systems Limited* v. *Carillion Construction Limited* (27 June 2001).
[17] *Rupert Morgan Building Services (LLC) Limited* v. *David Jervis and Harriet Jervis* [2003] EWCA Civ 1583, CA.

can withhold payment, that it is open to them by way of defence to prove that the items of work which go to make up the unpaid balance were not done at all, or were duplications of items already paid or were charged as extras when they were within the original contract, or represent "snagging" for works already done and paid for.

5. The rival arguments as presented in the courts below were for what can be termed "wide" and "narrow" constructions. The wide construction espoused by the builders is that once it is shown that there is a certificate and no withholding notice, the certified sum must be paid – it cannot be withheld. The narrow construction is roughly to the effect that if work has not been done there can be no "sum due under contract" and that accordingly section 111(1) simply does not apply. As HHJ Humphrey Lloyd put it in *KNS Industrial Services* v. *Sindall* (17th July 2000) "one cannot withhold what is not due".

6. Each construction has some basis in authority and learned writings. For the wide construction there is HHJ Bowsher QC in *Whiteways Contractors* v. *Impresa Castelli* (2000) 16 Const LJ 453, HHJ Humphrey Lloyd QC in *KNS*, HHJ Gilliland QC in *Millers Specialist Joinery* v. *Nobles* (3 August 2000) and *Keating on Building Contracts* (7th edn 2001, para15–15H). For (or apparently for) the narrow construction there is HHJ Thornton QC in *Woods* v. *Hardwicke* [2001] BLR 23, (arguably) HHJ Hicks QC in *VHE Construction* v. *RBSTB Trust* [2000] BLR 187 and Lord Macfadyen in *SL Timber Systems* v. *Carillion Construction* [2001] BLR 516 and Mr Wallace's article The views expressed in most of these cases are more or less oblique to the point directly in issue here. Moreover there were variants of the narrow construction, that is why I said "roughly to the effect". The variants were around the theme of whether the section merely prevented the raising of counterclaims or did it also cover matters of abatement and set-off? And what about a counterclaim based on work allegedly done badly?

7. I do not think these questions arise at all. This is because some of the debate seems to have been based upon an unspoken but mistaken assumption, namely that the provision is dealing with the ultimate position between the parties. That is not so as is pointed out by Sheriff J.A. Taylor in *Clark Contracts* v. *The Burrell Co* [2002] SLT 103. He casts a flood of light on the problem.

. . .

11. In this ASI contract, the sum is determined *by the certificate*. Clause 6.1 provides that "payments shall be made to the Contractor only in accordance with the Architects certificate". Clause 6.32 defines the sum – essentially the approved gross value of work done less retention and amounts previously paid. Clause 6.33 says when it is to be paid: "the employer shall pay to the Contractor the amount certified within 14 days of the date of the certificate, subject to any deductions and set-offs due under the Contract." So it is not the actual work done which either defines the sum or when it is due. The sum is the amount in the certificate. The due date is 14 days from certificate date. The certificate may be wrong – the architect may (though this is unlikely because he will be working from the builder's bill) have missed out work done (which would operate against the contractor) or he may have included items not in fact done or items already paid for (which would operate against the client). In the absence of a withholding notice, section 111(1) operates to prevent the client withholding the sum due. The contractor is entitled to the money right away. The fundamental thing to understand is that section 111(1) is a provision about cash-flow. It is not a provision which seeks to make any certificate, interim or final, conclusive. Analysed this way one sees that there is something inconsistent about the clients' argument here. Their duty to pay now and the sum they have to pay arise only because of the certificate. Yet they wish to ignore the certificate to reduce the amount they have to pay.

12. All this becomes blindingly clear following Sheriff Taylor's analysis. He was dealing with a case like this, one involving a system of architect's certificates. This is what he said:

"There was no dispute that the architect had issued an interim certificate. It therefore seems to me that the defenders became entitled to payment of the sum brought out in the interim certificate within 14 days of it being issued. In my opinion that is an entitlement to payment of a sum due under the contract. In order to reach the figure in the interim certificate one has made use of the contractual mechanism. To use the words deployed by Lord Macfadyen [in *SL Timber Systems*] in para. 20, the issue of an interim certificate was the occurrence of 'some other event on which a contractual liability to make payment depended'. This situation falls to be contrasted with the position in *SL Timber Systems* where, before the adjudicator, there had been no calculation of the sum sued for by reference to a contractual mechanism and which gave rise to an obligation under the contract to make payment. There had been no more than a claim by the pursuers which claim had not been scrutinised by any third party. Thus, in my opinion, if The Burrell Co (Construction Management) Ltd wished to avoid a liability to make such payment because the works did not conform to the contractual standard they would be withholding payment of a sum due under the contract. In order to withhold payment they would require to give notice in terms of section 111(1) of the Act. No such notice was given.

The interim certificate is not conclusive evidence that the works in respect of which the pursuers seek payment were in accordance with the contract (see clause 30.10). That however does not preclude the sum brought out in an interim certificate being a sum due under the contract. The structure and intent of the Act, as I understand it, and accepted by the solicitor for the defenders, is to pay now and litigate later.''

13. Sheriff Taylor earlier explained why Lord Macfadyen's case was different. The contract there had no architect or system of certificates. The builder simply presented his bill for payment. The bill in itself did not make any sums due. What, under that contract, would make the sums due is just the fact of the work having been done. So no withholding notice was necessary in respect of works not done – payment was not due in respect of them.'

The judgment of the Court of Appeal should be read in full.

A withholding notice on the grounds that nothing was due under the contract until the adjudicator made his decision, issued prior to enforcement, was held not to be a valid withholding notice in *The Construction Centre Group* v. *The Highland Council*[18] even in the circumstances of unusual contract provisions.

'22. Mr Currie founded his submission that a section 111 notice could competently be given after an adjudicator had issued his decision at least in part on the terms of the section which relate the notice to a "sum due under the contract". It is, no doubt, correct that the defender's obligation to pay the sum awarded by the adjudicator is of contractual nature, founded on the provisions made in the contract in order to comply with section 108(3). But that is not enough to persuade me that the language of section 111 is sufficiently wide to make it legitimate to give a withholding notice in respect of the adjudicator's award. What section 111 provides is that a party may not "withhold payment *after the final date for payment* of a sum due under the contract" (emphasis added). The final date for payment has a technical meaning in terms of section 110(1), which is reflected in this contract in clause 60(1). The context in which the phrase is used is ordinary payments becoming due at intervals during the course of the contract in consequence of the certification procedure. The structure of the contractual timetable is such that payments of that sort may not be withheld after the final date for payment unless a notice has been given. It follows, in my view, that section 111 is intended to apply only to the withholding of payments in respect of which the contract provides for a final date for payment. It was not intended to apply, and does not apply, to payments due in consequence of an adjudicator's decision.

23. If Mr Currie's construction of section 111 were correct, there would be nothing to confine the giving of a post-adjudication notice of intention to withhold the adjudicator's award to cases possessing the peculiarity which this case possesses, namely that there was no opportunity to give such a notice before the adjudication. If he were right that, because the sum awarded by the adjudicator is a sum due under the contract, section 111 permits the giving of a withholding notice in respect of the adjudicator's award, such a notice could be given in any case, whether the point had already been

[18] *The Construction Centre Group Limited* v. *The Highland Council* (23 August 2002).

argued before the adjudicator or not. There would, on that construction, be nothing in the section to restrict the scope of post-adjudication notices to cases in which the point taken in the notice could not have been, and had not been, taken before the adjudicator. It would, however, in my view be destructive of the effectiveness of the institution of adjudication if a responding party could decline to put forward an available defence in the course of the adjudication, then give a section 111 notice seeking to withhold on that ground the sum awarded by the adjudicator.

24. For these reasons I am of opinion that section 111 does not permit the giving of a withholding notice in respect of an adjudicator's award. I do not consider that that gives rise to any injustice.'

Although there is no specific mention in section 111 of the form the notice should take, the conclusion reached in the *Hestia Fireside* case[19] was that an effective notice needs to be issued in writing. The other interesting point in this case is that a section 111 is a notice in response to a particular payment application and certificate. A notice which is issued other than in response to a particular payment is not valid:

'13. There remains the pursuers' argument under the 1996 Act. Two matters of statutory interpretation were raised. The first was whether, as the pursuers contend, an effective notice within the meaning of section 111 requires to be in writing. Mr d'Inverno in the end effectively conceded that it did. Although the words "in writing" are not expressly used, I am satisfied that it unmistakably appears that writing in some form is required. This is so, in my view, having regard not only to the language of section 111 itself, including the use of the indefinite article ("an effective notice", "a notice") and the requirement to "specify" particular matters, but also to the language of section 115 and, in particular, section 115(6) which contemplates that a notice under Part II will be in some form of writing. A telephone message, even one referring to a particular letter of earlier date, will not suffice.

14. The second matter raised was whether a notice effective for the purposes of section 111 could be a communication in writing sent earlier than the making of the relevant Application. Mr d'Inverno pointed out that, while section 111(2) provided that any notice must be given not later than a particular time, it did not provide that it required to be given after any particular time, i.e. there was no terminus a quo. The letter of 17 August, albeit sent prior to the invoice of 27 November, was (or was arguably) a notice of intention to withhold payment within the meaning of section 111. I am unable to accept that argument. The purpose of section 111 is to provide a statutory mechanism on compliance with which, but only on compliance with which, a party otherwise due to make a payment may withhold such payment. It clearly, in my view, envisages a notice given under it being a considered response to the application for payment, in which response it is specified how much of the sum applied for it is proposed to withhold and the ground or grounds for withholding any amount. Such a response cannot, in my view, effactually be made prior to the application itself being made. It may, of course, be that the matter of withholding payment of any sum which might in the future be applied for has previously been raised. In such circumstances a notice in writing given after receipt of the application but which referred to or incorporated some earlier written communication might suffice for the purpose – though I reserve my opinion on that matter. But such an earlier written communication, whether alone or referred to subsequently in an oral communication, cannot, in my view, suffice. This is, as a matter of statutory interpretation, in my view, unmistakably the case.'

There have also been a number of cases where paying parties have issued withholding notices after an adjudicator's decision has been made, in an attempt to resist enforcement. These we deal with in Chapter 12.

---

[19] *Strathmore Building Services Limited* v. *Colin Scott Greig t/a Hestia Fireside Design* (18 May 2000).

Section 111(2)

(2) To be effective such a notice must specify–
   (a) the amount proposed to be withheld and the ground for withholding payment, or
   (b) if there is more than one ground, each ground and the amount attributable to it,
   and must be given not later than the prescribed period before the final date for payment.

For the notice to withhold payment to be effective it must specify the amount to be withheld and the grounds on which it is to be withheld. This is no more onerous than the requirement in many of the pre-Act sub-contract documents. This would also cover withholding a sum in respect of liquidated and ascertained damages under a main contract.

If there is more than one ground an effective notice must state each ground and the amount attributable to that ground. The notice where there are many grounds may become complex. If the intention is to set off sums from other contracts there would need to be an effective notice in respect of the other contract and the contract which is to bear the set-off.

The notice must not be given later than the prescribed period before the final date for payment. This gives a further facet to the period between the due date and the final payment date. The due date is set in section 109. The parties set the agreed (or default) period and the final payment date. Within five days of the due date section 110 requires a notice showing what the payment will be. If any part of the payment is to be withheld a further notice must be issued within an agreed number of days prior to the date for final payment.

Section 111(3)

(3) The parties are free to agree what that prescribed period is to be.
   In the absence of such agreement, the period shall be that provided by the Scheme for Construction Contracts.

The period before that final date for payment by which the notice of withholding money is to be issued is a matter of agreement between the parties. If the payer fails to issue the notice within that period the right to withhold monies against the particular payment is lost. There is nothing to prevent that notice being re-issued or given for the next payment due providing it is within the prescribed period. It is likely that this notice will be the 'trigger' for the offended party to give notice of adjudication. The notice of adjudication need not however be given immediately but can be given at any time. It would appear from the nature of many adjudications that parties are still trying to resolve their disputes during the contract without recourse to adjudication as very many references that are made when the parties have been unable to resolve the final account also allege underpayment in respect of interim applications or certificates.

Where the parties have failed to agree a period, the provisions of the Scheme provide the default position. All of the periods set by the Scheme are in days. For a definition of days see section 116.

Section 111(4)

(4) Where an effective notice of intention to withhold payment is given, but on the matter being referred to adjudication it is decided that the whole or part of the amount should be paid, the decision shall be construed as requiring payment not later than–
   (a) seven days from the date of the decision, or
   (b) the date which apart from the notice would have been the final date for payment, whichever is the later.

The payee who receives an effective notice may have a number of courses of action. He may accept the notice and withholding of the payment. He may seek to modify the notice by negotiation to secure withdrawal in its entirety or to a sum and grounds for which he does accept liability. He may issue the requisite notice that there is a dispute and that he requires adjudication of that dispute.

This sub-section deals with the position where the withholding of payment has been referred to an adjudicator for a decision. If the adjudicator decides that in part or whole the sum must be paid, payment must be made within seven days of the adjudicator's decision or by the date for final payment whichever is the later. Unless the period between the due date and the final date for payment is longer than the adjudication cycle in section 108, it is unlikely that the final date for payment provision will ever apply.

Section 112: Right to suspend performance for non-payment

There is no common law right to suspend the works for late or non-payment:

> 'Apart from suing for interim payments, or requiring arbitration where that is provided for, the remedy — and apparently the only remedy — which the contractor is recognised as having at common law is rescission if a sufficiently serious breach has occurred. If he chooses not to rescind, his own obligations continue. He is bound to go on with the work. All the available English and Commonwealth text books on building contracts state the law consistently with this view…'[20]

> 'It is well established that if one party is in serious breach, the other can treat the contract as altogether at an end; but there is not yet any established doctrine of English law that the other party may suspend performance, keeping the contract alive.'[21]

The Act now provides this right on a statutory basis.

Section 112(1)

> (1) Where a sum due under a construction contract is not paid in full by the final date for payment and no effective notice to withhold payment has been given, the person to whom the sum is due has the right (without prejudice to any other right or remedy) to suspend performance of his obligations under the contract to the party by whom payment ought to have been made ("the party in default").

The suspension provisions of the Act were initially thought to be a cause of great concern, there being nothing to prevent a party suspending performance at a critical stage of the works. In practice however this provision does not seem to have been used to a great extent or, if it has, it does not seem to have created any great difficulty as yet in that there have been no instances, to the authors' knowledge, of disputes relating to this provision having reached the courts.

Some contracts have contained contractual provisions which give a right to suspend performance when payment is not made in accordance with the contract. The Act now provides a statutory right in all construction contracts to suspend performance where payment is not made in full by the final date for payment and there is no effective notice of withholding payment. This right cannot be lost through any provision in the contract[22]. The

---

[20] *Canterbury Pipelines* v. *Christ Church Drainage* [1979] 2 NZLR 347.
[21] *Eurotunnel* v. *TML* [1992] CILL 754.
[22] See *Johnson* v. *Moreton* [1980] AC 37 p. 3–13.

right to suspend is without prejudice to any other rights or remedies available to the offended party. It is therefore conceivable that the suspension will be accompanied by some proceedings to collect the debt due.

In many contracts a certificate of the architect or engineer is condition precedent to any sum being due:

> 'In other words, it was a condition precedent to the contractor's entitlement to payment of a specified sum under clause 30(1) that the architect should have issued an interim certificate stating that such specified sum was due to the contractor from the employer.'[23]

In such contracts the absence of a certificate means that no sum is due and there are no grounds on which to suspend the works. The failure of a party to produce a certificate which was contractually due would be grounds on which to commence adjudication to seek the sum that should have been certified, but not to suspend performance. Similarly if a certificate has been issued and it represents a substantial undervaluation of the works, this would not be grounds for suspension of the works but would form the basis of a dispute that could be adjudicated.

Section 112(2)

(2) The right may not be exercised without first giving to the party in default at least seven days' notice of intention to suspend performance, stating the ground or grounds on which it is intended to suspend performance.

The seven days' notice is a condition precedent to the right to suspend the works. The notice must state the ground or grounds on which it is intended to suspend the works. These grounds by virtue of sub-section (1) above can only be the failure to make the payment at all or the payment of a lesser sum than is due. Before the Act applied, this would have presented a problem under a sub-contract because there was no mechanism to establish with certainty what sum was due. Whilst the sub-contract would have provided the method by which a payment should be calculated, in practice many sub-contractors would not know what they were to receive until they were actually paid. It is also unlikely they would have any concept of how the sum was calculated. Sub-section 110(2) of this Act resolves that problem.

Section 112(3)

(3) The right to suspend performance ceases when the party in default makes payment in full of the amount due.

As soon as the payment is made in full the right of suspension ceases and work must be commenced again. Conceivably there could be a stop/start situation throughout the contract. This warrants careful consideration concerning the extent to which resources are re-allocated or plant and equipment demobilised.

---

[23] *Lubenham Fidelities & Investments Co Ltd* v. *South Pembrokeshire District Council and Another; South Pembrokeshire District Council* v. *Lubenham Fidelities & Investments Co Ltd and Wigley Fox Partnership* (1986) 33 BLR 39 and 6 Con LR 15.

Section 112(4)

(4) Any period during which performance is suspended in pursuance of the right conferred by this section shall be disregarded in computing for the purposes of any contractual time limit the time taken, by the party exercising the right or by a third party, to complete any work directly or indirectly affected by the exercise of the right.
Where the contractual time limit is set by reference to a date rather than a period, the date shall be adjusted accordingly.

The period of suspension is ignored in calculating the time for performance of the contract. It is simply a matter that time does not count against a party who rightfully suspends.
If a party suspends the works and this in turn affects the performance of another party, that other party is also protected for the period of suspension. This is quite a likely occurrence when one considers the extent of sub-contracting in the industry and the interdependence of one trade on another.

The contract may set a period for completion or a definite date by which completion should occur. Where completion is set by a date, the date is adjusted to take account of the period of suspension. This is an automatic right and need not form part of the extension of time provisions in any contract.

There is no provision in the Act to cover the loss and expense or damages caused by a suspension of the works. Where there was a suspension provision in the pre-Act forms of sub-contract this was also the case. This is however not fatal to any claim. The Act covers the position on time almost on the basis that time stands still for the period of suspension. There is no liability on the suspending party. The loss and expense or damage arises from an event that is a breach of contract – failure to pay the sum due. Therefore the damages caused by the suspension would be claimable at common law by the suspending party, and the party in breach of the payment provisions of the contract suffers the consequences of that breach.

Section 113: Prohibition of conditional payment provisions

This section prohibits conditional payment provisions. It should be noted that this refers to conditional payment provisions and not simply to what are known as pay when paid clauses.

Section 113(1)

(1) A provision making payment under a construction contract conditional on the payer receiving payment from a third person is ineffective, unless that third person, or any other person pay-ment by whom is under the contract (directly or indirectly) a condition of payment by that third person, is insolvent.

The definition of a conditional payment provision is 'a payment under a construction contract conditional on the payer receiving payment from the third person'. These are known in the industry as pay when paid or pay if paid clauses. This provision in the Act may not be effective against pay when certified clauses. (For a debate on this point see *Durabella* v. *Jarvis*[24].) Any clause covered by section 113 in a contract is ineffective unless there is insolvency in the payment chain.

---

[24] *Durabella Limited* v. *J. Jarvis & Sons Limited* (19 September 2001).

A main contractor need not pay his sub-contractors where such a clause exists and the employer has become insolvent. This provision is to satisfy the *pari passu* rule (with equal step, equally without preference) in insolvency and not to create a new breed of preferential creditors. The eventuality of a third person who is to make the payment being dependent on other persons who have become insolvent is also dealt with here. In the payment cycles now imposed by sections 110 and 111, the importance of when the insolvency occurs will determine who in the chain is entitled to be paid before the pay when paid clause becomes effective. The key point is receipt of the payment. Therefore if the main contractor has received his payment by the time the employer becomes insolvent, the payment down the chain to the sub-contractors (and indeed the sub-sub-contractors) will have to be made and the main contractor cannot hide behind the pay when paid provision.

Section 113(2)–(5)

(2) For the purposes of this section a company becomes insolvent–
   (a) on the making of an administration order against it under Part II of the Insolvency Act 1986,
   (b) on the appointment of an administrative receiver or a receiver or manager of its property under Chapter I of Part III of that Act, or the appointment of a receiver under Chapter II of that Part,
   (c) on the passing of a resolution for voluntary winding-up without a declaration of solvency under section 89 of that Act, or
   (d) on the making of a winding-up order under Part IV or V of that Act.
(3) For the purposes of this section a partnership becomes insolvent–
   (a) on the making of a winding-up order against it under any provision of the Insolvency Act 1986 as applied by an order under section 420 of that Act, or
   (b) when sequestration is awarded on the estate of the partnership under section 12 of the Bankruptcy (Scotland) Act 1985 or the partnership grants a trust deed for its creditors.
(4) For the purposes of this section an individual becomes insolvent–
   (a) on the making of a bankruptcy order against him under Part IX of the Insolvency Act 1986, or
   (b) on the sequestration of his estate under the Bankruptcy (Scotland) Act 1985 or when he grants a trust deed for his creditors.
(5) A company, partnership or individual shall also be treated as insolvent on the occurrence of any event corresponding to those specified in subsection (2), (3) or (4) under the law of Northern Ireland or of a country outside the United Kingdom.

Sub-sections 113(2) to (4) inclusive give the definition of what constitutes insolvency (or bankruptcy in the case of an individual). These are largely drawn from the Insolvency Act 1986 for England and Wales. Sub-section 113(5) includes appropriate references to cover the position in Northern Ireland and Scotland.

Section 113(6)

(6) Where a provision is rendered ineffective by subsection (1), the parties are free to agree other terms for payment.
   In the absence of such agreement, the relevant provisions of the Scheme for Construction Contracts apply.

The parties are free to agree terms of payment where a pay when paid clause is rendered ineffective by the Act. Therefore where a contract is signed which includes a conditional payment clause outlawed by the Act, a new method of effecting payment can be agreed. This must necessarily comply with the provisions of section 110. If no such agreement is reached the provisions of the Scheme apply as the default position.

Chapter 5
The Secondary Legislation

The Scheme for Construction Contracts

The Scheme for England and Wales was introduced on 6 March 1998 as was the Scheme for Scotland. This was the required secondary legislation to make Part II of the Act operative. With this secondary legislation in place the whole of the adjudication and payment provisions could then be applied to construction contracts. Commencement Order 1998 No.649 made the Act and the Scheme operative from 1 May 1998. There was an equivalent Order in Scotland. On the same date an Exclusion Order was made under the powers conferred on the secretary of state to exclude certain types of contract from the definition of construction contracts in the Act.

The Scheme and the whole of the legislation for Northern Ireland became operative on 1 June 1999.

The Scheme is the default position where the parties have failed to address the requirements for construction contracts in the primary legislation. In the Act section 114 states:

(4) Where any provisions of the Scheme for Construction Contracts apply by virtue of this Part in default of contractual provision agreed by the parties, they have effect as implied terms of the contract concerned.

The terms of the Scheme are applied as if they were part of the contract. It has been argued that the Scheme does not actually comply with the Act. It does not have to; these are substitute provisions that stand in their own right.

The industry had adequate notice to change its contracts prior to the date of commencement of the legislation. Despite this many of the standard contract writing bodies were late in providing appropriate compliant amendments. It comes as no surprise therefore that in the first 40 months of the legislation the majority of the adjudications were carried out under the Scheme.[1] The authors suspect that this situation has not changed. Despite the publicity the legislation has had, there are still many who are ill-informed. Another possible explanation is that the industry is notoriously poor at addressing contractual relationships.

It has been argued that there should be a single set of rules for the industry and the way to do this is to adopt the Scheme wholesale. The JCT has recently adopted the Scheme as its adjudication provisions in the Major Projects Form. The authors do not agree with this. Where legislation is to be imposed, as is the case here, there must be a balance between freedom of contact and imposition. The legislation has addressed the balance correctly. Contract writing bodies should take the lead in providing compliant rules in their contracts.

Where there is a dispute as to whether the terms of a standard contract or the Scheme rules apply, the courts may well enforce an adjudicator's decision where the process has been

---

[1] *Adjudication – The First Forty Months*. A Report on Adjudication under the Construction Act. Construction Industry Council, 2002.

conducted under the Scheme despite it being argued that if the standard contract applied there would be no jurisdiction[2].

The Scheme is in two parts. There are provisions covering adjudication and provisions covering payment. This follows the pattern of the Act. If the contract does not comply fully with all of section 108(1) to (4) of the Act, Part I – Adjudication under the Scheme applies.

The payment provisions are not quite so straightforward. The Scheme applies on a section by section basis. Where the parties have failed to reach agreement under one or more of sections 109, 110, 111 or 113 of the Act, the Scheme will apply to each individual section where there is non-compliance. The Scheme imposes its own time periods where there has been failure to agree such time periods in accordance with the Act.

Parties can wholly avoid the Scheme by making contracts that comply with the Act. Experience has shown that adjudications which fall under the Scheme are those contracts made using simple orders, by exchange of letters, or work by small contractors, whose terms and conditions do not comply with the Act or do not refer to other terms which provide compliant clauses for both adjudication and payment.

The Exclusion Order regularises the definition of construction contracts and takes outside the regime of the Act certain arrangements that would have been unsuitable for the application of both the adjudication and payment provisions. The whole of the legislation now sets the minimum criteria for construction contracts. There is nothing to prevent any construction contract going beyond the minima required. What the legislation is doing is creating contractual terms either by voluntary amendments to the contracts or by imposition through the secondary legislation. If the parties include more in their contracts than is needed to comply with the legislation the courts will enforce those terms. There is no opportunity to opt back to the legislation when there is a dispute that falls to be adjudicated or the payment terms offer an improved position to that required by the Act.

The Scheme Part I – Adjudication

Notice of intention to seek adjudication

Paragraph 1

1. (1) Any party to a construction contract (the 'referring party') may give written notice (the 'notice of adjudication') of his intention to refer any dispute arising under the contract, to adjudication.

   (2) The notice of adjudication shall be given to every other party to the contract.

   (3) The notice of adjudication shall set out briefly–
      (a) the nature and a brief description of the dispute and of the parties involved,
      (b) details of where and when the dispute has arisen,
      (c) the nature of the redress which is sought, and
      (d) the names and addresses of the parties to the contract (including, where appropriate, the addresses which the parties have specified for the giving of notices).

Paragraph 1 deals with the notice of intention to refer any dispute arising under the contract to adjudication. It describes the basic requirements of the notice itself. If the notice does not comply with the minimum requirements it may be void. A notice that gives less than the

---

[2] *Pegram Shopfitters Limited* v. *Tally Wiejl (UK) Limited* (14 February 2003).

minimum requirements of the Scheme could give rise to arguments on jurisdiction when the adjudicator is appointed. The party seeking the adjudication is called the referring party. It will not be necessary to use this terminology in any decision made by an adjudicator; terms such as applicant or contractor, sub-contractor or employer should be adequate to describe the parties and will be more appropriate in the contractual context. The adjudicator becomes part of this contractual machinery. The importance of the notice is that it must be served on every other party to the contract. This envisages a situation where there can be more than two parties involved with the contract.

Paragraph 1(3) lists the minimum criteria for a valid notice. Brevity in the notice is envisaged. The matters to be listed really ought to be common sense. Even where notice is issued under adjudications that are not subject to the Scheme, it would be sensible to provide this minimum information. The details of the dispute and the nature of the redress sought are fundamental in any notice of adjudication.

The jurisdiction of the adjudicator comes from the notice. The notice has been described as the four corners of the claim[3]. It is important therefore to draft the notice carefully to identify properly the dispute to be decided and also to comply with any minimum criteria. This was explored in *K & D Contractors* v. *Midas Homes*[4]:

> 'In the context of any notice of adjudication it is essential to inform the other party and the adjudicator of the basis upon which such a claim is being made and which justifies these invoices, i.e. the statement of the nature of the redress, as required by the Scheme.'

Where the notice is inadequate it is possible that there is no dispute identified and therefore no jurisdiction to decide anything. There will therefore be nothing to enforce should the matter proceed through adjudication to a decision.

Paragraph 2

2. (1) Following the giving of a notice of adjudication and subject to any agreement between the parties to the dispute as to who shall act as adjudicator —

(a) the referring party shall request the person (if any) specified in the contract to act as adjudicator, or

(b) if no person is named in the contract or the person named has already indicated that he is unwilling or unable to act, and the contract provides for a specified nominating body to select a person, the referring party shall request the nominating body named in the contract to select a person to act as adjudicator, or

(c) where neither paragraph (a) nor (b) above applies, or where the person referred to in (a) has already indicated that he is unwilling or unable to act and (b) does not apply, the referring party shall request an adjudicator nominating body to select a person to act as adjudicator.

(2) A person requested to act as adjudicator in accordance with the provisions of paragraph (1) shall indicate whether or not he is willing to act within two days of receiving the request.

(3) In this paragraph, and in paragraphs 5 and 6 below, an 'adjudicator nominating body' shall mean a body (not being a natural person and not being a party to the dispute) which holds itself out publicly as a body which will select an adjudicator when requested to do so by a referring party.

---

[3] *Whiteways Contractors (Sussex) Limited* v. *Impresa Castelli Construction UK Limited* (9 August 2000).
[4] *Ken Griffin & John Tomlinson, t/a K & D Contractors* v. *Midas Homes Limited* (21 July 2000).

Paragraph 1 deals with 'the notice'. This paragraph deals with 'the request' for the adjudicator to act.

Paragraph 2(1) envisages that the parties may agree on who shall act as adjudicator. Sub-paragraph (a) deals with the position where there is an adjudicator named in the contract. Although the provisions of the contract may not comply with the Act, and therefore the Scheme applies, the naming of the adjudicator in the contract is recognised by the Scheme. The naming of the adjudicator in the contract may not survive where the scheme for adjudication in the contract is non-compliant[5].

While the Scheme envisages the naming of an adjudicator in the contract, this ought to be used with caution. There is always a question as to the possibility of bias where an adjudicator is named widely across a number of contracts by the party proposing the contract. It is wise to name an individual rather than a practice or some other form of collective. Although not doing so is not fatal, it may be a cause of difficulty at a time when the adjudicator is needed[6].

The referring party requests the adjudicator to act. It should be noted that where an adjudicator is named in the contract, the request to that adjudicator is a mandatory requirement.

Sub-paragraph (b) deals with two situations. If no person is named in the contract, this sub-paragraph applies. It also applies if there is a person named in the contract but, for whatever reason, that person has already indicated that he is unwilling or unable to act.

This sub-paragraph applies where a nominating body is named in the contract. The Scheme seeks to preserve those aspects of the contract where the parties have reached agreement, notwithstanding the fact that the contract clause itself does not comply with the Act and therefore the Scheme applies by default. The request to the nominating body is mandatory in this situation.

Sub-paragraph (c) covers the situation where no nominating body is named in the contract. The referring party can then request any adjudicator nominating body to select a person to act. In the earlier drafts of the Scheme adjudicator nominating bodies had a different status. There were 16 bodies who were to be designated as adjudicator nominating bodies and who were to be named in the legislation. This was dropped at a late stage on the grounds of avoiding the need to take the legislation back to parliament each time it was necessary to add further names or to remove names from the list.

Paragraph 2(2) requires that a person requested to act as adjudicator shall indicate whether or not he is willing to act within two days of receiving the request. This request can come from one of the parties, usually the referring party, or from the nominating body. Days include non-working days such as weekends but exclude bank holidays. This duty on the prospective adjudicator or nominating body is therefore onerous.

Paragraph 2(3) gives the definition of an adjudicator nominating body. The body cannot be a natural person or a party to the dispute but firms, companies and partnerships will fit the definition. The body simply needs to hold itself out publicly as a body that will select an adjudicator when requested to do so by a referring party.

This leaves the task open to any commercial concern that wishes to enter the market place to make money from such a selection. It has no regard to any training, quality or indeed whether the body holds a list or has any adjudicators of its own. They could almost be a labour agency-type arrangement. This total lack of control must be unsatisfactory. Having

---

[5] *R.G. Carter Limited* v. *Edmund Nuttall Limited* TCC Judge Thornton (21 June 2000).
[6] *John Mowlem & Company Plc* v. *Hydra-Tight Ltd (t/a Hevilifts)*, TCC Judge Toulmin (6 June 2000).

said this, there are non-professional bodies who act as adjudicator nominating bodies (ANBs) who are reputable. At least in the original proposals the organisations selected had been vetted, had given undertakings as to their practice and procedure and were to be monitored.

This monitoring process has taken place independently to a limited degree. The Construction Industry Council has monitored ANBs in order that it may form its own panel of adjudicators. Some of the professional institutions have carried out examination of other ANBs in order to allow exemptions on training courses. Unfortunately none of the findings of any monitoring have been published.

Paragraph 3

3. The request referred to in paragraphs 2, 5 and 6 shall be accompanied by a copy of the notice of adjudication.

A copy of the notice referred to in paragraph 1 must accompany the request for a person to act. This is only the notice referred to in paragraph 1. There are two notices required in the adjudication process under the Scheme: the notice of adjudication referred to in paragraph 1 and the notice that refers the dispute to the adjudicator under paragraph 7. These are known as the 'notice of adjudication' and the 'referral notice'. Most of the standard form contracts also make the distinction between the notice of adjudication and the notice that refers the dispute to the adjudicator.

Paragraph 4

4. Any person requested or selected to act as adjudicator in accordance with paragraphs 2, 5 or 6 shall be a natural person acting in his personal capacity. A person requested or selected to act as an adjudicator shall not be an employee of any of the parties to the dispute and shall declare any interest, financial or otherwise, in any matter relating to the dispute.

This paragraph requires that the adjudicator is an individual, acting in his personal capacity, and thus bars companies from being adjudicators under the Scheme. It does not bar the employee of a company from acting as an adjudicator providing he acts in his personal capacity. However he cannot be an employee of any of the parties to the dispute.

If a company were to be named in a contract as the adjudicator or as the provider of an adjudicator by agreement and subsequently it was found that the contract did not comply with the Act, that company could not act as adjudicator under the Scheme.

It must be questionable in any event as to whether there are any benefits in naming a company as an adjudicator in a contract. The risks must be high that you will be provided with a person from that company to carry out the adjudication purely on the grounds of availability with no regard to their suitability. This must be compared with the lists of adjudicators of professional institutions that are the basis for selection on the grounds of suitability as well as availability.

The declaration required in the last part of this paragraph would appear not to bar a person with an interest, financial or otherwise, in any matter relating to the dispute from acting as adjudicator, providing such interest was declared. In such circumstances the independence and possibly the impartiality of the individual must be questionable. If the declaration is made, it must be open to one or both of the parties to refuse the proposed adjudicator. What happens if one party seeks to refuse the proposed adjudicator and the other wants him to proceed? Paragraph 10 would not permit any objection by a party to invalidate an appointment or the decision made by the adjudicator. The adjudicator can resign at any time under the provisions of paragraph 9. This is unsatisfactory. A potential adjudicator should consider carefully his relationship with either party and/or the dispute at the time of the original enquiry and before saying he is willing to act. If there is a remote possibility that the acceptance may lead to the possibility of having to resign if there is an objection, the appointment should not be taken. The option to resign at any time should not be taken lightly. Adjudicators who are too ready to resign at the first difficulty will bring the whole system into disrepute.

Paragraph 5

5. (1) The nominating body referred to in paragraphs 2(1)(b) and 6(1)(b) or the adjudicator nominating body referred to in paragraphs 2(1)(c), 5(2)(b) and 6(1)(c) must communicate the selection of an adjudicator to the referring party within five days of receiving a request to do so.

   (2) Where the nominating body or the adjudicator nominating body fails to comply with paragraph (1) the referring party may –
   (a) agree with the other party to the dispute to request a specified person to act as adjudicator, or
   (b) request any other adjudicator nominating body to select a person to act as adjudicator.

   (3) The person requested to act as adjudicator in accordance with the provisions of paragraphs (1) or (2) shall indicate whether or not he is willing to act within two days of receiving the request.

This paragraph refers to a nominating body and an adjudicator nominating body to cover the differentiation in paragraphs 2 and 6. We shall refer to a 'nominating body' for the purposes of discussion where the Scheme uses these two alternate terms. This paragraph deals with the timing of a nomination. The selection of an adjudicator must be communicated to the referring party within 5 days of receiving the request to do so. This is not difficult with modern means of communication. The requirement in the Scheme is only to communicate to the referring party. This illustrates the point that adjudication is a process that can be commenced unilaterally. In practice the nominating bodies communicate their selection to the responding party as well as the referring party.

It is the general practice of nominating bodies not to proceed with a selection until an application form is completed and received accompanied by the fee that they require for making the nomination. The Scheme obviously does not take account of any rules that the nominating bodies may have or of the procedures they adopt before they make a selection. The request can therefore be received before the fee and as far as the Scheme is concerned this is the starting point of the five days. The five days could be lost before the fee is received but in practice it appears rare that the nominating body exceeds the allotted five days except for good reason, probably because the referring party generally pays the fee on making the application.

The forms required for a nomination of an adjudicator will often contain much more information than just the notice of adjudication. This information is useful to the nominating body in determining a suitable adjudicator. It is also useful to the adjudicator in deciding whether the dispute is within his experience, and where it contains full contact information he will have all the necessary data he needs to make contact as soon as he is nominated.

Paragraph 5(2) covers the situation if there is a failure on the part of the nominating body to comply with paragraph 5(1). It requires that the parties almost start again. This must be frustrating for a referring party. Paragraph 5(2)(a) is unhelpful because the odds are that there already has been a failure to reach an agreement on the name of an adjudicator and it is therefore unlikely that agreement will be reached at this stage. It is more likely that help will have to be sought from another nominating body as envisaged in paragraph 5(2)(b). It is however of little use to make a fresh application if a reason that the first nominating body has not nominated anyone is the failure to provide the fee.

Paragraph 5(3) requires that the person requested to act as adjudicator confirms whether or not he is willing to act within two days of receiving the request. Clearly the periods of five and two days equal the seven days envisaged by the Act. However there can be a gap between the expiry of five days and the commencement of the two days where the adjudicator confirms his willingness or otherwise to act. This could simply be caused by postal communications. It would not invalidate the selection of an adjudicator. Unusually the governing date is receipt of the request and not the date on which the request was made.

Paragraph 6

6. (1) Where an adjudicator who is named in the contract indicates to the parties that he is unable or unwilling to act, or where he fails to respond in accordance with paragraph 2(2), the referring party may —
        (a) request another person (if any) specified in the contract to act as adjudicator, or
        (b) request the nominating body (if any) referred to in the contract to select a person to act as adjudicator, or
        (c) request any other adjudicator nominating body to select a person to act as adjudicator.
    (2) The person requested to act in accordance with the provisions of paragraph (1) shall indicate whether or not he is willing to act within two days of receiving the request.

Paragraph 6 simply deals with the eventuality of an adjudicator named in the contract confirming that he is unwilling to act or failing to respond in accordance with paragraph 2(2).

Paragraph 6(1)(a) contemplates the possibility that there could be more than one adjudicator named in the contract.

Paragraphs 6(1)(b) and (c) follow a similar pattern to paragraph 5(2), the first choice being to seek selection from a nominating body if named in the contract, and where there is none to seek selection from an adjudicator nominating body.

As with paragraphs 2 and 5 the person selected to act as the replacement for the adjudicator named in the contract must indicate whether or not he is willing to act within two days of receiving the request. In practice the request will be the standard form of enquiry from the nominating body to the potential adjudicator. This enquiry must seek to eliminate any conflicts of interest. Notwithstanding the liberal rules of the Scheme, the nominating body can impose such rules as it wishes on those it is prepared to nominate.

In practice the nominating bodies have all the matters required by the scheme off to a fine art. The telephone and facsimile are used widely at this stage.

Paragraph 7

7. (1) Where an adjudicator has been selected in accordance with paragraphs 2, 5 or 6, the referring party shall, not later than seven days from the date of the notice of adjudication, refer the dispute in writing (the 'referral notice') to the adjudicator.

    (2)   A referral notice shall be accompanied by copies of, or relevant extracts from, the construction contract and such other documents as the referring party intends to rely upon.

    (3)   The referring party shall, at the same time as he sends to the adjudicator the documents referred to in paragraphs (1) and (2), send copies of those documents to every other party to the dispute.

The referral notice required by paragraph 7(1) sets the whole process in motion. The adjudicator is selected or where named in the contract confirms willingness to act. The adjudication is then under way. The notice referred to here must not be confused with the notice in paragraph 1. The notice in paragraph 1 is the notice of adjudication. The notice required by paragraph 7 is the referral notice. It must be issued not later than seven days from the date of the notice of adjudication. It must be issued to the adjudicator. This could be difficult if there is a breakdown of the procedures envisaged in paragraphs 2, 5 and 6 and no adjudicator is selected within seven days of the notice of adjudication. This may lead to an argument from a reluctant responding party that the adjudication cannot proceed because the referral notice was too late. This may result in the referring party having to commence the whole process again. It is likely that many adjudicators will ignore these technicalities and proceed until a substantial point on jurisdiction is raised. What is certain is that the referring party must really have everything in place before the notice of adjudication is issued. This almost invites the 'ambush' situation that many of the pundits in the trade press feared.

    Disappointingly there has been no reported case law on the use of the mandatory language 'shall' in paragraph 7(1). This is stronger than 'with the object of' in section 108(2)-(b) of the Act. It is thought the courts will not apply this technicality strictly in the absence of sanction when faced with an action to enforce a decision.

    Paragraph 7(2) requires that copies of all relevant extracts from the construction contract shall accompany the referral notice, together with such other documents as the referring party intends to rely upon. There may be some difficulty here having regard to section 107 of the Act. That section gives a very wide definition of what constitutes a construction contract that is in writing. Among other matters an oral contract which refers to terms which are in writing makes an agreement in writing. The referring party will have to do the best it can in these circumstances.

    At the same time that the documents are sent to the adjudicator by the referring party, they must also be sent to every other party to the dispute.

Paragraph 8

8. (1)   The adjudicator may, with the consent of all the parties to those disputes, adjudicate at the same time on more than one dispute under the same contract.

    (2)   The adjudicator may, with the consent of all the parties to those disputes, adjudicate at the same time on related disputes under different contracts, whether or not one or more of those parties is a party to those disputes.

    (3)   All the parties in paragraphs (1) and (2) respectively may agree to extend the period within which the adjudicator may reach a decision in relation to all or any of these disputes.

    (4)   Where an adjudicator ceases to act because a dispute is to be adjudicated on by another person in terms of this paragraph, that adjudicator's fees and expenses shall be determined in accordance with paragraph 25.

During the consultation period on the Scheme there was a question as to whether there should be any provision for joinder. There is always an argument with multi-party disputes that decisions will be inconsistent if more than one person examines the same facts and matters across the disputes with differing parties. This paragraph has not really assisted in achieving true joinder provisions.

Paragraph 8(1) permits the adjudicator to deal with more than one dispute under the same contract. It requires consent of the parties. The other disputes therefore are not those contained within the original notice of adjudication. This paragraph is not needed at all where a number of matters are included in the original notice. The adjudicator would in any event have the jurisdiction to deal with all of the disputes in the original notice. See Chapter 3 for the debate on single and multiple disputes under section 108(1) of the Act.

Paragraph 8(2) is an attempt at joining related disputes. This could apply where there is a dispute between the main contractor and a sub-contractor and a related dispute on the same facts between the main contractor and the employer. However what is needed for this to apply is the consent of all the parties to the related disputes. All parties will see the benefit of this procedure but unless the contractor (for example) persuades the sub-contractor that it is not a delaying tactic, it is unlikely that this will be readily achieved. The sub-contractor will no doubt consider that he can get an adjudicator's decision in less time than it takes to sort out the joinder.

There will inevitably be a difficulty with the timescales where more than one dispute is being adjudicated or disputes are added to an adjudication which has already commenced.

The provisions of paragraph 8(2) have proved problematical in practice. They were first examined in *Grovedeck* v. *Capital Demolition*.[7] The comments by the judge were remarks he made without making a decision on what was a second line of defence in the case. Judge Bowsher's remarks are as follows:

> **'One contract, not more than one to be referred.**
> It is not necessary for me to make any decision on the second line of defence. However, it is a matter of practical importance so I shall say something about it.
> In part II of the Act, wherever there is reference to contract it is in the singular, "a contract" or "the contract". Mr Royce relies on sections 5 and 6 of the Interpretation Act 1978, "Unless the contrary intention appears, 'words in the singular include words in the plural'". So does a contrary intention appear? I do not think that a contrary intention does appear from the Act. Reading the Act alone, I see nothing to prevent more than one contract being included in one referral. If there is to be any restriction on the number of contracts, or the number of disputes under one contract, to be referred, one has to look to the terms of the contract or the statutory scheme. But the restriction, if any, is to be derived from the contract or the statutory scheme. The statute is not to be construed by reference to the statutory instrument made under it.
> Paragraph 8 of the statutory scheme indicates that it is only with the consent of the parties that an adjudicator can adjudicate at the same time on more than one dispute under the same contract or adjudicate at the same time on related disputes under different contracts. (What is one dispute may raise interesting philosophical questions. In *Fastrack Construction* v. *Morrison Construction* (unreported, 4 January 2000) Judge Thornton at paragraph 20 said that "the dispute which may be referred to adjudication is all or part of whatever is in dispute at the moment that the referring party first intimates an adjudication reference".) So where the statutory scheme applies, that is the position. But I see no reason why a construction contract in writing which sufficiently complied with section 108 of the Act as to avoid the application of the Scheme should not provide for the referral of more than one dispute or more than one contract without the consent of the other party. Parties might be unwise to agree to such a term, but I do not see why they should not do so. Section 108(2)(a)

---

[7] *Grovedeck Limited* v. *Capital Demolition Limited* [2001] BLR 181.

of the Act requires that a construction contract shall "enable a party to give notice at any time of his intention to refer *a* dispute to adjudication" but I do not read that as showing any intention that the singular does not include the plural.

In the present case, if the Act had applied to the contracts, the Scheme would have applied and the claimants would have had no right to refer more than one dispute or more than one contract except with the consent of the defendants. That is another ground for refusing to enforce this adjudication.'

This was revisited in *Pring & St Hill* v. *Hafner*[8]:

'15. It was thus submitted that the purpose of paragraph 8(2) was to ensure that the parties knew that the adjudicator might acquire knowledge or hear submissions in relation to the instant dispute which would have to be considered in the light of what he might learn or be told or find out, carrying out his investigative powers, on the other dispute.

16. This is a risk which the parties might well wish to take. As the adjudicator himself pointed out in the correspondence, and as suggested by Mr Scott Holland, that risk is minimised by conducting the proceedings not in parallel but in tandem, or by ensuring in some way that what is learnt in the one is not revealed in the other. Of course, that latter course cannot take place without the consent of all the parties involved. In my judgment paragraph 8(2) is intended to cover, and does cover, a variety of circumstances. It is intended to cover all the situations in which there may be related disputes under different contracts, whether or not the parties are the same and whether or not there may be [sic] permissibly be consolidation of the two proceedings. It applies whenever where one party needs to know or may need to know, before allowing the adjudication to proceed in that way, whether the adjudicator is going to have to pass on information or may acquire information which would not be available in the other adjudication to which it is not a party. In other words they are all circumstances where, as a matter of principle, a party's rights to the resolution of a dispute, privately and con-fidentially, would or might be infringed by the introduction of a third party, either in the same proceedings or by having the dispute determined by a person who would or could acquire knowledge from the other proceedings but which could not be used in the resolution of the dispute, yet might either consciously or unconsciously influence its outcome. A party must give a real and informed consent to any reduction in such rights.'

The courts are expecting conscionable effort on the party or the parties to agree to any joinder or consolidation. The conduct of the parties may not be sufficient; where a party does not object to the consolidation during the currency of the adjudication this is one of the few instances where the right to object on grounds of want of jurisdiction may be preserved. The situation envisaged in *Pring & St Hill* could cover four different parties under two separate contracts with disputes that are nevertheless related.

Paragraph 8(3) attempts to deal with this situation but requires agreement of all parties involved in the disputes to agree to an amended timetable. It is unlikely that all parties will agree a timetable where they all have separate but connected disputes. The disputes may be based on the same facts but the parties' needs and expectations all differ. The adjudicator should therefore agree the timetable with the parties before he is willing to take on the related disputes. This will ensure there is no difficulty with the 28-day period for reaching a decision.

Paragraph 9

9. (1) An adjudicator may resign at any time on giving notice in writing to the parties to the dispute.

---

[8] *Pring & St Hill Limited* v. *C.J. Hafner t/a Southern Erectors*, TCC, Judge Humphrey Lloyd QC (31 July 2002).

(2) An adjudicator must resign where the dispute is the same or substantially the same as one which has previously been referred to adjudication, and a decision has been taken in that adjudication.

(3) Where an adjudicator ceases to act under paragraph 9(1) —
(a) the referring party may serve a fresh notice under paragraph 1 and shall request an adjudicator to act in accordance with paragraphs 2 to 7; and
(b) if requested by the new adjudicator and insofar as it is reasonably practicable, the parties shall supply him with copies of all documents which they had made available to the previous adjudicator.

(4) Where an adjudicator resigns in the circumstances referred to in paragraph (2), or where a dispute varies significantly from the dispute referred to him in the referral notice and for that reason he is not competent to decide it, the adjudicator shall be entitled to the payment of such reasonable amount as he may determine by way of fees and expenses reasonably incurred by him. The parties shall be jointly and severally liable for any sum which remains outstanding following the making of any determination on how the payment shall be apportioned.

The whole of paragraph 9 deals with circumstances under which an adjudicator can or should resign.

The first premise in paragraph 9-(1) is that an adjudicator may resign at any time on giving notice in writing to the parties to the dispute. Generally speaking this is undesirable. Adjudicators should only take on work which is within their competence and that they can carry out within the time-frame required by the legislation. Having regard to the primary intent of the adjudication process, to expedite cash flow, an adjudicator should not consider resignation lightly.

Where an adjudicator resigns in accordance with paragraph 9(1), this may be a breach of contract. This will certainly lead to a situation where the parties are unlikely to pay him and he is unlikely to be in a position where he could enforce any right to payment. It is unlikely that such a resignation will constitute an act or omission in bad faith. It is questionable therefore if an adjudicator resigns, whether or not the parties would get very far in an action against him for breach of contract. The only prejudice they will have suffered in most circumstances will be a short delay while they appoint a new adjudicator.

There is a perfectly legitimate circumstance in which an adjudicator should consider resignation. This is when the process is unfair due to some circumstance that has arisen during the adjudication. The remedy in this situation is for the parties to agree more time to allow the process to be conducted fairly. In practice most adjudicators will not have any difficulty obtaining more time when the reasons are properly explained to the parties. In *Balfour Beatty* v. *Lambeth*[9] Judge Humphrey Lloyd QC stated the following:

'That is not something which could necessarily and practicably be done within the time allowed. On the other hand, if, as Mr Richards himself recognised right from the outset, the nature of BB's case was likely to make it extremely difficult for himself (or for Lambeth) to complete a reasonably fair investigation of it within the original 28 days, or even the further period agreed by the parties, then an adjudicator would have to say that it would not be possible to carry out such an investigation and to arrive at a decision, even if it was "coarse" (to quote paragraph 6.4 of his decision) within the time available. An adjudicator does not act impartially or fairly if he arrives at a decision without having given a party a reasonable opportunity of commenting upon the case that it has to meet (whether presented by the other party or thought to be important by the adjudicator) simply because there is

---

[9] *Balfour Beatty Construction Limited* v. *The Mayor And Burgesses Of The London Borough of Lambeth*, TCC, Judge Humphrey Lloyd QC (12 April 2002).

not enough time available. An adjudicator, acting impartially and in accordance with the principles of natural justice, ought in such circumstances to inform the parties that a decision could not properly reasonably and fairly be arrived at within the time and invite the parties to agree further time. If the parties were not able to agree more time then an adjudicator ought not to make a decision at all and should resign.'

There is also some helpful comment in *Ballast* v. *Burrell*[10]:

'It also appears to me to be necessary to remember that, although both parties to the contract undoubtedly have a strong interest in the enforceability, without delay, of adjudicators' decisions, they also have an interest in being protected against decisions which are unjust. . . . Notwithstanding the ephemeral and subordinate character of an adjudicator's decision, and the deemed intention that adjudication should be an expeditious procedure rooted in commercial common sense, I would be slow to attribute to the parties an intention that the adjudicator's decision should always be binding notwithstanding errors of law, procedural unfairness or lack of consideration of relevant material submitted to him by the parties, no matter how fundamental such a breach of the adjudicator's obligations might be.'

Correctly adjudicators have been trained not to resign simply because they are challenged or because the matter they are dealing with is difficult. There are limited circumstances, primarily when what they are faced with is unjust, where they should consider resignation as a last resort.

Paragraph 9(2) is sensible. It seems there would be nothing to prevent a party who is dissatisfied with an adjudicator's decision, from framing his dispute in a slightly different manner and seeking a further adjudication. In such circumstances this paragraph makes it mandatory that the adjudicator must resign. This will only occur when the similarity in the disputes is readily apparent. There is nothing to prevent the respondent in the second adjudication arguing as a defence that the dispute has already been decided in a previous adjudication. This will ensure, if this is the case, that the dispute will not be adjudicated again.

This was examined in *Sherwood & Casson* v. *McKenzie*[11]. There are some important and practical matters which arise from this judgment:

'27. However, unlike the question of whether or not there is an underlying contract in existence, the adjudicator is given jurisdiction to determine whether or not the two disputes are substantially the same. The jurisdiction is analogous to that given to arbitrators for the first time by the Arbitration Act 1996 to determine their own jurisdiction. This jurisdiction has long held by arbitrators acting under Civil Law systems, a power often characterised in such systems as kompetenz kompetenz. It might well be thought that if the adjudicator is given the power to determine the jurisdictional question of substantial overlap, he also has the power to make an error in determining that question which is not open to challenge, save perhaps on grounds of perversity or unreasonableness.

28. However, I do not accept that the adjudicator's powers are not open to challenge. Since the scheme provides that the adjudicator has an obligation to furnish reasons for his decision, if these are requested, it can readily be ascertained whether any jurisdictional challenge is being mounted on reasonable and bona fide grounds and whether the adjudicator has correctly determined that challenge. It makes no sense, as I see it, to impose on an adjudicator a mandatory requirement to resign if there is a substantial overlap between the dispute referred to him and one already decided

---

[10] *Ballast Plc* v. *The Burrell Company (Construction Management) Limited*, Outer House, Court Of Session, Lord Reed (21 June 2001).
[11] *Sherwood & Casson Ltd* v. *McKenzie*, TCC, Judge Thornton (30 November 1999).

by an earlier adjudication decision but then to make such an obligation unenforceable. This would be the effect of making an adjudicator's jurisdiction decision unchallengeable.

29. I draw attention to the fact that my decision on this issue is only, strictly speaking, one affecting scheme adjudications. Most adjudications are conducted under Institutional Rules and many place no obligation on the adjudicator to resign where the dispute has already been the subject of an earlier adjudication. Indeed, I was provided with information that "none of seven particular sets of Institutional Rules reviewed contained a provision similar to paragraph 9(3) of the Scheme". However, the same jurisdictional question would arise under these Institutional Rules since, if a dispute had already been substantially decided by an adjudicator, there would not remain in existence a "dispute or difference" capable of being referred to a second adjudicator in relation to the matters decided in that adjudication. Any second appointment would probably be one without jurisdiction.'

There is authority here for the adjudicator to decide his own jurisdiction. Where he takes the route of mandatory resignation, because the two disputes are substantially the same, that decision becomes binding, at least for the purposes of that adjudication and that adjudicator. There is nothing that prevents the referring party trying again with another adjudicator. Where the adjudicator decides to proceed, his decision is open to challenge on the grounds of lack of jurisdiction because the disputes were substantially the same.

In *Holt* v. *Colt*[12] the adjudicator was asked to decide whether the referring party was entitled to a certain sum. When he decided no, the same party referred the same dispute in terms of the same sum but added the words 'or such other sum as the adjudicator may determine'. The judge held that this was different enough for the two disputes not to be substantially the same.

As a general rule an adjudication concerning an interim account followed by an adjudication concerning a final account will always be different disputes because they apply to a different stage in the contract[13].

Paragraph 9(3) deals with the consequences of resignation under paragraph 9(1). This simply allows the referring party to serve a fresh notice and then, if requested by the new adjudicator, the parties will as far as possible make available all the documents given to the previous adjudicator.

Paragraph 9(4) deals with two situations. First, when the adjudicator resigns in the circumstances referred to in paragraph 9(2) where the dispute is the same as one in which a 'decision has been taken'. We have looked at paragraph 9(2) above. Second, where after the referral is made the dispute develops to 'vary significantly' from the dispute that was referred and turns out to be outside his competence. This would be a valid reason to resign and in an adjudication under the Scheme would not be a breach of contract. This relates to changes to the case during the adjudication after the referral. It does not deal with a change of case between the notice of adjudication and the referral notice; such a change would mean absence of jurisdiction which is not a matter for resignation. We suspect that the decision of Judge Seymour QC in *Carter* v. *Nuttall* and later cases that we consider in Chapter 9 mean that a change in the case of the nature envisaged by paragraph 9(4) also takes the matter out of the adjudicator's jurisdiction even if he is competent to continue. If there is no jurisdiction there is nothing to resign from.

Where an adjudicator has resigned for good reason as in paragraph 9(2) or for the further reason given in paragraph 9(4), he is entitled to a reasonable amount in respect of his fees

---

[12] *Holt Insulation Limited* v. *Colt International Limited*, Liverpool District Registry, TCC, Judge Mackay.

[13] *Skanska Construction UK Ltd* v. *(First) The ERDC Group Ltd and (Second) John Hunter*, Outer House, Court Of Session, Opinion Of Lady Paton (Scotland) (28 November 2002); also *Sherwood & Casson Limited* v. *Mackenzie* (30 November 1999).

and expenses. That amount is to be determined by the adjudicator and he should also determine the apportionment of those fees and expenses. The parties remain jointly and severally liable for these fees as they do for the fees of the new adjudicator.

Paragaph 10

10. Where any party to the dispute objects to the appointment of a particular person as adjudicator, that objection shall not invalidate the adjudicator's appointment nor any decision he may reach in accordance with paragraph 20.

It is essential that the adjudicator be allowed to proceed unfettered. This is therefore a sensible provision. Although this allows the adjudicator to proceed despite any objection to him as a particular person and to reach a decision without the danger of that decision being invalidated by the objection, this provision will not support decisions made where the adjudicator has no jurisdiction at all. The position on this is clear from *Pring & St Hill* v. *Hafner*[14]:

'In my judgment, paragraph 10 of the Scheme is concerned about the consequences of an objection to the appointment of a particular person to be the adjudicator – for that is what it says – and it has nothing to do with whether that person, if otherwise validly chosen and appointed, has jurisdiction.'

Paragraph 11

11. (1) The parties to a dispute may at any time agree to revoke the appointment of the adjudicator. The adjudicator shall be entitled to the payment of such reasonable amount as he may determine by way of fees and expenses incurred by him. The parties shall be jointly and severally liable for any sum which remains outstanding following the making of any determination on how the payment shall be apportioned.

(2) Where the revocation of the appointment of the adjudicator is due to the default or misconduct of the adjudicator, the parties shall not be liable to pay the adjudicator's fees and expenses.

There is nothing to prevent the parties ceasing the proceedings at any time. It is only likely that they will do this by agreement and the most likely reason is that they have reached an agreement on the matters in dispute. In these circumstances paragraph 11(1) makes the parties jointly and severally liable for the adjudicator's fees. The adjudicator still makes a decision on how the fees will be apportioned. This does not go as far as the concept of an agreed award under section 51 of the Arbitration Act 1996. The adjudicator only has the authority to determine the apportionment of fees. There is nothing to prevent the parties from seeking the equivalent of an agreed award in a form of a consent decision which deals with both the substantive matters settled in respect of the dispute and the apportionment of fees. However there would need to be agreement for this to be done.

Paragraph 11(2) causes some concern. In drafting the Arbitration Act 1996 effort was made to avoid the use of the term misconduct and the stigma which it carried. There is no clue or guidance here as to what might constitute a default or misconduct of the adjudicator. The scope to argue that the adjudicator has misconducted himself must be very limited in that the powers available to the adjudicator are so wide. There must be very little he can do that is outside these powers save for failing to act impartially. The adjudicator may only find out that this is the line that the parties wish to adopt when he seeks his fees.

---

[14] *Pring & St Hill Limited* v. *C.J. Hafner t/a Southern Erectors*, TCC, Judge Humphrey Lloyd QC (31 July 2002).

This paragraph sets out the only grounds on which the adjudicator's fees and expenses can be challenged. This matter was visited briefly in *Paul Jensen* v. *Staveley*[15]. The adjudicator had been asked to deal with his own jurisdiction as a preliminary issue. He did so and ruled he had no jurisdiction and his appointment came to an end. The court quickly supported the claimant for payment of his fees remarking that there had been no suggestion of any default or misconduct. In *Stubbs Rich* v. *Tolley*[16] the paying party lost a challenge to the adjudicator's fees on the basis that these were excessive and the court gave the following reasons for its decision:

> 'I regret to say that in applying that criterion the Deputy District Judge fell into grave error. First, he was not comparing like with like. I accept Mr Stead's submissions. The role of an adjudicator is wholly different from that of a solicitor who is preparing a client's case for trial. The solicitor prepares a one-sided case for argument in court. The adjudicator, on the other hand, had to read the files – in this case a total of 834 pages – interview the parties, visit the sites, and then prepare his decisions. He was acting in the role of both investigator and judge.
>
> Second, if comparative evidence was relevant and admissible then it should have been in the form of an expert architect/adjudicator; as the court had earlier made provision for, and no doubt by implication wished to hear from.
>
> Thirdly, a court must be very slow indeed to substitute its own view of what constitutes reasonable hours. I quote again from *Mustill and Boyd*, at page 240, "The court does not substitute its own view for that of the arbitrator. In order to make good an allegation of misconduct, very clear evidence is required and it is not enough to show that the amount demanded is more than the court would have considered appropriate if it had been approaching the matter afresh."
>
> For these reasons I reject the allegation that the hours, which were in fact truly done by Mr Smart, were excessive...'

Powers of the adjudicator

Paragraph 12

12. The adjudicator shall—
    (a) act impartially in carrying out his duties and shall do so in accordance with any relevant terms of the contract and shall reach his decision in accordance with the applicable law in relation to the contract; and
    (b) avoid incurring unnecessary expense.

This paragraph has some similarities to section 33 of the Arbitration Act 1996. It is not a word for word reproduction of that section nor would it have been appropriate to adopt the Arbitration Act 1996 for the purposes of adjudication. Adjudication under the Scheme or under the statutory right to adjudication only deals with disputes that arise under the contract. It is therefore appropriate in the Scheme that reference is made both to the relevant terms of the contract and the applicable law in relation to the contract. The obligations here are mandatory. The use of the word 'shall' in three places in this paragraph places onerous duties on the adjudicator. There are requirements to apply both the terms of the contract and the law in relation to the contract. There is no doubt that adjudication here is a judicial process involving the determination of the parties' legal rights. There are occasions when arbitrators and judges get the law wrong, which is why we have a system of appeals. It is no

---

[15] *Paul Jensen Limited* v. *Staveley Industries PLC* (27 September 2001).
[16] *Stubbs Rich Architects* v. *W.H. Tolley & Son Ltd* (8 August 2001).

less likely, particularly with the time restraints, that adjudicators will make decisions that do not comply with the law.

The law is simply difficult in some areas and gives rise to the individual making an error. This should not be something that should occur as a matter of course.

For a commentary on the duty to act impartially see Chapter 9 on jurisdiction, powers and duties.

It was hoped that adjudication would be both quick and relatively cheap. This is an aspect that has come under scrutiny with anecdotal reports of high total fee levels. These should not be read out of context and not treated as the norm.

There is nevertheless an obligation imposed on the adjudicator to avoid incurring unnecessary expense.

Paragraph 13

13. The adjudicator may take the initiative in ascertaining the facts and the law necessary to determine the dispute, and shall decide on the procedure to be followed in the adjudication. In particular he may —

(a) request any party to the contract to supply him with such documents as he may reasonably require including, if he so directs, any written statement from any party to the contract supporting or supplementing the referral notice and any other documents given under paragraph 7(2),

(b) decide the language or languages to be used in the adjudication and whether a translation of any document is to be provided and if so by whom,

(c) meet and question any of the parties to the contract and their representatives,

(d) subject to obtaining any necessary consent from a third party or parties, make such site visits and inspections as he considers appropriate, whether accompanied by the parties or not,

(e) subject to obtaining any necessary consent from a third party or parties, carry out any tests or experiments,

(f) obtain and consider such representations and submissions as he requires, and, provided he has notified the parties of his intention, appoint experts, assessors or legal advisers,

(g) give directions as to the timetable for the adjudication, any deadlines, or limits as to the length of written documents or oral representations to be complied with, and

(h) issue other directions relating to the conduct of the adjudication.

What is clear in any event is that if the adjudicator is going to succeed in reaching a decision within the restraints of a 28-day timetable, he must take the initiative. The process may be an investigation as opposed to a reaction to what the parties provide. The widest single power is in the first sub-paragraph of paragraph 13. The adjudicator may take the initiative in ascertaining the facts and the law necessary to determine the dispute. There are two powers given here. The first is to take the initiative in ascertaining the facts and the law, which is permissive; the second is to decide on the procedure to be followed in the adjudication, which is mandatory.

The adjudicator must therefore make up his mind from the outset if he is to exercise inquisitorial powers and to what extent. He must also set the procedure, part of which will be the timetable he sets from the outset.

The other powers listed in sub-paragraphs (a) to (h) really do no more than enlarge on the overall power that the adjudicator has. These are debated more widely in Chapter 9. It is not thought that this list will restrain the adjudicator in any way. The list is not exhaustive. The approach taken here has similarities to section 34 of the Arbitration Act 1996.

Paragraph 14

14. The parties shall comply with any request or direction of the adjudicator in relation to the adjudication.

This paragraph imposes a duty on the parties. They are obliged to comply with any request or direction of the adjudicator. Compliance is mandatory. However such requirements are to no avail if the adjudicator does not have sanction against the party who fails to comply. The sanctions are given in paragraph 15 below.

Paragraph 15

15. If, without showing sufficient cause, a party fails to comply with any request, direction or timetable of the adjudicator made in accordance with his powers, fails to produce any document or written statement requested by the adjudicator, or in any other way fails to comply with a requirement under these provisions relating to the adjudication, the adjudicator may —
(a) continue the adjudication in the absence of that party or of the document or written statement requested,
(b) draw such inferences from that failure to comply as circumstances may, in the adjudicator's opinion, be justified, and
(c) make a decision on the basis of the information before him attaching such weight as he thinks fit to any evidence submitted to him outside any period he may have requested or directed.

These powers are essential if the adjudicator is to reach a decision within the timetable required. Obviously the Scheme spells out the powers but it is questionable as to whether this is entirely necessary. The Act itself allows the adjudicator to take the initiative in ascertaining the facts and the law, and reading the legislation as a whole the courts ought to be unsympathetic to an uncooperative party who does not participate or comply with whatever is necessary to assist the adjudicator in reaching his decision.

The three sanctions in sub-paragraphs (a) to (c) are helpful. However these need to be exercised with caution. Having regard to that overall timetable the adjudicator must be able to continue in the absence of a party or documents or written statement requested. However, the time given to a party to supply information required must be realistic having regard to the timetable itself. It is a matter of balance and the adjudicator should endeavour to allow the party who has the most to do an appropriate time in which to do it.

When drawing inferences from failure to comply, the adjudicator cannot simply find against a party. He must consider the grounds for the failure to comply with any request. If the grounds are reasonable they will be of less weight against the party failing to comply than when a party deliberately withholds or conceals documents.

There is no actual power here to ban a document if it arrives late, particularly given the provisions of paragraph 17. It will simply get the appropriate consideration in the remaining time available.

It is inevitable that a point will be reached when decisions will be made on the basis of the information before the adjudicator. Some consideration will have to be given to evidence that arrives late in order to comply with paragraph 17, but the time available for such consideration will of course relate to the overall time available. Very late information will only get the scantiest regard if the 28-day timetable or any extended period to that timetable is not to be jeopardised.

Paragraph 16

16. (1) Subject to any agreement between the parties to the contrary, and to the terms of paragraph (2) below, any party to the dispute may be assisted by, or represented by, such advisers or representatives (whether legally qualified or not) as he considers appropriate.

(2) Where the adjudicator is considering oral evidence or representations, a party to the dispute may not be represented by more than one person, unless the adjudicator gives directions to the contrary.

This paragraph gives the parties the right to utilise such representation as they may wish. There may well be situations where parties choose a lawyer, particularly where the adjudication turns on a legal point. There may be situations when no representation is necessary and the adjudication is carried out solely on the investigations undertaken by the adjudicator, principally with the parties themselves.

It is important that the adjudicator has the powers to limit the number of representatives that a party may have. It can be difficult to control any oral hearing necessary with several individuals wishing to make their own representations and put forward their own arguments, although a pointed comment from the adjudicator that 'you are not helping me' generally has a salutary effect.

Paragraph 17

17. The adjudicator shall consider any relevant information submitted to him by any of the parties to the dispute and shall make available to them any information to be taken into account in reaching his decision.

There are two important points under this paragraph. The adjudicator need only consider information that is relevant. This follows the principles of the rules of evidence. There is nothing in the legislation to suggest whether the rules of evidence apply at all. What is important here is not the application of the rules of evidence themselves but the common sense approach of taking into account only those materials that are relevant and the right to reject and not take into account those materials that are not relevant. The other point is that the adjudicator shall make available to either party the information that is to be taken into account in reaching a decision, a point made quite clear in *RSL* v. *Stansell*[17]. This part of the paragraph is written in the future tense, indicating that the adjudicator must tell the parties what he will take into account before the decision is made. This would include any legal or technical advice that the adjudicator has obtained and intends to use in reaching his decision. This does not necessarily put the adjudicator under pressure. It may mean that he will be unable to take into account relevant evidence that arrives too close to the end of the 28-day period unless the time is extended. If the whole process is conducted as an investigation there will be matters that the adjudicator may discover which are privy to only one party or indeed not within the knowledge of either party. Where there are key matters that are discovered by the adjudicator, these ought to be drawn to the attention of the parties before the decision is made.

Paragraph 18

18. The adjudicator and any party to the dispute shall not disclose to any other person any information or document provided to him in connection with the adjudication which the party

---

[17] *RSL (South West) Ltd* v. *Stansell Ltd* (16 June 2003).

supplying it has indicated is to be treated as confidential, except to the extent that it is necessary for the purposes of, or in connection with, the adjudication.

This paragraph does not deal with privilege. It only deals with the matter of privacy and then only insofar as documents supplied have been indicated to be treated as confidential. The parties to the dispute cannot keep documents confidential from one another. The adjudicator can use those documents in making his decision. If at some later date the parties finally decide to settle their dispute by arbitration or litigation, all documents except those which are privileged would be discoverable in any event. At that stage the adjudicator's decision should be ignored and the issues dealt with again from new (*de novo*).

Paragraph 19

19. (1) The adjudicator shall reach his decision not later than —
(a) twenty eight days after the date of the referral notice mentioned in paragraph 7(1), or
(b) forty two days after the date of the referral notice if the referring party so consents, or
(c) such period exceeding twenty eight days after the referral notice as the parties to the dispute may, after the giving of that notice, agree.

(2) Where the adjudicator fails, for any reason, to reach his decision in accordance with paragraph (1)
(a) any of the parties to the dispute may serve a fresh notice under paragraph 1 and shall request an adjudicator to act in accordance with paragraphs 2 to 7; and
(b) if requested by the new adjudicator and insofar as it is reasonably practicable, the parties shall supply him with copies of all documents which they had made available to the previous adjudicator.

(3) As soon as possible after he has reached a decision, the adjudicator shall deliver a copy of that decision to each of the parties to the contract.

Paragraph 19(1)(a) to (c) deals with the timetable. Not surprisingly it reflects exactly the periods given in the Act and permits the extensions to the 28-day period given in the Act.

Paragraph 19(2) deals with the situation where for any reason the adjudicator fails to reach his decision in accordance with the timetable. This could conceivably be due to the default of one of the parties. This is why the adjudicator must keep full control of the proceedings and ensure that a reticent party does not succeed in disrupting the timetable to such an extent that the decision cannot be made in due time. There is no provision here which deals with the decision that is a day or two late. Conceivably if the parties were in agreement they could accept a late decision and agree to be bound by it. If this is to be the case the agreement must be recorded in writing. The court judgment in *St Andrews Bay* v. *HBG Management*[18] should be read in this respect.

In the absence of any such agreement the whole process must commence again. This is done by serving a fresh notice. The new adjudicator is entitled to have all documents from the previous adjudication insofar as it is reasonably practicable. There can be nothing that prevents the new adjudicator in his investigatory role talking to the previous adjudicator to obtain such documents as he had obtained.

Paragraph 19(3) requires that the decision be delivered to each of the parties as soon as possible after it is made. The Act does not make any reference as to whether or not the decision needs to be in writing. What is clear in the Scheme is that the decision must be recorded in some manner, otherwise it would not be possible to deliver a copy of the

---

[18] *St Andrews Bay Development Limited* v. *HBG Management Limited* (20 March 2003).

decision to each of the parties. This opens up the concept of speaking decisions either on audio or video tape.

There was some debate during the drafting stages of the Scheme as to whether or not the adjudicator should hold a lien on the decision until his fees are paid. This was dropped from this final version of the Scheme. Equally there is no obligation in law to extend credit to the parties. There is no reason why the adjudicator is not entitled to payment of his fees and expenses in full on delivery of the decision. We consider the question of liens in Chapter 8.

Adjudicator's decision

Paragraph 20

> **20.** The adjudicator shall decide the matters in dispute. He may take into account any other matters which the parties to the dispute agree should be within the scope of the adjudication or which are matters under the contract which he considers are necessarily connected with the dispute. In particular, he may –
>
> (a) open up, revise and review any decision taken or any certificate given by any person referred to in the contract unless the contract states that the decision or certificate is final and conclusive,
>
> (b) decide that any of the parties to the dispute is liable to make a payment under the contract (whether in sterling or some other currency) and, subject to section 111(4) of the Act, when that payment is due and the final date for payment,
>
> (c) having regard to any term of the contract relating to the payment of interest decide the circumstances in which, and the rates at which, and the periods for which simple or compound rates of interest shall be paid.

The adjudicator will only have jurisdiction to decide the matters in dispute. These are the matters that are identified in the notice of adjudication. Parties often limit or restrict those matters that can be dealt with by poorly worded notices. There is some leeway here on strict jurisdiction points. The parties may agree that the adjudicator can take further matters into account within the scope of the adjudication. The adjudicator can also unilaterally take into account matters under the contract that he considers are necessarily connected with the dispute. This was explored in *A & D Maintenance* v. *Pagehurst*[19]:

> '24. These payments and such matters as set off under the contract or abatement may properly be within his remit, as matters arising under the contract. Given that the adjudicator has been properly appointed under the Scheme and the timetable laid down has been properly observed, he would have the jurisdiction to consider the types of issues raised as to the payment of these invoices. The correctness of his decision is not a matter that falls to be considered at this time by this court which is considering the limited issue arising out of the claimant's claim, namely the enforceability of the adjudicator's decision. For this court to review the adjudicator's decision given that he has been properly appointed under the Scheme and was considering matters arising under the contract, properly within in his remit would be to go behind the intention of Parliament that his decision should be binding. The correctness of the decision may be reviewed, revised, challenged where appropriate in subsequent arbitration proceedings or legal proceedings or by way of an agreement. In the instant case there are the pending legal proceedings commenced by the defendant under 1999 TCC 170, where the disputes between the parties, now provisionally adjudicated by the adjudicator, will be finally determined by the court.'

---

[19] *A & D Maintenance & Construction Limited* v. *Pagehurst Construction Services Limited* (23 June 1999).

Repudiation arises under the contract. This was explored in *Northern Developments* v. *Nichol*[20]:

> 'Accordingly, if there was in this case a repudiation and an acceptance of repudiation (which has not been established) the performance of the contract was terminated but any rights arising under the contract remained to be enforced under the contract. Such rights would include rights enforceable in adjudication. The repudiation issues were matters arising "under the contract", and, if they had been mentioned in the notice of intention to withhold payment, the adjudicator would have had a discretion under paragraph 20 of the Scheme to take them into account if he considered them to be necessarily connected with the dispute. Paragraph 20 says that the adjudicator *may* take such matters into account. If he had the discretion, it would be a wrongful exercise of his discretion to refuse to exercise the discretion. If he did exercise such a discretion, it would almost certainly be impossible to challenge the exercise of that discretion whichever way he decided the discretion, in favour of or against considering the other matters. But in this case, the repudiation not having been mentioned in the notice of intention to withhold payment, the adjudicator did not have a discretion and his refusal to consider exercising a discretion was not a denial of a jurisdiction which had any existence.'

A dispute arising under the contract is not the same as connected with the contract. Frequently arbitration clauses are drafted widely permitting arbitrators to examine matters under or in connection with the contract. Some permit the investigation of whether or not there is a contract at all. This is not so in adjudication under the Scheme. There is only express authority to deal with disputes which arise under the contract and to take into account those things necessarily connected with that dispute. Contract clauses may however be drafted with a wider remit.

There is no doubt in English law that damages for breach of contract or professional negligence arose under the contract. That this concept applied in Scots law was clarified in *Gillies Ramsay* v. *PJW Enterprises*[21]:

> '31. In relation to the terms of paragraph 20(2)(b) of the Scottish adjudication regulations (SI 1998 No. 687), counsel contended that the list in paragraph 20(2) (prefaced by the words "and, in particular, he may") was not exhaustive. In any event, "payment under the contract" could mean damages arising under the contract: *Heyman* v. *Darwins Ltd.* [1942] AC 356, and the additional authorities above cited. To give the adjudication scheme and paragraph 20 a more restricted meaning as contended by Diamond would hamper the operation of the scheme, and would result in fragmentation of claims and tactical behaviour. Professionals would go to adjudication to recover fees, knowing that the other party could not defend on the basis of poor professional conduct. Disputes involving demands for payment and notices to withhold payment could not be determined. Such consequences could not have been intended by parliament.'

The list of matters, which the adjudicator can consider, is not exhaustive. Express power is required to open up, revise and review any decision or certificate and it is in paragraph 20(a). This was explored in *David McLean* v. *Swansea HA*[22]:

> 'The Scheme (and, so as far as I am aware, other standard forms of contract) does not confer on an adjudicator a right to adapt, vary or otherwise modify a contract. Under the statutory Scheme an adjudicator has to decide a dispute under the contract (and in other schemes, disputes arising out of or in connection with the contract). It is a decision about the rights and liabilities of the contract which are questioned. Thus paragraph 20 of the Scheme expressly provides for the review of a certificate that has been issued (sub-para (a)) and for the adjudicator to decide a person "is liable to

[20] *Northern Developments (Cumbria) Limited* v. *J & J Nichol* (24 January 2000).
[21] *Gillies Ramsay Diamond* v. *PJW Enterprises Limited*, Outer House, Court Of Session, Lady Paton (27 June 2002).
[22] *David McLean Housing Contractors Limited* v. *Swansea Housing Association Limited* (27 July 2001).

make a payment *under the contract* ... [emphasis supplied] and, subject to section 111(4) of the Act, when that payment is due and the final date for payment''.'

The courts no longer need such express powers. The decision in *Beaufort Developments* v. *Gilbert-Ash*[23] has overruled the decision in *Northern Regional Health* v. *Derek Crouch*[24] which restrained the courts' jurisdiction in such matters. The courts have inherent powers in this respect. Adjudicators and arbitrators need express terms to permit them to open up, review or revise any certificate or decision.

A late revision to the draft of the Scheme introduced the restriction 'unless the contract states that the decision or certificate is final and conclusive'. This could have disastrous consequences. It is arguable that there is need for some certainty when final certificates are issued under contracts. There is a fall-back at that point with the arbitration provisions in most contracts. However, there is a limited time for the arbitration to be commenced after the issue of the final certificate. If the arbitration is not commenced within that time, leave of the courts must be sought to extend time to commence arbitration at all. It is obviously sensible that adjudication is not used to flout these provisions or to run in parallel with an arbitration that is properly commenced following a final certificate. It is also possible that adjudication could be commenced long after the time allowed in the contract for arbitration, thus avoiding the need to seek any extension to the time stipulated for commencement of the arbitration.

However there is nothing to prevent divisive parties amending their contracts to make all decisions and certificates final and conclusive. The only means of then dealing with such certificates is on the grounds that they are procured in breach of contract, i.e. the contract has been applied incorrectly. See *John Barker* v. *London Portman Hotel*[25] 50 Con LR in which the courts used this approach. Whether it can be used successfully by an adjudicator without challenge will be seen as jurisprudence is developed in the field of adjudication. There are no reported cases on this point at the time of writing.

The adjudicator can decide that payment be made, when the payment is due and the final date for payment. This aligns the Scheme with the Act on payment provisions. There is a question as to whether set-off or abatement can be used as a defence to not pay against an adjudicator's decision. We consider this in Chapter 12 when looking at enforcement. Paragraph 20(b) deals with the position where an effective notice to withhold payment has been issued and the whole matter is then subject to adjudication. Section 111(4) of the Act refers to the position where there is still a sum due notwithstanding that there is a valid withholding notice. This then deals with the timing of when the residual payment becomes due.

There is a strongly held view, accepted on a technical basis by the authors but not on a practical level, that an adjudicator has no power to order anything, his duty being limited to declaring what the contract says. Thus the adjudicator identifies that the contract entitles the receiving party to be paid, states the date due and the final date for payment. These dates are often some time before the decision is made. He does no more and it is then for the parties to comply with his determination without being told specifically to do so. On a practical level however we take the view that there is nothing wrong with actually telling the parties what they have to do.

Paragraph 20(c) gives the adjudicator powers to deal with interest, in wording that is open to interpretation. Some commentators suggest that the words 'having regard to any term of

---

[23] *Beaufort Developments (NI) Limited* v. *Gilbert-Ash NI Limited and Others* (1988) 88 BLR 1, HL.
[24] *Northern Regional Health Authority* v. *Derek Crouch Construction Co.* [1984] QB 644.
[25] *John Barker Construction Ltd* v. *London Portman Hotel* 50 Con LR 43.

the contract relating to the payment of interest' mean that the power to award interest under the Scheme is predicated on the existence of interest provisions in the contract. In our view this cannot be right as this would mean the effective exclusion of any right to interest as it is unlikely that such a provision would be included in the type of contract that the Scheme is designed to assist. We interpret the wording as giving the adjudicator the discretionary power to award interest but if he does so he must consider any related provisions in the contract.

This is now pretty much obsolete with the coming into force of the Late Payment of Commercial Debts (Interest) Act 1998. This Act applies to all contracts entered into after 1 August 2002. Where the contract is entered into prior to that date, the adjudicator will need to work through the commencement orders to determine if and to what extent that legislation applies. From 1 August 2002 the contract will always have an interest provision. This is because the Late Payment of Commercial Debts (Interest) Act 1998 imports its provisions into the contract. It is possible that the adjudicator may be faced with deciding whether the contract provision on interest is a substantial remedy sufficient to oust the application of the Late Payment of Commercial Debts (Interest) Act 1998. The JCT contracts and the related sub-contracts now have provisions for contractual interest, as do other standard forms, in recognition of the provisions of the Late Payment of Commercial Debts (Interest) Act 1998.

The question still remains: should the adjudicator award interest if it is not claimed? There are again differing views on this. On the one hand the discretionary power is there to be exercised and the parties are, or should be, aware that the Scheme gives this power to the adjudicator. On the other hand it can be said that the adjudicator could be exceeding his jurisdiction if he did so in the absence of a claim. More importantly is the question: is the adjudicator being fair if he does not give the opposing party the opportunity to make submissions resisting such an award and particularly to deal with the amount of interest? Whatever view the adjudicator has, it is suggested that the principles of fairness suggest that any award of interest should not be made without the adjudicator acquainting the parties beforehand that he is minded to make such an award.

Paragraph 21

21. In the absence of any directions by the adjudicator relating to the time for performance of his decision, the parties shall be required to comply with any decision of the adjudicator immediately on delivery of the decision to the parties in accordance with this paragraph.

Where a payment is involved the adjudicator must have regard to the payment provisions both of the contract and the Act. The adjudicator should direct in his decision when the payment is due. Whether payment is involved or not, short of any direction in the decision, this paragraph leaves no doubt that the decision is operative forthwith on delivery. It is both a breach of the contract and these statutory provisions not to comply with a decision.

Paragraph 22

22. If requested by one of the parties to the dispute, the adjudicator shall provide reasons for his decision.

This was a late addition to the draft that went before Parliament in December 1997. There are arguments that favour the issuing of reasons for any decision taken by a third party. The default position under the Arbitration Act 1996 is for a reasoned award to be given. It is argued that parties find it more acceptable and are more likely to comply with a decision if they know why they have won or lost. However we have a system here which is restrained to a strict timetable. There is nothing to prevent a party insisting on reasons at the last minute and thus placing the decision date in jeopardy. Reasons can be requested after the decision has been made. The adjudicator cannot refuse to provide reasons for the decision where so requested. However there is nothing to prevent the adjudicator providing the decision on time and then providing the reasons after the decision date if the request is delayed. It is advisable for the adjudicator in any event to set a deadline for such requests to be made. See Chapter 11 for a more detailed discussion.

On balance, reasons given with the decision are the favoured approach. This was explored in *Joinery Plus* v. *Laing*[26]:

'40. Indeed, although adjudicators are, under some adjudication rules, permitted to issue a decision without reasons, it is usually preferable for the parties for reasons to be given so that they can understand what has been decided and why the decision has been taken and so as to assist in any judicial enforcement of the decision. A silent decision is more susceptible to attack in enforcement proceedings than a reasoned one. Moreover, judicial enforcement, being a mandatory and compulsory exercise imposed by the state, should only be ordered by a court once it has been satisfied that the underlying adjudication decision is valid, is in accordance with law and complies with all applicable contractual and statutory procedures. For that purpose, a court will always be greatly assisted by cogent, albeit succinct, reasons.'

Effects of the decision

Paragraph 23

23. (1) In his decision, the adjudicator may, if he thinks fit, order any of the parties to comply peremptorily with his decision or any part of it.
    (2) The decision of the adjudicator shall be binding on the parties, and they shall comply with it until the dispute is finally determined by legal proceedings, by arbitration (if the contract provides for arbitration or the parties otherwise agree to arbitration) or by agreement between the parties.

There were concerns about enforcement of adjudicators' decisions. The first part of paragraph 23 gives a decision a peremptory status if so ordered. The effect of this is discussed below under paragraph 24. The purpose of this was probably to give some status to the decision to assist enforcement. The word peremptorily means so as to preclude debate, discussion or opposition. This does little more than strengthen paragraph 21.

Paragraph 23(2) follows section 108(3) of the Act. This is more important than the preceding sub-paragraph. The basic nature of adjudicators' decisions is that they are to be complied with until some substitute process has taken place.

Paragraph 24

24. Section 42 of the Arbitration Act 1996 shall apply to this Scheme subject to the following modifications—
    (a) in subsection (2) for the word 'tribunal' wherever it appears there shall be substituted the word 'adjudicator',

---

[26] *Joinery Plus Limited (in administration)* v. *Laing Limited*, TCC, Judge Thornton QC (15 January 2003).

(b) in subparagraph (b) of subsection (2) for the words 'arbitral proceedings' there shall be substituted the word 'adjudication',

(c) subparagraph (c) of subsection (2) shall be deleted, and

(d) subsection (3) shall be deleted.

This is an attempt to use a part of the Arbitration Act 1996 to provide a mechanism for enforcement of adjudicators' decisions. To understand the effect of this paragraph the amended version of section 42 of the Arbitration Act 1996 is reproduced here:

Enforcement of peremptory orders of tribunal

42. (1) Unless otherwise agreed by the parties, the court may make an order requiring a party to comply with a peremptory order made by the tribunal.

(2) An application for an order under this section may be made —

(a) By the ~~tribunal~~ adjudicator (upon notice to the parties)

(b) By a party to the ~~arbitral proceedings~~ adjudication with the permission of the ~~tribunal~~ adjudicator (and upon notice to the other parties)~~, or~~

~~(c) Where the parties have agreed that the powers of the court under this section shall be available.~~

~~(3) The court shall not act unless it is satisfied that the applicant has exhausted any available arbitral process in respect of failure to comply with the tribunal's order.~~

(4) No order shall be made under this section unless the court is satisfied that the person to whom the tribunals order was directed has failed to comply with it within the time prescribed in the order or, if no time was prescribed, within a reasonable time.

(5) The leave of the court is required for any appeal from a decision of the court under this section.

Unfortunately there are some drafting errors which remain as a result of the instructions here to amend section 42 of the Arbitration Act 1996. There are instances where the word tribunal remains in place and the word adjudicator should have been substituted.

The clumsiness of the device presented in paragraphs 24 and 23 was quickly revealed in *Macob* v. *Morrison*[27]:

'38. It is not at all clear why section 42 of the Arbitration Act 1996 was incorporated into the Scheme. It may be that Parliament intended that the court should be more willing to grant a mandatory injunction in cases where the adjudicator has made a peremptory order than where he has not. Where an adjudicator has made a peremptory order, this is a factor that should be taken into account by the court in deciding whether to grant an injunction. But it seems to me that it is for the court to decide whether to grant a mandatory injunction, and, for the reasons already given, the court should be slow to grant a mandatory injunction to enforce a decision requiring the payment of money by one contracting party to another.'

The possibilities under the Scheme were that if a party fails to comply with the decision an application can be made to the court to order compliance with the decision. It is unlikely that the adjudicator himself will apply to the court. It is more likely that the party seeking enforcement will apply. To do this the adjudicator must give permission and notice must be given to the other parties. Section 42(4) also requires that the adjudicator sets a time limit for compliance with his decision if this provision is to be used, and for this provision to be effective the decision needs to be drafted with these points incorporated from the outset.

In essence, from *Macob* the peremptory order and the application of section 42 of the Arbitration Act 1996 is dead. The way to enforce a decision is to seek summary judgment.

---

[27] *Macob Civil Engineering Ltd* v. *Morrison Construction Ltd* (1999) 1 BLR 93.

Paragraph 25

> **25.** The adjudicator shall be entitled to the payment of such reasonable amount as he may deter-
> mine by way of fees and expenses reasonably incurred by him. The parties shall be jointly and
> severally liable for any sum which remains outstanding following the making of any deter-
> mination on how the payment shall be apportioned.

The adjudicator is entitled to a reasonable sum and he may determine what the amount is.
The expenses must be reasonably incurred, i.e. he has spent the money and it was reasonable
to do so in the execution of his duties. In practice it is unlikely that any adjudicator will
proceed without at least giving the parties notice of his terms, including the way in which
expenses will be incurred and charged. The adjudicator can determine the way in which fees
and expenses will be apportioned.

There is no provision in the Scheme to deal with the parties' costs. They must bear their
own costs in any adjudication unless they agree otherwise. There is nothing to prevent a
party spending inordinate and disproportionate sums in attempting to win in the adjudication, or in attempting to increase unnecessarily the costs in an attempt to 'break' a
financially weaker party. These costs may be revisited in the final determination of the
dispute in arbitration or litigation but their admissibility in such later proceedings is a
question that has not to our knowledge been considered.

Most adjudicators apportion their own fees and expenses on the basis that costs follow the
event. The party who wins will then not have to bear the adjudicator's charges. Some
adjudicators split their fee on a 50/50 basis, seemingly on the basis that the adjudication is a
temporary contractual process. This can cause a winning party to feel extremely upset but if
the adjudicator considers that such a conclusion is appropriate there is nothing to stop him
from doing it. If he does so however when reasons have been requested, he will need to be
sure that he gives his reasons for this decision as well.

Whatever the approach, until the fee is paid in full the parties are jointly and severally
liable for the adjudicator's fees and expenses. This applies notwithstanding that he may have
decided that one party will bear all of those costs. A winning party may consider itself
obliged to pay the adjudicator's fee in full, but this is generally only when the responding
party resists payment of the amount awarded and court proceedings have to be taken to
enforce the decision and recover the adjudicator's charges from the other party.

Paragraph 26

> **26.** The adjudicator shall not be liable for anything done or omitted in the discharge or purported
> discharge of his functions as adjudicator unless the act or omission is in bad faith, and any
> employee or agent of the adjudicator shall be similarly protected from liability.

This paragraph repeats section 108(4) of the Act and thus provides the contractual immunity
as an implied term where the contract itself fails to do so. The immunity is not however any
different from that which would be provided were the immunity written into the contract
itself. The parties cannot hold the adjudicator liable for anything that he does in the course of
the adjudication, unless it is in bad faith, but he has no immunity in respect of third parties.

The Scheme: Part II – Payment

The Construction Industry Council discovered in its survey[28] that 73% of the disputes reported involved payment under sections 109 to 112 of the Act. The payment provisions are slightly more complex than those for adjudication. With adjudication it is a simple matter. The contract either complies with all of the requirements of section 108(1) to (4) inclusive of the Act, or it does not. Where it does not the Scheme applies.

In respect of payment the Scheme is referred to in sections 109, 110, 111 and 113. Reversion to the Scheme is on an individual basis and any one or more non-compliant clause is replaced and the compliant parts remain in force.

The payment provisions apply to contracts that exceed a duration of 45 days.

Entitlement to and amount of stage payments

Paragraph 1

1. Where the parties to a relevant construction contract fail to agree –
   (a) the amount of any instalment or stage or periodic payment for any work under the contract, or
   (b) the intervals at which, or circumstances in which, such payments become due under that contract, or
   (c) both of the matters mentioned in sub-paragraphs (a) and (b) above
   the relevant provisions of paragraphs 2 to 4 below shall apply.

The Act leaves the parties free to agree their payment regime provided it complies with the specific requirements in each section of the Act. The construction industry is the only industry as a matter of statute that has the right to payment by instalments. Paragraph 1 of the Scheme deals with those matters required to be agreed in accordance with section 109(2) of the Act. If the matters in paragraph 1 are not agreed, paragraphs 2 to 4 apply.

Paragaph 2

2. (1) The amount of any payment by way of instalments or stage or periodic payments in respect of a relevant period shall be the difference between the amount determined in accordance with sub-paragraph (2) and the amount determined in accordance with sub-paragraph (3).
   (2) The aggregate of the following amounts –
   (a) an amount equal to the value of any work performed in accordance with the relevant construction contract during the period from the commencement of the contract to the end of the relevant period (excluding any amount calculated in accordance with sub-paragraph (b)),
   (b) where the contract provides for payment for materials, an amount equal to the value of any materials manufactured on site or brought onto site for the purposes of the works during the period from the commencement of the contract to the end of the relevant period, and
   (c) any other amount or sum which the contract specifies shall be payable during or in respect of the period from the commencement of the contract to the end of the relevant period.

---

[28] *Adjudication – The First Forty Months.* A Report on Adjudication under the Construction Act. Construction Industry Council, 2002.

(3)   The aggregate of any sums which have been paid or are due for payment by way of instalments, stage or periodic payments during the period from the commencement of the contract to the end of the relevant period.

(4)   An amount calculated in accordance with this paragraph shall not exceed the difference between
(a) the contract price, and
(b) the aggregate of the instalments or stage or periodic payments which have become due.

The method for determining the sum of any payment is less sophisticated than the provisions in the standard form contracts used in the industry. It will ensure that some money flows by way of instalments where the contract is silent on any method to determine such payments. The constituent parts of the gross figure to be calculated (there is no retention or discount to consider) are determined in accordance with paragraph 2(2) and there is deducted from this sum previous payments in accordance with paragraph 2(3).

The gross sum due is calculated at the end of the relevant period. The relevant period, if no period is specified, is 28 days in accordance with paragraph 12 of the Scheme. This sets a 28-day cycle for the period covering the instalment to be calculated. Paragraph 2(2) then describes the component parts to be aggregated to form the gross payment.

Paragraph 2(2)(a) requires that an amount be calculated for the total value of work from commencement date to the end of the relevant period. This amounts to carrying out a gross valuation of all the works from commencement every 28 days. The work to be valued is all the work performed in accordance with the relevant construction contract. 'Work' means any of the work or services mentioned in section 104 of the Act as defined in paragraph 12. This will include any varied or additional work. It excludes materials; these are dealt with separately. There does not need to be a contract sum for there to be a contract. There simply needs to be a mechanism to determine price. A simple example of this would be a contract let on a schedule of rates. Where there is no mechanism at all in the contract to determine the value, the basis required in the Scheme is the cost of the work performed plus an amount equal to any overhead or profit that may be included in the contract price. The contract price means the entire sum payable under the construction contract in respect of the work as defined in paragraph 12.

Paragraph 2(2)(b) only applies where the contract provides for payment for materials. This must not be confused solely with the standard form-type arrangement for payment for unfixed materials on site or even off site, although this could apply also in those situations. In the first instance it applies where payment for materials is part of the mechanism for determining the payment. All of the materials brought on to the works from commencement date to the end of the relevant period are valued and included in the gross sum. Obviously if the contract provides for the traditional payment for materials on site, those terms will operate and there is no danger of a grossed-up value of materials resulting in an over-payment. It does not apply to any materials off site. If the contract is silent on payment for materials, paragraph 2(2)(b) does not fall to be considered at all.

Paragraph 2(2)(c) deals with any other sum which is specified to be payable in accordance with the contract. This really is a catch-all provision. It is unlikely that the contract will be sophisticated enough to make such provisions and also be lacking in other areas sufficient for it to fall within this paragraph.

Paragraph 2(3) simply requires the previous payments, including sums that are calculated as due but not yet paid, to be deducted from the gross sum calculated in accordance with paragraph 2(2). This is simply a method of calculating gross sums due and deducting previous payments, and in principle it follows the practices already prevalent in the industry.

Paragraph 2(4) could lead to some confusion on first reading. The maximum payment due is the contract price less the aggregate of all previous payments. Contract price is not to be confused with contract sum. The contract price is defined in paragraph 12 as being the entire sum payable under the construction contract in respect of the work. This is actually the final sum payable under the construction contract and should not be confused with the contract sum, which is the starting point. The contract price is the total value of the work including all varied and additional work.

Dates for payment

Paragraph 3

3. Where the parties to a construction contract fail to provide an adequate mechanism for determining either what payments become due under the contract, or when they become due for payment, or both, the relevant provisions of paragraphs 4 to 7 shall apply.

Paragraph 3 simply sets the scene for the application of paragraphs 4 to 7 inclusive.

Paragraph 4

4. Any payment of a kind mentioned in paragraph 2 above shall become due on whichever of the following dates occurs later —
    (a) the expiry of 7 days following the relevant period mentioned in paragraph 2(1) above, or
    (b) the making of a claim by the payee.

Every payment under a construction contract must now have a due date. The due date is the date on which a liability at law to pay is established. This does not coincide with the date on which payment must actually be made. This actual final date on which payment can be made is also determined by this legislation. This is known as the final date for payment (see paragraph 8).

The due date is the later of either seven days after the expiry of the relevant period (payment cycle) or of the making of a claim by the payee. It is imperative in these circumstances that those wishing to be paid always lodge their claim for payment at the end of each payment cycle or slightly earlier than that date. A claim by the payee is defined in paragraph 12. It means a written notice given by the party carrying out work under a construction contract to the other party specifying the amount of any payment or payments which he considers to be due and the basis on which it is, or they are, calculated. If no claim is made it must follow that no payment becomes due.

Paragraph 5

5. The final payment payable under a relevant construction contract, namely the payment of an amount equal to the difference (if any) between —
    (a) the contract price, and
    (b) the aggregate of any instalment or stage or periodic payments which have become due under the contract,
    shall become due on the expiry of —
    (a) 30 days following completion of the work, or
    (b) the making of a claim by the payee,
    whichever is the later.

Paragraph 5 deals with the final payment payable under a construction contract. These are the payments which normally fall under final certificates. The terminology here is not to be confused with final date for payment. The demands here on the paying party are much higher than anything contained in the standard forms of contract. If a claim is made relating to the final payment under the contract within the 30-day period, the final payment then becomes due on the thirtieth day. In these circumstances the final account must be settled no later than 30 days following the completion of the work. This does give the paying party a bit of leeway but when these provisions apply the final account itself must be agreed or determined in a much shorter period than is customary in the industry.

Paragraph 6

6. Payment of the contract price under a construction contract (not being a relevant construction contract) shall become due on
    (a) the expiry of 30 days following the completion of the work, or
    (b) the making of a claim by the payee,
    whichever is the later.

This paragraph almost repeats paragraph 5. The difference here is that it covers construction contracts which are not relevant construction contracts as defined in the Act. These are contracts which specify that the duration of the work is to be less than 45 days or where the parties have agreed that the duration of the work is estimated to be less than 45 days.

Paragraph 7

7. Any other payment under a construction contract shall become due
    (a) on the expiry of 7 days following the completion of the work to which the payment relates, or
    (b) the making of a claim by the payee,
    whichever is the later.

This paragraph deals with the situation where the construction contract has a mechanism for arriving at the final sum which is due but fails to state the date on which the payment actually becomes due.

Final date for payment

Paragraph 8

8. (1) Where the parties to a construction contract fail to provide a final date for payment in relation to any sum which becomes due under a construction contract, the provisions of this paragraph shall apply.
    (2) The final date for the making of any payment of a kind mentioned in paragraphs 2, 5, 6 or 7, shall be 17 days from the date that payment becomes due.

Every construction contract must provide a final date for payment. This final date for payment applies to all payments, both interim and final certificates or payments. This sets the date on which the money must actually be paid. For the purpose of the Scheme this is 17 days from the due date. The Scheme does not need to deal with the right to suspend performance for non-payment under section 112 of the Act. What the Scheme simply does is determine the final date for payment and this is the point at which the necessary notice can be issued to suspend performance. This also gives a date from which any entitlement to interest will run.

Notice specifying amount of payment

Paragraph 9

> 9. A party to a construction contract shall, not later than 5 days after the date on which any payment—
> (a) becomes due from him, or
> (b) would have become due, if—
> (i) the other party had carried out his obligations under the contract, and
> (ii) no set-off or abatement was permitted by reference to any sum claimed to be due under one or more other contracts,
> give notice to the other party to the contract specifying the amount (if any) of the payment he has made or proposes to make, specifying to what the payment relates and the basis on which that amount is calculated.

This paragraph covers requirements of section 110(2) of the Act. The notice must be given even where the payment nets down to nothing due to set-off or abatement. The statement must show the gross amount due on the basis that the obligations under the contract had been properly carried out and without any set-off or abatement by reference to any sum claimed to be due under one or more other contracts. Set-off or abatement are then dealt with separately under the notice requirements in accordance with paragraph 10 below. The notice must specify to what the payment relates and the basis on which the amount is calculated. This has changed the practice of the industry. Payments used to be made with little or no detail as to how they were arrived at. The provisions of the Scheme and the Act require much more than this. The basis on which the amount is calculated requires full detail of the build-up to the payment.

There is a possibility of some difficulty here when stage payments or a payment mechanism dependent on the construction programme cover the works. If the contractor has not performed to the programme or achieved a stage, is the sub-contractor still entitled to his money? On the basis of the wording in the Act he probably is.

There has been some argument as to the effect of failing to issue the section 110(2) or the paragraph 9 notice required here. The argument is that by default, if no section 110(2) notice is issued the amount applied for becomes the amount due. This is not the correct premise under the Scheme. The matter was dealt with in *SL Timber v. Carillion*[29]. See the discussion under section 110(2) in Chapter 4.

There is no doubt that the failure to issue a section 110(2) notice (paragraph 9) does not make the sum applied for or claimed, the sum due. There is nothing to prevent a responding party questioning the sum due in adjudication. The whole of the *SL Timber* judgment is worth reading.

The position is different under some of the standard contracts, particularly the JCT with Contractor's Design suite in which the sum applied for may become the sum due if there is no section 110(2) notice.

---

[29] *S L Timber Systems Limited* v. *Carillion Construction Limited*, Outer House, Court of Session, Lord Macfadyen (27 June 2001).

Notice of intention to withhold payment

Paragraph 10

10. Any notice of intention to withhold payment mentioned in section 111 of the Act shall be given not later than the prescribed period, which is to say not later than 7 days before the final date for payment determined either in accordance with the construction contract, or where no such provision is made in the contract, in accordance with paragraph 8 above.

This paragraph imports section 111 of the Act. The content of the notice must conform with all the requirements of section 111 of the Act. What this paragraph specifically deals with is the prescribed period for giving the notice when no such period is contained in the contract. Notice must be given not later than seven days before the final date for payment. The final date for payment will be that stated in the contract, or if no provision is made in the contract it will be the date calculated in accordance with paragraph 8 of the Scheme.

Prohibition of conditional payment provisions

Paragraph 11

11. Where a provision making payment under a construction contract conditional on the payer receiving payment from a third person is ineffective as mentioned in section 113 of the Act, and the parties have not agreed other terms for payment, the relevant provisions of —
     (a) paragraphs 2, 4, 5, 7, 8, 9 and 10 shall apply in the case of a relevant construction contract, and
     (b) paragraphs 6, 7, 8, 9 and 10 shall apply in the case of any other construction contract.

The Act bars conditional payment provisions. The 'pay when paid', 'pay if paid' and 'pay what paid' clauses can no longer be applied in construction contracts. The only time when they do become operative is when the person paying the payer becomes insolvent. Where a contractor is relying on the employer to pay him, a conditional payment clause would apply in any sub-contract should the employer become insolvent.

At all other times conditional payment provisions are inoperative. Where there is a relevant construction contract, paragraphs 2, 4, 5, 7, 8, 9 and 10 apply to determine the payments which are due. In the case of any other construction contract, paragraphs 6, 7, 8, 9 and 10 apply to determine the payments which are due.

Interpretation

Paragraph 12

12. In this Part of the Scheme for Construction Contracts —
     'claim by the payee' means a written notice given by the party carrying out work under a construction contract to the other party specifying the amount of any payment or payments which he considers to be due and the basis on which it is, or they are calculated;
     'contract price' means the entire sum payable under the construction contract in respect of the work;
     'relevant construction contract' means any construction contract other than one —
     (a) which specifies that the duration of the work is to be less than 45 days, or
     (b) in respect of which the parties agree that the duration of the work is estimated to be less than 45 days;

'relevant period' means a period which is specified in, or is calculated by reference to the construction contract or where no such period is so specified or is so calculable, a period of 28 days;

'value of work' means an amount determined in accordance with the construction contract under which the work is performed or where the contract contains no such provision, the cost of any work performed in accordance with that contract together with an amount equal to any overhead or profit included in the contract price;

'work' means any of the work or services mentioned in section 104 of the Act.

This paragraph gives definitions and does not warrant any further explanation.

The Exclusion Order

The purpose of the exclusion order is to exclude further contracts from the definition of construction contracts in the Act. Technically the contracts that this order excludes would have been within the definition of construction contracts in the Act. In some instances they may have been borderline. The premise simply is that the Act was not designed to apply to these contracts. The exclusion order is exhaustive and covers contracts with very specific criteria rather than in broad categories. If the criteria are not met there is no room for argument that the contract in question closely resembles an excluded contract. If the criteria are not met in every detail the contract is still a construction contract. There is nothing however to prevent the parties to a contract covered by this order making a contractual adjudication arrangement.

1998 No. 648

CONSTRUCTION, ENGLAND AND WALES

The Construction Contracts (England and Wales) Exclusion Order 1998

Made   *6th March 1998*

Coming into force in accordance with article 1(1)

The Secretary of State, in exercise of the powers conferred on him by sections 106(1)(b) and 146(1) of the Housing Grants, Construction and Regeneration Act 1996 and of all other powers enabling him in that behalf, hereby makes the following Order, a draft of which has been laid before and approved by resolution of, each House of Parliament:

Citation, commencement and extent

1   (1) This Order may be cited as the Construction Contracts (England and Wales) Exclusion Order 1998 and shall come into force at the end of the period of 8 weeks beginning with the day on which it is made ('the commencement date').

(2) This Order shall extend to England and Wales only.

Interpretation

2.   In this Order, 'Part II' means Part II of the Housing Grants, Construction and Regeneration Act 1996.

The order came into operation on the same date as the Scheme and Part II of the Act on 1 May 1998. There is a similar order that covers Scotland. Northern Ireland also has its own Exclusion Order which became effective from 1 June 1999.

Agreements under statute

3. A construction contract is excluded from the operation of Part II if it is—

    (a) an agreement under section 38 (power of highway authorities to adopt by agreement) or section 278 (agreements as to execution of works) of the Highways Act 1980;

    (b) an agreement under section 106 (planning obligations), 106A (modification or discharge of planning obligations) or 299A (Crown planning obligations) of the Town and Country Planning Act 1990;

    (c) an agreement under section 104 of the Water Industry Act 1991 (agreements to adopt sewer, drain or sewage disposal works); or

    (d) an externally financed development agreement within the meaning of section 1 of the National Health Service (Private Finance) Act 1997 (powers of NHS Trusts to enter into agreements).

It was obviously sensible to exclude such matters as adoption agreements. Although these involve construction works they really are on the fringe of the mischief that the legislation was seeking to correct. The most significant exclusion is work in connection with private finance initiatives. It is only the finance element that is excluded and not the construction works procured using those finances.

Private finance initiative

4. (1) A construction contract is excluded from the operation of Part II if it is a contract entered into under the private finance initiative, within the meaning given below.

    (2) A contract is entered into under the private finance initiative if all the following conditions are fulfilled—

    (a) it contains a statement that it is entered into under that initiative or, as the case may be, under a project applying similar principles;

    (b) the consideration due under the contract is determined at least in part by reference to one or more of the following—

    (i) the standards attained in the performance of a service, the provision of which is the principal purpose or one of the principal purposes for which the building or structure is constructed;

    (ii) the extent, rate or intensity of use of all or any part of the building or structure in question; or

    (iii) the right to operate any facility in connection with the building or structure in question; and

    (c) one of the parties to the contract is—

    (i) a Minister of the Crown;

    (ii) a department in respect of which appropriation accounts are required to be prepared under the Exchequer and Audit Departments Act 1866(a);

    (iii) any other authority or body whose accounts are required to be examined and certified by or are open to the inspection of the Comptroller and Auditor General by virtue of an agreement entered into before the commencement date or by virtue of any enactment;

    (iv) any authority or body listed in Schedule 4 to the National Audit Act 1983(b) (nationalised industries and other public authorities);

    (v) a body whose accounts are subject to audit by auditors appointed by the Audit Commission;

    (vi) the governing body or trustees of a voluntary school within the meaning of section 31 of the Education Act 1996(c) (county schools and voluntary schools), or

    (vii) a company wholly owned by any of the bodies described in paragraphs (i) to (v).

For a contract to be excluded under the Private Finance Initiative it must satisfy all of the criteria listed in paragraph 4. If it were not for this exclusion, PFI contracts would constitute construction contracts. This matter however is not as simple as it may first seem. Any construction contract made under the PFI contract itself will be within the Act. It is only the PFI concession contract that is excluded from the ambit of the Act. The risks inherent in construction contracts will eventually find their way back to the concessionaire. Such risks in previous arrangements have been on the basis that the contractor would only be able to claim extra monies from the concessionaire to the extent that the concessionaire is able to recover extra from the government. Such an arrangement now constitutes a pay when paid clause which is vetoed under section 113 of the Act. The concessionaire will therefore have to pick up the risk for extra work and delay that it cannot recover from the government. It remains to be seen whether the banks who finance these arrangements are prepared to accept risks which were not previously found in construction contracts under PFI schemes.

Finance agreements

5. (1) A construction contract is excluded from the operation of Part II if it is a finance agreement, within the meaning given below.
(2) A contract is a finance agreement if it is any one of the following—
(a) any contract of insurance;
(b) any contract under which the principal obligations include the formation or dissolution of a company, unincorporated association or partnership;
(c) any contract under which the principal obligations include the creation or transfer of securities or any right or interest in securities;
(d) any contract under which the principal obligations include the lending of money;
(e) any contract under which the principal obligations include an undertaking by a person to be responsible as surety for the debt or default of another person, including a fidelity bond, advance payment bond, retention bond or performance bond.

Under the heading of finance agreements, the types of contract excluded are obvious and self-explanatory. The only area of doubt concerns bond arrangements. Where the principal obligation is to act as surety there is now no doubt that this contract will be outside the provisions of the Act. Where the arrangement involves stepping into the shoes of the contractor to have the work completed, these still may be construction contracts. This will very much depend on the terms of the arrangement.

Development agreements

6. (1) A construction contract is excluded from the operation of Part II if it is a development agreement, within the meaning given below.
(2) A contract is a development agreement if it includes provision for the grant or disposal of a relevant interest in the land on which take place the principal construction operations to which the contract relates.
(3) In paragraph (2) above, a relevant interest in land means—
(a) a freehold; or
(b) a leasehold for a period which is to expire no earlier than 12 months after the completion of the construction operations under the contract.

To be excluded, the development agreement must involve a grant or disposal of the relevant interest in the land on which the construction operations take place. The relevant interest in the land will mean a freehold or leasehold which has more than 12 months to run after completion of the construction operations under the contract. If the primary contract includes the disposal of the relevant interests in the land other contracts for professional services, building contracts and sub-contracts will not be excluded.

Chapter 6
Adjudication Clauses, Rules and Procedures

Adjudication clauses

All the standard forms of contract include adjudication provisions that comply with the Act. There are two distinct ways in which the contract drafting bodies have approached the content of the adjudication clause included in their contract forms. One approach is to do little more than cover the matters that must be included in order to comply with the Act and to set out procedures in a separate set of rules. The other approach is to do away with any separate rules and include all procedural matters in the adjudication clause itself.

The drafting bodies that have decided to avoid the use of separate rules are the Joint Contracts Tribunal and the Property Advisers to the Civil Estate (PACE) which is responsible for the government contracts (GC/Works/1 and its fellows). The Construction Confederation (previously the BEC) contracts, DOM/1 and its fellows, which follow JCT very closely, likewise have no separate adjudication rules. The adjudication clauses in the contracts produced by these bodies provide the procedure which governs the conduct of the adjudication.

The forms of contract produced by the Institution of Civil Engineers, apart from, interestingly, the Engineering and Construction Contract (ECC), refer to the Adjudication Procedure that the ICE also publishes. The adjudication provision of the ECC has the strange omission of not referring to the ICE Adjudicator's Contract. The Institution of Chemical Engineers also produces separate rules. The Association of Consulting Engineers has a minimalist approach to its adjudication provisions complying with the Act by an express incorporation of, and relying entirely upon, the Construction Industry Council Model Adjudication Procedure.

The approach to the adjudication provision in each 'family' of contracts is the same throughout all the contracts encompassed in it and we review the following as examples:

- JCT 98 Private Edition
- DOM/1 (this has now been superseded by the JCT DSC/C)
- ICE 6th Edition
- GC/Works/1 (1998)
- ACE Conditions of Engagement
- The IChemE contracts.

Joint Contracts Tribunal

Part 4 of JCT 98 includes clauses 41A, 41B and 41C covering adjudication, arbitration and legal proceedings. Clause 41A, Adjudication, comprises eight sub-clauses, many of which have sub-sub-clauses, a natural result of the JCT's decision that there should be no accompanying adjudication rules.

Clause 41A.1 confirms that clause 41A as a whole applies where either party refers a dispute or difference to adjudication. It uses the words 'arising under this Contract', which echoes the provisions of section 108 of the Act and excludes disputes which arise other than under the contract. This clause does not however limit the adjudicable matters to those set out in the Act. If parties include work that is not defined as a 'construction contract' by the Act within the works governed by this contract, the adjudication provisions will apply as a term of the contract rather than as a statutory right.

Clause 41A.2 allows the adjudicator to be either an individual agreed by the parties or, on application by a party, an individual to be nominated by the person named in the appendix to the contract. There are two provisos to this provision in sub-sub-clauses 41A.2.1 and 41A.4.2. The first of these is to the effect that no adjudicator shall be agreed or nominated who will not execute the JCT Adjudication Agreement. The second is that any agreement between the parties on the appointment of the adjudicator must be reached, or any application to the nominator must be made, with the object of securing the appointment of, and the referral of the dispute to, the adjudicator within seven days of the date of the notice of intention to refer. This all echoes the objectives of the Act. Clause 41A.2.3 requires the parties to execute the JCT Adjudication Agreement with the adjudicator upon his nomination or appointment but failure to do this is of no consequence – see clause 41A.5.6.

Clause 41A.3 deals with the situation if the adjudicator dies, becomes ill or is otherwise unavailable, allowing the parties to agree a replacement or to seek the nomination of another adjudicator from the named person in the contract.

Clause 41A.4.1 deals with the notice of intention to refer the dispute to adjudication and the notice of referral. In each case it sets out how the notice is to be structured, which is necessary in the absence of rules. This sub-clause requires the notice of referral to deal with the particulars of the dispute, a summary of the referring party's contentions, a statement of the relief or remedy sought and any material that the party wishes the adjudicator to consider. The court has made it clear that it is only a crystallised dispute that can be referred to adjudication and, unless the parties agree otherwise, the referring party has to put its complete case in the referral.

Clause 41A.4.2 deals in detail with the mechanics of delivering the notice of referral, citing requirements relating to fax transmission and postal delivery. This procedure relates to the applicant's referral, similar requirements being contained in clause 41A.5.2 in respect of the initial response of the party not making the referral. It is of note that the detail of this procedure is not extended to cover any other exchange of information or documents during the subsequent course of the adjudication. Customarily this is done by fax and increasingly by e-mail.

Clause 41A.5.1 requires the adjudicator to confirm the date that he receives the referral and its accompanying documentation to the parties immediately on its receipt. This requirement relates to clause 41A.5.3, which sets the 28-day period for the completion of the decision from the date of that receipt.

The JCT have decided to provide for a formal response by the party not making the referral as a standard requirement in clause 41A.5.2. This is to be sent to the adjudicator and to the referring party within seven days of the date of the referral. Note that this is not seven days from the date that the referral is received, it is seven days from the date it is sent. It is, as a result, conceivable that if the referral is sent by post the response may have to be produced in no more than five days. This response is required to be in the form of a written statement of the contentions relied on and any material that the party not making the referral wants the adjudicator to consider. There are, as mentioned previously, specific requirements as to delivery of this statement.

Clause 41A.5.2 does, however, create what may be seen as a conflict with the provision set out in Clause 41A.5.5, of which more later, to the effect that the adjudicator may set his own procedure. It is entirely possible that the adjudicator may consider that the time allowed for the written response should be more than seven days. Clause 41A.5.2 provides that the responding party 'may' send a written statement within seven days; there is nothing to suggest that it must do this and it is our view that there is nothing to prevent the adjudicator from allowing a greater time than that stated in the contract.

In addition to defining the 28-day period as starting from the receipt of the referral, clause 41A.5.3 requires the adjudicator to act as 'an Adjudicator for the purposes of section 108 of the Act and not as an expert or an arbitrator'. It does not, however, define what the JCT means by 'an Adjudicator' and how this differs from an expert or an arbitrator. The authors understand that it is the intention of the JCT in using these words to do no more than accentuate that adjudication is a different process from arbitration and expert determination.

There is, however, one possible problem that arises from the words used. It is not the intention of the JCT to limit the adjudication matters under any JCT contract. For example, any dispute arising under any contract let under the JCT Minor Works form is intended to be adjudicable. In particular this will mean that a contract in this form with a residential occupier will include full adjudication provisions. It is possible to argue that the use of the specific words '. . . and acting as an Adjudicator for the purposes of section 108 of the Housing Grants, Construction and Regeneration Act 1996' in clause 41A.5.3 restricts the adjudicator to dealing only with the precise matters that are the subject of the statutory right to adjudication under the Act. As noted above, we understand that this is not the intention of the JCT and we are not aware that this has been taken as an issue to resist enforcement of a decision that deals with matters not within those adjudicable under the Act.

It is noteworthy in this respect that the drafters of GC/Works/1 have also sought to define how the adjudicator is to act by making a similar comparison, but they chose not to refer to section 108 when doing so.

Clause 41A.5.3 also requires the adjudicator to reach his decision within 28 days and forthwith send it to the parties. This clause also deals with the extension of the 28-day period in compliance with the requirements of the Act.

Clause 41A.5.4 states that the adjudicator shall not be obliged to give reasons for his decision but this does not of course prevent him from doing so if he so chooses, subject of course to any agreement otherwise by the parties.

Clause 41A.5.5 requires the adjudicator to act impartially, to set his own procedure and at his own discretion take the initiative in ascertaining the facts and the law. This clause then sets out eight particular actions that the adjudicator may utilise. It is noteworthy that this list, which appears at first glance to be along similar lines to lists of powers enjoyed by the adjudicator that appear in other adjudication procedures, relates to the making of this ascertainment only and does not include general procedural matters. These are entirely at the discretion of the adjudicator, as they are in other adjudication procedures, but the JCT has decided not to delve into this area.

Clause 41A.5.6 states that a failure of a party to complete the JCT Adjudication Agreement or to comply with any requirement of the adjudicator or the adjudication provisions of JCT clause 41A does not invalidate the adjudicator's decision.

Clause 41A.5.7 requires that the parties meet their own costs of the adjudication except that the adjudicator may direct how costs fall in respect of any test or opening up that he may have required.

Clause 41A.6.1 allows the adjudicator to direct how his fee and expenses are to be apportioned but does not require him to do so. Should he not do so these costs fall equally between the parties. Clause 41A.6.2 confirms that the parties have joint and several liability for the adjudicator's fees and expenses.

Clause 41A.7.1 covers the Act's requirement relating to the binding nature of the decision.

Clauses 41A.7.2 and 41A.7.3 are drafted to formalise the effect of the binding nature of the decision and require that the adjudicator's decision is complied with and that legal proceedings can be taken to ensure such compliance pending final determination of the dispute.

Clause 41A.8 provides the adjudicator and employees and agents with the immunity required by the Act.

Construction Confederation

The Construction Confederation produced the suite of contracts which included DOM/1 and DOM/2. The JCT has now produced the DSC/C contract and its fellows and the Construction Confederation is now to stop producing its own forms. DOM/1 will no doubt however remain in use for a considerable time to come.

These contracts are aimed specifically at sub-contracts formed in respect of works being carried out under the Joint Contracts Tribunal contracts. This being so the provisions are identical to those in the main contracts.

The drafting includes reference to the JCT Adjudication Agreement and has the same requirement on the parties and the adjudicator to complete this Agreement and also the provision that failure to complete it will not invalidate the adjudicator's decision.

Institution of Civil Engineers Conditions 6th Edition

The ICE 6th edition is still much in use and there are no changes of any significance in the 7th edition other than the incorporation of the adjudication provisions within the contract, rather than as a supplement as is the case with the 6th edition. These contracts introduce the concept of a 'matter of dissatisfaction' – in the 6th edition this is clause 66(2) – in an endeavour to avoid the issue of a notice to refer a dispute to adjudication before the engineer or employer's representative has had the opportunity to address the problem that is identified by the matter of dissatisfaction. We consider this in detail elsewhere. At the time of writing we understand that the ICE is considering modifying this to reflect the recent clarifications set out by the court regarding the existence of a dispute by the use of an 'advance warning' procedure.

Clause 66 also includes a provision for conciliation in clause 66(5), which allows the dispute to be considered under the Institution of Civil Engineers' Conciliation Procedure at any time before service of a notice to refer to arbitration. This clause is in no way obligatory but does offer an alternative way of resolving the dispute. Note that this procedure will, if it is successful, resolve the dispute and will be deemed to have done so unless a notice of adjudication or arbitration is issued within one month. The statutory right to adjudication is in no way interfered with by this provision.

This contract then proceeds to deal with adjudication. It does no more than repeat the eight compliance points that a contract is required to include in order to avoid the Scheme for Construction Contracts coming into play. This contract, as with the JCT, does not limit the matters that are adjudicable; anything that forms a part of the contract works is subject to adjudication even if it is not within those matters covered by the statutory right to adjudication.

Institution of Civil Engineers – NEC Engineering and Construction Contract

It is of interest that this form of contract approaches the adjudication provisions in a significantly different way from that used in the ICE 6th edition. There is an adjudicator's contract, which we consider in Chapter 8, but surprisingly this is not mentioned in the main contract adjudication clauses although it does appear in respect of the appointment of a replacement adjudicator. There are no rules. The ICE Adjudication Procedure is not referred to. This form of contract, as a result, contains all procedural matters that the drafters considered necessary for the conduct of an adjudication.

In the same way as the ICE 6th edition, a procedure relating to a notice of dissatisfaction is interposed between the event that might otherwise create a dispute and be immediately adjudicable and the commencement of the adjudication process itself. Consequently, as with the 6th edition, a 'dispute' is defined as a matter of dissatisfaction that has not been resolved within four weeks. In this contract it is a requirement of the notice of dissatisfaction procedure that the notice is issued within four weeks of the party that is dissatisfied becoming aware of the matter of dissatisfaction. Taking the contract literally there is no dispute if this procedure is not followed. It is, of course, as discussed elsewhere, arguable that in defining a dispute in this way by agreement the parties have side-stepped the requirement of the Act that there be a statutory right to have a dispute adjudicated at any time.

There is no provision for the optional conciliation procedure in this form of contract.

The procedure for referring the dispute is in accordance with the Act, requiring a notice of intent to refer and a referral of the dispute to the adjudicator within seven days thereof. The referring party includes with the referral the information that he wishes the adjudicator to consider. Each party is then allowed 14 days to provide the adjudicator with any further information. The adjudicator then makes his decision.

This form of contract differs from all others, except GC/Works, in one particular aspect. In most adjudication clauses, or the rules associated with them, there is a provision that empowers the adjudicator to set the procedure for the adjudication, which is accompanied by a non-exhaustive list of actions that he may take. The ECC adjudication does none of this. All it does is to include the basic Act requirement that he may use his initiative in ascertaining the facts and the law. This would appear to put the procedure into a total straitjacket. The referral notice with accompanying information is sent to the adjudicator, and presumably, although the contract does not spell this out in detail, to the other party as well. Both parties then have 14 days to make further submissions and it is down to the adjudicator, if he so wishes, to investigate the facts and the law. There is no provision for anything else and whilst undoubtedly the adjudicator will be able to bring in many of the procedural matters that are spelt out in other forms of contract under the heading 'ascertaining the facts and the law', there may well be occasions where a little more flexibility might assist.

We note in particular in this respect that the adjudication provisions in other forms of contract include such matters as the adjudicator being allowed to use his own knowledge and/or experience, and that he may obtain legal and technical advice. In the absence of confirmation of such matters it is conceivable that a challenge might arise should the adjudicator proceed along these lines. We are sure that adjudicators will on occasion have done just this, if only to make the process work, but we are not aware of any challenge to the enforcement of a decision that has been based upon allegations of excess of jurisdiction or procedural unfairness for this reason.

One useful piece of guidance for the parties in this contract at clause 90.7 is as follows: 'Unless and until the *Adjudicator* has given his decision on the dispute, the Parties proceed as if the action, failure to take action or other matters were not disputed' (italics in original).

This makes the requirements on the parties clear in a situation that could cause problems. What do the parties do while awaiting an adjudicator's decision? The subject matter of the dispute might be covered up by later activities on site. The initial reaction might be to stop doing anything that might affect the subject matter of the adjudication or that might be affected by it. There is certainly no intention to allow extensions of time for the period of an adjudication. This contract makes it quite clear that the work proceeds as the adjudication progresses.

There is one possible downside to this wording. It is arguable that if the parties must proceed as if there were no dispute until the adjudicator has given his decision, there is an absolute requirement to go through the adjudication process before any other dispute resolution procedure can be entered into. Does this mean that a party cannot go directly to arbitration or the courts in respect of a dispute? We await any testing of this situation with interest.

The remaining provisions of this adjudication agreement cover the compliance points required by statute.

The Government Contracts

GC/Works/1 has a Model Form of Adjudicator's Appointment attached to it, as considered in Chapter 8. There are no equivalent rules so, as with JCT, the whole procedure is contained in the contract clauses, which are called 'Conditions' in this contract.

The adjudication provisions in this contract (in condition 59) include many aspects that are rather different from those included in other contracts and it is worth setting them out in full and commenting on each provision:

Condition 59: Adjudication

(1)   The Employer or the Contractor may at any time notify the other of intention to refer a dispute, difference or question arising under, out of, or relating to, the Contract to adjudication. Within 7 Days of such notice, the dispute may by further notice be referred to the adjudicator specified in the Abstract of Particulars.

This wording not only covers the notice of intent to refer the dispute to adjudication, but also allows 'a dispute, difference or question arising under, out of, or relating to the Contract' to be referred to adjudication. This immediately widens the adjudication provision beyond that required by statute, which does, of course, only require disputes arising under the contract to be referred. This is in fact far wider than even the JCT and ICE contracts.

(2)   The notice of referral shall set out the principal facts and arguments relating to the dispute. Copies of all relevant documents in the possession of the party giving the notice of referral shall be enclosed with the notice. A copy of the notice and enclosures shall at the same time be sent by the party giving the notice to the PM, the QS and the other party.

This contract, as does the ECC, brings in the project manager (PM) and the quantity surveyor (QS) as having a detailed involvement in the adjudication.

(3)   (a) If the person named as the adjudicator in the Abstract of Particulars is unable to act, or ceases to be independent of the Employer, the Contractor, the PM and the QS, he shall be substituted as provided in the Abstract of Particulars.

This provision sets a procedure for the replacement of an adjudicator who is unable to act or ceases to be independent of the parties or their agents. As noted in Chapter 8 where the Adjudicator's Agreement is discussed relating to this contract, there is a requirement that the adjudicator be independent of the parties. There is of course no requirement of independence in the Act, only impartiality.

> (b) It shall be a condition precedent to the appointment of an adjudicator that he shall notify both parties that he will comply with this Condition and its time limits.

This notification is done automatically by completion of the Model Form of Adjudicator's Appointment, which includes this provision, but, as there is no requirement that the Model Form be completed, this needs to be in the adjudication clause as well.

It is of note that there is no fall-back provision to cover the status of the decision if the adjudicator does not do this; note how the JCT deals with a failure to enter into the JCT Adjudicator's Agreement as a comparison. The authors are sure that failure to give this notice could not be held to be any bar to the validity of a decision.

> (c) The adjudicator, unless already appointed, shall be appointed within 7 Days of the giving of a notice of intention to refer a dispute to adjudication under paragraph (1). The Employer and the Contractor shall jointly proceed to use all reasonable endeavours to complete the appointment of the adjudicator and named substitute adjudicator. If either or both such joint appointments has not been completed within 28 Days of the acceptance of the tender, either the Employer or the Contractor alone may proceed to complete such appointments. If it becomes necessary to substitute as adjudicator a person not named as adjudicator or substitute adjudicator in the Abstract of Particulars, the Employer and Contractor shall jointly proceed to use all reasonable endeavours to appoint the substitute adjudicator. If such joint appointment has not been made within 28 Days of the selection of the substitute adjudicator, either the Employer or Contractor alone may proceed to make such appointment. For all such appointments, the form of adjudicator's appointment prescribed by the Contract shall be used, so far as is reasonably practicable. A copy of each such appointment shall be supplied too [sic] each party. No such appointment shall be amended or replaced without the consent of both parties.

This contract requires the adjudicator and a named substitute adjudicator to be appointed within 28 days of the acceptance of the tender for the contract. If this is not done either party may proceed to do this unilaterally. There is also provision for appointing a replacement for the adjudicator or for the substitute adjudicator in a similar way. All things being equal these provisions ought to mean that there will always be an adjudicator available, but as there is no obligation on the parties to complete the appointments unilaterally, only an option, there has to be a provision for the adjudicator to be appointed within seven days of the giving of a notice of intention to refer a dispute to adjudication in order to comply with the Act.

> (4) The PM, the QS and the other party may submit representations to the adjudicator not later than 7 Days from the receipt of the notice of referral.

This adjudication provision approaches the submissions to the adjudicator in the same way as the ECC, making specific provision for the project manager and quantity surveyor to make submissions.

> (5) The adjudicator shall notify his decision to the PM, the QS, the Employer and the Contractor not earlier than 10 and not later than 28 Days from receipt of the notice of referral, or such longer period as is agreed by the Employer and the Contractor after the dispute has been referred. The adjudicator may extend the period of 28 Days by up to 14 Days, with the consent of the party by whom the dispute was referred. The adjudicator's decision shall nevertheless be valid if issued after the time allowed. The adjudicator's decision shall state how the cost of the adjudicator's

fee or salary (including overheads) shall be apportioned between the parties, and whether one party is to bear the whole or part of the reasonable legal and other costs and expenses of the other, relating to the adjudication.

There are a number of interesting and unique provisions in this clause. The adjudicator cannot notify his decision earlier than ten days from the notice of referral. This seems to us to create unnecessary difficulties if the matter is one that would be best thrashed out very quickly on site so that the contract works can proceed in the knowledge of the adjudicator's decision. To have to wait ten days for a decision on the quality of works whilst those works proceed could mean that at least a week's worth of work could be done that is of the same standard as the earlier work, which may be found to be inadequate. An immediate decision would mean that, at the very least, this further work would not have been done to the standard found to be unacceptable by the adjudicator. This problem might be addressed by the adjudicator issuing a preliminary provisional decision and a formal ratification being issued on the tenth day.

Another matter of specific note is that the adjudicator's decision is still valid if it is issued after the time allowed. Unfortunately, there is no indication of any time restriction on this and one wonders where this leaves the parties if the adjudicator still has not produced his decision weeks after the original due date. It must be noted that there is no requirement that the adjudicator performs within the timescales in the appointment associated with this form.

The last of the matters covered by this clause relates to the allocation of not only the adjudicator's fees and expenses as between the parties but also the costs of the parties themselves. This is unusual as it is generally the case that the parties to an adjudication bear their own costs whatever the result. If there is such an agreement the court will enforce it but this more usually occurs where both parties have claimed their own costs during the course of the adjudication, and an agreement that such costs are recoverable is deemed to have been reached.

(6) The adjudicator may take the initiative in ascertaining the facts and the law, and the Employer and the Contractor shall enable him to do so. In coming to a decision the adjudicator shall have regard to how far the parties have complied with any procedures in the Contract relevant to the matter in dispute and to what extent each of them has acted promptly, reasonably and in good faith. The adjudicator shall act independently and impartially, as an expert adjudicator and not as an arbitrator. The adjudicator shall have all the powers of an arbitrator acting in accordance with Condition 60 (Arbitration and choice of law), and the fullest possible powers to assess and award damages and legal and other costs and expenses; and, in addition to, and notwithstanding the terms of, Condition 47 (Finance charges), to award interest. In particular, without limitation, the adjudicator may award simple or compound interest from such dates, at such rates and with such rests as he considers meet the justice of the case —

(a) on the whole or part of any amount awarded by him, in respect of any period up to the date of the award;

(b) on the whole or part of any amount claimed in the adjudication proceedings and outstanding at the commencement of the adjudication proceedings but paid before the award was made, in respect of any period up to the date of payment;

and may award such interest from the date of the award (or any later date) until payment, on the outstanding amount of any award (including any award of interest and any award of damages and legal and other costs and expenses).

This sub-clause adds some notable points to the required provision that the adjudicator may take the initiative in ascertaining the facts and the law. It uses the term 'expert adjudicator' and identifies the powers that the adjudicator enjoys. One small criticism is the use of the word 'award' in the final paragraph. This should in the authors' view be 'decision' for the sake of consistency.

(7) Subject to the proviso to Condition 60(1) (Arbitration and choice of law), the decision of the adjudicator is binding until the dispute is finally determined by legal proceedings, by arbitration (if the Contract provides for arbitration, or the parties otherwise agree to arbitration), or by agreement: and the parties do not agree to accept the decision of the adjudicator as finally determining the dispute.

(8) In addition to his other powers, the adjudicator shall have power to vary or overrule any decision previously made under the Contract by the Employer, the PM or the QS, other than decisions in respect of the following matters—

(a) decisions by or on behalf of the Employer under Condition 26 (Site admittance);

(b) decisions by or on behalf of the Employer under Condition 27 (Passes) (if applicable);

(c) provided that the circumstances mentioned in Condition 56(1)(a) or (b) (Determination by Employer) have arisen, and have not been waived by the Employer, decisions of the Employer to give notice under Condition 56(1)(a), or to give notice of determination under Condition 56(1);

(d) decisions or deemed decisions of the Employer to determine the Contract under Condition 56(8) (Determination by Employer);

(e) provided that the circumstances mentioned in Condition 58A(1) (Determination following suspension of Works) have arisen, and have not been waived by the Employer, decisions of the Employer to give notice of determination under Condition 58A(1); and

(f) decisions of the Employer under Condition 61 (Assignment).

In relation to decisions in respect of those matters, the Contractors's [sic] only remedy against the Employer shall be financial compensation.

The exclusions in this sub-clause appear to run in the face of the statutory requirement that any dispute is adjudicable. It will however be noted that all the matters covered by this clause relate to decisions of the Employer, the PM or the QS that are expressed by the contract as being final and in those terms these exclusions would be the same if the Scheme applied. The Scheme by paragraph 20(a) precludes the opening up of a decision or certificate which the contract states is final and conclusive. It does not however prevent a dispute of this nature from being adjudicated, merely restricting the remedy to financial compensation should the adjudicator be persuaded that the matter has adversely affected the contractor.

(9) Notwithstanding Condition 90 (Arbitration and choice of law), the Employer and the Contractor shall comply forthwith with any decision of the adjudicator; and shall submit to summary judgment and enforcement in respect of all such decisions.

The immediate compliance term of this sub-clause clarifies the binding nature of the decision, and the points on summary judgment and enforcement clarify any possible misapprehensions that may exist about these matters.

(10) If requested by one of the parties to the dispute, the adjudicator shall provide reasons for his decision. Such requests may only be made within 14 Days of the decision being notified to the requesting party.

This is a unique method of approaching reasons and may well cause some confusion. The adjudicator notifies the parties of his decision. A party may request reasons for 14 days thereafter. It seems that this clause prevents any earlier request for reasons but common sense suggests that if an earlier request is made there is no prejudice of any sort and the procedure must be eased if this is done. Should reasons be requested for the first time after the decision has been notified the adjudicator will be embarrassed unless he has them to hand already. If he does not do this, he may find that the request comes at an inconvenient time and the reasons take a considerable time to produce. This is not really a problem for the adjudicator as it is only the request that has to be made within the 14 day period, not the giving of the reasons.

If this provision is drafted in the hope of reducing the adjudicator's fee we do not believe that it works. We suggest that an adjudicator proceeding under this provision should enquire of the parties as to whether they want reasons at an early stage or, alternatively, if he is so minded, tell them that he intends to provide reasons unless they both agree that he should not.

> (11) The adjudicator is not liable for anything done or omitted in the discharge or purported discharge of his functions as adjudicator, unless the act or omission is in bad faith. Any employee or agent of the adjudicator is similarly protected from liability.

This is the standard immunity provision.

Association of Consulting Engineers

The Association of Consulting Engineers (ACE) has published the 2nd edition (1998) of the ACE Conditions of Engagement 1995.

The ACE is extremely economical in its use of words, with no more than two clauses within the section entitled 'Disputes and Differences'. The first of these, clause 9.1, relates to an encouragement to mediate. The second, clause 9.2, provides for the express incorporation of the Construction Industry Council Model Adjudication Procedure into the Consulting Engineer's Agreement. The CIC Procedure, which is considered in detail elsewhere, itself contains all the requirements of the Act and by this incorporation the Engineer's Agreement is compliant with the Act.

Having achieved compliance with the Act in this way there is no need for anything else.

The Institution of Chemical Engineers Forms of Contract 2001

The IChemE publishes the following forms of contract: the Red Book for lump sum contracts; the Green Book for reimburseable contracts; the Yellow Book for subcontracts; the Brown Book for subcontracts for civil engineering works; and the Orange Book for minor works. Each of these contracts includes adjudication provisions which are prefaced as follows:

> This Clause 46 shall only apply to disputes under a construction contract as defined in the Housing Grants, Construction and Regeneration Act 1996 or any amendment or re-enactment thereof.

Clause 46 must be put into the context of the submissions made by the IChemE and others involved in the process industries when the Act was under consideration, that their contracts were of a nature that made adjudication an unsuitable procedure and that those industries should be exempted from the statutory right to adjudication. The success of this lobbying is reflected in section 105 of the Act. It is also of interest to note the preface to the IChemE Adjudication Rules, the Grey Book, which includes what appears to be an apologia for adjudication. This sets out a brief history of the development of the process and puts forward the suggestion that adjudication is only suitable for disputes of a relatively minor nature. This preface also identifies the fact that works relating to specific activities in the 'process industries' are excluded from the definition of construction contracts, the main thrust of this point being set out in the preface in terms that 'most disputes will be complex, firstly because of the sheer size of potential disputes (possibly involving massive quantities of information/documentation) and secondly because of the multi-disciplinary nature of the process industries'.

This clause states very clearly that the adjudication procedures apply only to contracts to which the Act applies and confirms that there is no unilateral right to adjudication under the IChemE forms unless the work from which the dispute arises comes within the provisions of the Act. This does not of course mean that the parties cannot by agreement adjudicate a dispute over matters otherwise excluded by section 105 if they so wish but it does prevent the use of a unilateral application for the appointment of an adjudicator as a tactical ploy to move a dispute forward quickly. Our experience in the construction industry suggests that the use of such tactical ploys has been one of the great successes of adjudication and that perhaps parts of the process industries have missed out as a result of section 105 but we understand that this is not perceived to be a problem. That said, in our view the IChemE has made a superb job of addressing the procedural aspects of adjudication in these rules.

Adjudication rules and procedures

The Act does no more than require that the contract includes the eight compliance points that have been discussed in earlier chapters. It requires nothing more. As far as the procedure for the adjudication is concerned, provided that the contract includes these points, all the adjudicator has to do is to reach his decision within 28 days of referral or any extended period and act impartially.

In such a case the Scheme would not apply and the procedure for the adjudication will be totally at the discretion of the adjudicator. In practice, however, this is unlikely to happen. There are two common scenarios. The first is that the parties may know nothing about, or decide to ignore, the statutory requirement for adjudication, in which case their contract will not comply with the requirements of the Act and the Scheme will apply. Alternatively, they will use some standard form or write their own ad hoc contract which will have compliant adjudication provisions within it. These may have pet 'bolt-ons' included by the party who originally drafts the contract terms (which will need agreement by the other) or it will include the Scheme as the procedure for any adjudication.

The adjudicator will naturally have to comply with the requirements of the contract and any rules in order to fulfil the obligations that he takes on when accepting the position of adjudicator.

At the time of writing the following bodies have published adjudication rules:

- The Technology and Construction Solicitors Association (2002 Version 2.0 Procedural Rules for Adjudication)
- The Construction Industry Council (Model Adjudication Procedure Third Edition)
- The Centre for Dispute Resolution (2002 edition)
- The Institution of Civil Engineers (ICE Adjudication Procedure 1997)
- The Institution of Chemical Engineers (Adjudication rules Second edition 2001).

The JCT has decided that it will not produce any adjudication rules, nor will it adopt any of the rules produced by other bodies. Instead it has produced clause 41A in JCT 80, with similar clauses in its other forms, which deal with adjudication and include all procedural matters, running to nearly three pages of closely spaced type. There is some indication at the time of writing that the JCT may be considering moving to the position of using the Scheme as the adjudication rules for its contracts.

The Government Construction Contracts (GC/Works 1998), which comprise ten separate publications, do not include any rules.

The Engineering and Construction Contract (ECC) also includes all the procedural matters that the drafters consider to be necessary in the contract clauses.

The Scheme has been looked at in detail elsewhere. The Scheme is itself a set of rules for the conduct of adjudication for parties to contracts, for ANBs and for adjudicators. The principles of the Scheme and of the various rules are very similar. There are certain differences, however, and set out below in tabular form is a comparison of the points of similarity and difference between the Scheme and the various rules in respect of those areas that the authors consider to be of particular interest. The reader should refer to the text of the various rules in order to make a full comparison of all the provisions. The nature of similar provisions, if any, in the contracts that do not rely on rules is also noted.

Comparison of the provisions

(The DSC/C and DOM/1&2 follow the procedures in the JCT forms)

The notice of adjudication

The Scheme	The nature and a brief description of the dispute and of the parties involved. Details of when and where the dispute has arisen. The nature of the redress which is sought. The names and addresses of the parties.
The CIC Procedure 3rd edition	A brief statement of the issue or issues which it is desired to refer and the redress sought.
The TeCSA Rules 2002 version 2.0	A written notice identifying in general terms the dispute in respect of which adjudication is required.
The ICE Procedure	The details and date of the contract between the parties. The issues which the adjudicator is being asked to decide. Details of the nature and extent of redress sought.
The CEDR Rules (2002)	The names and addresses and full contact details of the parties and any representatives. A copy of the relevant provisions of the contract providing for adjudication. Brief details of the dispute. Details of the remedy sought.
JCT Standard Forms	Notice must briefly identify the dispute or difference.
GC Works	No requirements.
Engineering and Construction Contract	No requirement as to the nature of the notice.
The IChemE Rules	Give notice with details of the matter in dispute.

The nomination of the adjudicator

The various adjudication rules, the contractual adjudication procedures and the Scheme comply with the requirements of the Act by including a timetable with the intention of referring the dispute to the adjudicator within seven days of the notice of adjudication being given.

The referral notice

The Scheme	A reference in writing to be accompanied by copies of or relevant extracts from the construction contract. Such other documents as the referring party intends to rely upon.
The CIC Procedure 3rd edition	A statement of case. A copy of the notice of adjudication. The contract. Details of the circumstances giving rise to the dispute. The reasons why it is entitled to the redress sought. The evidence upon which it relies. The statement of case shall be confined to the issues in the Notice (of adjudication). (The parties can agree to include additional matters subsequently)
The TeCSA Rules 2002 version 2.0	No requirements as to a referral notice.
The ICE Procedure	A copy of the notice of adjudication. A copy of the adjudication provision in the contract. The information upon which the referring party relies including supporting documents.
The CEDR Rules (2002)	A concise statement of case including a copy of the notice of adjudication. A copy of the conditions of contract and other provisions in the contract on which the referring party intends to rely. Details of the circumstances giving rise to the dispute. The reasons for entitlement to the remedy sought. The evidence, including relevant documentation, in support of its case.
JCT Standard Forms	Particulars of the dispute or difference. Summary of the contentions on which the referring party relies. A statement of the relief or remedy which is sought. Any material the referring party wishes the adjudicator to consider.
GC Works	Principal facts and arguments relating to the dispute. Copies of all relevant documents in the possession of the party giving the notice.
Engineering and Construction Contract	Information to be considered by the adjudicator.
The IChemE Rules	Adjudicator directs the parties each to provide a statement within a specified time of his appointment which shall be reasonable. A view of the dispute in summary and detail, with supporting documentary evidence and a copy of the contract. Names and organisational position of representatives at any meeting. Address of any place which the adjudicator may need to visit. The name and address etc. of anyone the party intends shall present any evidence or opinion and the subject to be covered by each.

The date of referral

The Scheme	Not later than seven days after the notice of adjudication.
The CIC Procedure 3rd edition	The date the adjudicator receives the statement of case.
The TeCSA Rules 2002 version 2.0	The date the adjudicator receives the Referral Notice.
The ICE Procedure	The date the adjudicator receives the documents.
The CEDR Rules (2002)	The date on which both the adjudicator and the other party receive the concise statement of case
JCT Standard Forms	Upon the adjudicator receiving the referral. Specific provisions as to this receipt are given and as far as postal delivery is concerned this is deemed to have taken place two days after posting, Sundays and public holidays excepted, subject to proof to the contrary.
GC Works	The date the adjudicator receives the referral notice.
Engineering and Construction Contract	Within seven days of notice of adjudication.
The IChemE Rules	Not specified.

Adjudicator's agreement

The Scheme	No agreement form.
The CIC Procedure 3rd edition	Includes agreement form.
The TeCSA Rules 2002 version 2.0	No agreement form.
The ICE Procedure	Includes agreement form.
The CEDR Rules (2002)	No agreement form.
JCT Standard Forms	Two forms, for use either on letting the contract or when the dispute arises.
GC Works	Form intended for use at time of letting contract.
Engineering and Construction Contract	Adjudicator's contract. One for use when Scheme applies, the other for when it does not.
The IChemE Rules	Draft letter provided.

Adjudicator's fees

The Scheme	Adjudicator may determine and apportion between the parties.
The CIC Procedure 3rd edition	Allowed reasonable fees and expenses. Adjudicator may direct who pays. (If he does not – equal shares.) Not entitled to fees and expenses if decision late and replacement adjudicator appointed. (Other than the cost of legal or technical advice – if the parties have received that advice.)
The TeCSA Rules 2002 version 2.0	Adjudicator may not require advance payment or security. Adjudicator may direct who pays. Not to exceed £1,250.00 per day plus expenses plus VAT.
The ICE Procedure	Adjudicator may direct who pays. If no direction parties pay in equal shares.
The CEDR Rules (2002)	Adjudicator has discretion to apportion liability.
JCT Standard Forms	Adjudicator directs how apportioned.
GC Works	Adjudicator directs how apportioned.
Engineering and Construction Contract	Equal shares unless otherwise agreed (if adjudicator's contract applies).
The IChemE Rules	Payable in equal parts.

Adjudicator's jurisdiction

The Scheme	No provision for adjudicator to determine own jurisdiction. Adjudicator may, with consent of all parties, adjudicate at the same time on one or more disputes under the same contract. Adjudicator may, with consent of all parties, adjudicate at the same time on related disputes under the different contracts.
The CIC Procedure 3rd edition	No provision for adjudicator to determine own jurisdiction.
The TeCSA Rules 2002 version 2.0	Adjudicator may decide upon his own substantive jurisdiction. More than one notice of adjudication may be given in respect of disputes arising out of the same contract.
The ICE Procedure	No provision for adjudicator to determine own jurisdiction.
The CEDR Rules (2002)	No provision for adjudicator to determine own jurisdiction.
JCT Standard Forms	No provision for adjudicator to determine own jurisdiction.
GC Works	No provision for adjudicator to determine own jurisdiction.
Engineering and Construction Contract	No provision for adjudicator to determine own jurisdiction.
The IChemE Rules	No provision for adjudicator to determine own jurisdiction.

The parties' costs

The Scheme	Silent.
The CIC Procedure 3rd edition	Parties bear own costs.
The TeCSA Rules 2002 version 2.0	Adjudicator can award costs to the successful party if the parties so agree. Notwithstanding anything in the contract to the contrary the adjudicator has no jurisdiction to require a party which referred the dispute to adjudication to pay the costs of any other party solely by reason of having referred the dispute to adjudication.
The ICE Procedure	Parties bear own costs.
The CEDR Rules (2002)	Parties bear own costs.
JCT Standard Forms	Parties bear own costs save that the adjudicator may direct who should pay the cost of any test or opening up procedure.
GC Works	Adjudicator can award costs against a party.
Engineering and Construction Contract	Silent.
The IChemE Rules	Parties bear own costs.

Adjudicator's ability to resign

The Scheme	At any time. Must resign where dispute has already been referred to adjudication and decision 'taken'.
The CIC Procedure 3rd edition	At any time upon giving notice.
The TeCSA Rules 2002 version 2.0	Silent.
The ICE Procedure	Silent.
The CEDR Rules (2002)	Adjudicator must resign if: The dispute already referred to adjudication and a decision made. The adjudicator is not competent to decide because the nature of the dispute is significantly different to that in the notice of adjudication. The adjudicator becomes unable to give a decision by the due date.
JCT Standard Forms	Silent.
GC Works	Silent.
Engineering and Construction Contract	If adjudicator's contract completed, may terminate if conflict of interest, unable to act, not been paid within five weeks of date by which payment should have been made.
The IChemE Rules	Silent.

Termination by parties

The Scheme	At any time by agreement.
The CIC Procedure 3rd edition	Silent.
The TeCSA Rules 2002 version 2.0	Silent, but Chairman of TeCSA may replace one adjudicator with another if and when it appears necessary for him to do so.
The ICE Procedure	Silent.
The CEDR Rules (2002)	Silent.
JCT Standard Forms	By agreement at any time.
GC Works	Silent.
Engineering and Construction Contract	If adjudicator's contract completed, by agreement at any time.
The IChemE Rules	Silent.

Joinder provisions

The Scheme	Adjudicator may, with consent of all parties.
The CIC Procedure 3rd edition	Subject to agreement of adjudicator and existing and additional parties.
The TeCSA Rules 2002 version 2.0	No joinder provisions.
The ICE Procedure	Subject to agreement of adjudicator and existing and additional parties.
The CEDR Rules (2002)	Subject to agreement of adjudicator and existing and additional parties.
JCT Standard Forms	No joinder provisions.
GC Works	No joinder provisions.
Engineering and Construction Contract	No joinder provisions.
The IChemE Rules	No joinder provisions.

Procedure and timetable (all within the 28-day limit)

The Scheme	Adjudicator decides.
The CIC Procedure 3rd edition	Adjudicator decides. Subject to any limitation of the contract or the Act.
The TeCSA Rules 2002 version 2.0	Adjudicator decides.
The ICE Procedure	Adjudicator decides.
The CEDR Rules (2002)	Adjudicator decides timetable and procedure. Adjudicator may not take into consideration any statement unless it has been made available to the parties for consideration.
JCT Standard Forms	Adjudicator decides. Responding party may respond within seven days
GC Works	Procedure set. PM, QS and other party may make submissions within seven days of notice of referral. Other than the required provision that the adjudicator may take initiative to ascertain the facts and the law, there are no other provisions.
Engineering and Construction Contract	Procedure set. Both parties have 14 days after referral to provide further information. Other than the required provision that the adjudicator may take initiative to ascertain the facts and the law there are no other provisions.
The IChemE Rules	Adjudicator decides subject to any limitation on the timetable in the contract or the Act.

Power of adjudicator to open up, review and revise

The Scheme	Any decision or certificate unless stated to be final and conclusive.
The CIC Procedure 3rd edition	Any certificate, decision, direction, instruction, notice, opinion, requirement or valuation made in relation to the contract.
The TeCSA Rules 2002 version 2.0	Any certificate or other things issued or made pursuant to the contract as would an arbitrator and/or a court.
The ICE Procedure	Any decision, opinion, instruction, direction, certificate or valuation made under or in connection with the contract and which is relevant to the dispute.
The CEDR Rules (2002)	Any certificate, decision, direction, instruction, notice, requirement or valuation made under the contract except where the contract precludes this.
JCT Standard Forms	Any certificate, opinion, decision, requirement or notice as if no such had been issued, given or made.
GC Works	Adjudicator has all the powers of an arbitrator. In certain defined areas the adjudicator cannot vary or overrule decisions made by the employer, project manager or quantity surveyor, the contractor's only remedy being financial compensation.
Engineering and Construction Contract	Silent.
The IChemE Rules	Adjudicator may open up, review and revise any decision taken (other than that of an adjudicator unless agreed by both parties) or certificate given under or in connection with the contract and which is relevant to the dispute unless the contract states that the decision or certificate is final and conclusive.

Expert assistance

The Scheme	The parties must agree.
The CIC Procedure 3rd edition	At discretion of the adjudicator provided he notifies the parties first.
The TeCSA Rules 2002 version 2.0	One party must consent or request.
The ICE Procedure	At discretion of adjudicator provided he notifies the parties.
The CEDR Rules (2002)	Parties must be informed beforehand with reasons.
JCT Standard Forms	On prior notice to parties.
GC Works	No prior notice needed.
Engineering and Construction Contract	If adjudicator's contract completed. Adjudicator may obtain help. No notice needed.
The IChemE Rules	After prior notice and where possible an indication of likely costs.

Matters for decision

The Scheme	The matters in dispute and any other matters that the parties agree should be within the scope of the adjudication or which are matters that the adjudicator considers are necessarily connected with the dispute.
The CIC Procedure 3rd edition	The matters in the notice of adjudication and other matters which the parties and the adjudicator agree. (But the statement of case must be limited to the matters included in the notice.)
The TeCSA Rules 2002 version 2.0	The matters identified in the notice of adjudication, any further matters which all parties agree should be within the scope of the adjudication and any further matters that the adjudicator determines must be included to make the adjudication effective and/or meaningful.
The ICE Procedure	The matters in the notice of adjudication and any other matters which the parties and the adjudicator agree shall be within the scope.
The CEDR Rules (2002)	The adjudicator shall decide the dispute and any other matters which the adjudicator determines should be taken into account in deciding the dispute.
JCT Standard Forms	The dispute.
GC Works	The dispute.
Engineering and Construction Contract	The dispute
The IChemE Rules	The matters in the notice of adjudication together with any other matters which the parties and the adjudicator agree should be within the scope of the adjudication.

Method of decision

The Scheme	In accordance with the applicable law in relation to the contract.
The CIC Procedure 3rd edition	In accordance with the law of the contract.
The TeCSA Rules 2002 version 2.0	Wherever possible to reflect the parties' legal entitlements. Where this is not possible the decision shall represent a fair and reasonable view in light of the facts and law insofar as they have been ascertained by the adjudicator.
The ICE Procedure	Silent.
The CEDR Rules (2002)	The Rules are governed by English Law.
JCT Standard Forms	Silent.
GC Works	The law of the contract is in the adjudicator's agreement. Official Secrets Act applies. Atomic Energy Act may apply.
Engineering and Construction Contract	If adjudicator's contract completed. In accordance with the law of the contract. Contract itself is silent.
The IChemE Rules	Silent.

Reasons

The Scheme	If required by the parties.
The CIC Procedure 3rd edition	Adjudicator required to give reasons unless parties agree otherwise.
The TeCSA Rules 2002 version 2.0	If any or all parties so request within seven days of the Referral. If requested by one party reasons shall be delivered to all parties.
The ICE Procedure	Adjudicator shall not be required to give reasons.
The CEDR Rules (2002)	Adjudicator shall give reasons unless the parties agree to the contrary.
JCT Standard Forms	Adjudicator not obliged to give reasons.
GC Works	If requested by a party.
Engineering and Construction Contract	Reasons to be given.
The IChemE Rules	Reasons to be given unless otherwise agreed by the parties.

Status of decision if not reached in 28 days or any extended period

The Scheme	Silent.
The CIC Procedure 3rd edition	Effective if reached before the referral of the dispute to any replacement adjudicator but not otherwise.
The TeCSA Rules 2002 version 2.0	Silent.
The ICE Procedure	Effective if reached before the referral of the dispute to any replacement adjudicator.
The CEDR Rules (2002)	Silent.
JCT Standard Forms	Silent.
GC Works	Valid if issued late. No time limit.
Engineering and Construction Contract	Silent.
The IChemE Rules	A party may give adjudicator and other party seven days' notice of intention to refer the dispute to another adjudicator. Adjudicator's decision valid if given within the seven-day notice period. If decision notified after expiry of seven-day notice period it is not valid.

Adjudicator's lien

The Scheme	Adjudicator must deliver decision as soon as possible after it is reached.
The CIC Procedure 3rd edition	Adjudicator shall reach his decision within the time limits in paragraph 16. (The 2nd edition provision that the adjudicator is not required to release decision until paid is now deleted but has not been replaced with any requirement for delivery of the decision.)
The TeCSA Rules 2002 version 2.0	Silent. Adjudicator not required to do any more than reach a decision in 28 days.
The ICE Procedure	Adjudicator need not release decision until paid if advance notice given.
The CEDR Rules (2002)	Adjudicator shall communicate the decision in writing to the parties by the 28th day or any extended date.
JCT Standard Forms	No lien. Adjudicator must send decision to parties.
GC Works	No lien. Adjudicator must notify decision to parties not later than 28 days from receipt of referral.
Engineering and Construction Contract	No lien but no requirement to do more than reach a decision in 28 days.
The IChemE Rules	Adjudicator may apply a lien if he so notifies the parties up to seven days before the decision is due.

Correction of decision

The Scheme	Silent.
The CIC Procedure 3rd edition	Correction of any error arising from an accidental error or omission or to clarify or remove any ambiguity. The adjudicator has five days after the delivery of his decision to the parties to correct errors.
The TeCSA Rules 2002 version 2.0	The adjudicator may on his own initiative or on application by a party, correct his decision. Parties have up to five days to apply for a correction Adjudicator to make correction as soon after application received as possible or, if made on the adjudicator's own initiative, as soon as possible after he becomes aware of the need to make a correction.
The ICE Procedure	The adjudicator may correct any clerical mistake, error or ambiguity. Adjudicator has up to 14 days to correct his decision on his own initiative. Parties have 14 days to request a correction, adjudicator then has seven days to make the correction.
The CEDR Rules (2002)	The adjudicator has five days after the communication of his decision to the parties to correct errors.
JCT Standard Forms	Silent.
GC Works	Silent.
Engineering and Construction Contract	Silent.
The IChemE Rules	Silent.

Destruction of documents

The Scheme	Silent.
The CIC Procedure 3rd edition	The adjudicator may destroy the documents after six months provided that he enquires beforehand as to whether the parties want them returned.
The TeCSA Rules 2002 version 2.0	Silent.
The ICE Procedure	The adjudicator shall inform the parties if he intends to destroy the documents which have been sent to him in relation to the adjudication and he is required to retain the documents for a further (unspecified) period at the request of either party.
The CEDR Rules (2002)	Silent.
JCT Standard Forms	Silent.
GC Works	Silent.
Engineering and Construction Contract	Adjudicator keeps documents until termination.
The IChemE Rules	Silent.

Chapter 7
The Appointment

Once a party to a construction contract has reached what he considers to be the point of no return in his dealings with the other party and has concluded that there is no alternative but to implement his right to adjudication, he issues a notice of adjudication.

Before he issues this notice it is essential that he is certain that there is a dispute and that it is sufficiently crystallised so that problems can be avoided should there be a need for enforcement proceedings. In the five years since the statutory right to adjudication came into force, this has become one of the common challenges to adjudicators' jurisdiction. Winning parties often have to commence proceedings in the court to enforce the adjudicator's decision and in these proceedings the other party may argue that there was no dispute at all or that it has insufficiently crystallised and thus the adjudicator had no jurisdiction to make his decision. We deal with this aspect in detail in Chapter 9 where we consider the adjudicator's jurisdiction.

One other principal matter that will exercise the mind of a party considering adjudication relates to the extent of the matters that he wants to refer to adjudication. In some instances a party may take the view that he wishes to refer only a discrete part of the overall matters that he considers to be in dispute to adjudication. Alternatively the parties may agree that it is in both their interests to refer specific generic issues to the adjudicator in the anticipation that once a decision is made it may be possible for them to apply that decision to other aspects of the overall dispute. If they do not succeed in this, there is of course no bar to starting an adjudication at any time, and to refer part of the dispute in this way will not prevent other parts being referred later, provided of course that they are different disputes.

Prior to 1 May 1998, adjudication was a consensual process and was exclusively the result of provisions in certain standard form contracts. These are now, by the definition in the Act, construction contracts. Then the parties provided for adjudication in their contracts by agreement. The right to adjudication was solely a contractual one. If there was no adjudication provision in the contract, a party had no right to adjudication. The only possible exception was an agreement to adjudicate reached subsequently to the formation of the contract, but in the absence of any culture relating to adjudication this was probably never done.

Since 1 May 1998 every party to a construction contract has a statutory right to refer their disputes arising under the contract to adjudication under a procedure complying with section 108 of the Act. As considered in earlier chapters, the Act requires that the right to adjudication must be included as part of the contract in the form of a compliant provision. If it is not, the Act by section 114 creates that right by requiring that the provisions of the Scheme are implied as terms in the contract.

At the time when a dispute occurs, a single party can therefore set the whole process in motion; it does not require the agreement of both the parties for an adjudication to commence. It is a unilateral right which arises from provisions in the contract that comply with the Act or from the implied terms of the Scheme. It is in fact very often the case that an adjudication is brought by one party to a construction contract where the other party is extremely reluctant.

Adjudication can and does apply to contracts that are outside the provisions of the Act. Contracts in the JCT Minor Works Form for example include adjudication provisions. This form of contract is often used for works that would otherwise not be affected by the statutory right to adjudication. One example is a contract with a residential occupier which is excluded from the ambit of statutory adjudication by the provisions of section 106 of the Act. Another instance occurs where a contract in any of the standard forms used for construction work that includes adjudication provisions, which they all now do, includes work that is excluded from the definition of construction contracts in section 105 of the Act and to which adjudication would not otherwise apply. A party then applies its contractual right to adjudication and should it have difficulty in getting the other party to honour the decision the court will enforce the contractual right – see *Parsons Plastics* v. *Purac*[1].

Where there are contractual arrangements for adjudication that do not comply with the requirements of the Act, the Scheme normally applies under the provisions of section 108(5) of the Act. It is however not compulsory to exercise the right to imply the terms of the Scheme. The parties could choose to ignore the Scheme and adjudicate under their non-complying contractual arrangement. In so doing they would be bound by the results of that arrangement. The enforcement of any decision reached by an adjudicator acting under such a contractual arrangement would be the enforcement of a contractual rather than a statutory right to have an adjudicator make a decision on the dispute in question.

It is also possible that the parties have agreed provisions in relation to the adjudication process that are in addition to those that comply with the Act. This could be either where there are compliant provisions in the contract or where the Scheme applies in default. Examples would be a provision that requires a response from the responding party and sets a timetable for that response to be provided, or a requirement limiting the submissions to a certain number of pages. Where such provisions are in addition to the eight compliance points, there would appear to be no difficulty in applying the additional provisions on the basis of a contractual agreement between the parties.

In a situation where the contract provisions are not compliant, things are possibly a little more complicated. There are two possible interpretations that can be put on section 108(5) of the Act which reads: 'If the contract does not comply with the requirements of subsections (1) to (4) [of section 108] the adjudication provisions of the Scheme for Construction Contracts apply.' The first interpretation is that this means that all procedures in the contract, even those that do not conflict with the Act, go out of the window and only the provisions of the Scheme apply. The second is that the Scheme provisions apply to replace those that are non-compliant, but the parties' agreement as to supplementary provisions that do not conflict remains in place. We do not see this as creating any real problem for the adjudicator. If there is no objection from the parties to the additional provisions, it seems wise for the adjudicator to comply with the parties' agreement and follow them. If he does not, he might create some difficulty if the parties are expecting him to do so. To avoid possible problems later it would be sensible for the adjudicator to reflect the agreed provisions in his directions given at the outset of the adjudication. If a party objects to or fails to fulfil an additional provision the adjudicator should follow the same course as he does in any situation where there is a difference, or possible difference, between the parties, that is to obtain submissions from each party and decide between them.

---

[1] *Parsons Plastics (Research & Development) Limited* v. *Purac Limited* (13 August 2001) unreported; [2002] BLR 334, CA.

There is a view, strongly held by some adjudicators, that where the contract is non-compliant, the adoption of the Scheme supersedes all the contractual provisions regarding adjudication, and any additional agreements that do not conflict with the Scheme should be ignored. If this is the route chosen by the adjudicator, he should be careful to confirm in his directions at the outset of the adjudication that this is the way he intends to proceed.

If the parties are agreed that they wish to proceed under a non-complying contractual arrangement, the adjudicator is duty bound to proceed in that way. He cannot insist that the parties operate in accordance with the Scheme. It is however essential for the smooth running of the adjudication process that the adjudicator is aware of the parties' agreement, as the last thing anyone wants in the somewhat pressurised situation of an ongoing adjudication is for the adjudicator to try to apply procedures which are not agreed.

The Institution of Civil Engineers' standard forms include a provision for a matter of dissatisfaction to be followed before an adjudication can commence. We consider this provision elsewhere in detail. The courts have not, as far as we know, been asked to decide whether such a provision complies with the Act; the nearest to it has been the case of *Mowlem* v. *Hydra-Tight*[2], where the parties had agreed before the matter reached court that the clause did not comply. This agreement was endorsed by the court but it was given no specific judicial scrutiny. We suggest that this is probably the right answer given the statutory right to adjudication on demand. It is likely that any attempt to delay the commencement of an adjudication by citing this provision will fail, as it rightly should, so that the party desiring the adjudication is not improperly delayed from having his dispute decided. Whatever the rights and wrongs of this provision we are, however, of the opinion that this can be an admirable means of ensuring that the dispute has properly crystallised before commencing the adjudication.

A dispute that has already been adjudicated cannot be re-opened save in arbitration or litigation. It is vital when preparing a notice of adjudication where there has already been one or more adjudications, that the party preparing the notice identifies with utmost clarity what it is that is being referred so that it is clear that what is being referred has not been decided already. The adjudicator clearly has no jurisdiction to decide a dispute that has already been the subject of an adjudicator's decision, but see the discussion in Chapter 5 relating to paragraph 9 of the Scheme in which we consider this point.

Another problem arises from the provisions of paragraph 8 of the Scheme. Paragraph 8(1) provides that the adjudicator may, with the consent of all the parties to those disputes, adjudicate at the same time on one or more disputes under the same contract. Paragraph 8(2) applies similar requirements to related disputes under different contracts. This provision was interpreted strictly in the case of *Grovedeck* v. *Capital Demolition*[3]. The losing party in *Grovedeck* was upset at having lost and was successful in using the provisions of the Scheme to avoid the consequences of the adjudicator's decision. It is generally the case however that there are considerable benefits in respect of time and cost if one adjudicator deals with related disputes, and it appears generally to be the case that parties accept this. This restriction does in any event only occur where the Scheme applies.

---

[2] *John Mowlem & Company plc* v. *Hydra-Tight Limited (t/a Hevilifts)* [2000] CILL 1650.
[3] *Grovedeck Limited* v. *Capital Demolition Limited* [2000] BLR 181.

Selecting the adjudicator

The time for the appointment of the adjudicator

The Act requires, at section 108(2)(b), that the construction contract must 'provide a time-table with the object of securing the appointment of the adjudicator and referral of the dispute to him within 7 days of such notice [of adjudication]'. If it does not so provide, the Scheme applies.

It is a peculiarity of the Scheme that it contains a provision that in itself may mean that the seven-day period is not achieved. The Scheme provides that communication of the selection of an adjudicator by a nominating body must be made within five days of receiving the request to select. If the request for that selection is received by the nominating body more than two days after the notice of adjudication, the nominating body may do what the Scheme requires even though the selection [nomination] may still be outside the required seven days.

The timetable of seven days is not thought to be strict. The contract must only provide a timetable with the object of securing the appointment and making the referral within seven days. The Scheme itself uses the word 'shall' in the context of the seven-day period; it does however include no sanction if the nomination is outside the seven-day period. Arguments may be put up by the responding party to the effect that the person nominated lacks jurisdiction as adjudicator, but the majority of adjudicators now appear prepared to proceed in these circumstances if the referring party so desires, the challenge being seen to be the sham that it is. In the case of a referring party who is not prepared to take the risk, however remote, of the court refusing enforcement due to a late referral, it is a simple matter to make a further application to the nominating body and experience seems to be that in general such re-nominations will be made without further charge.

Who are adjudicators?

Adjudication as brought in by the Act is not new. What is new is the much wider remit the adjudicator is now given by the legislation. Over the 20 years or so prior to the Act coming into force, adjudication was available under the DOM/1 and DOM/2 type sub-contract arrangements, as they were then, and also under the JCT with Contractor's Design main contract. This process was however limited in its application and dealt principally with questions relating to payment. The statutory regime extends adjudication to all disputes arising under construction contracts, but perhaps unsurprisingly the majority of the disputes that are referred relate to money.

In essence, adjudication is a judicial process under severe time restraints. The powers of an adjudicator are great; he has, to all intents and purposes, total discretion as to the way in which he conducts the adjudication and the effects of his decision can be as far reaching as an arbitrator's award and, in some cases, more so. He makes decisions on the facts and law without anything like the in-depth investigation of the arbitrator. The adjudicator has to be very good at what he does.

When drafting the first edition of this book we said that it was hardly surprising that the initial trainees for adjudication had their roots and experience in arbitration. After five years we can say that many of the successful adjudicators are just those people. They understand the industry, they understand how construction contracts work and they can apply that knowledge in producing decisions that the parties can either accept as resolving their dispute or use as a basis for such resolution. We accept that this is a rather different concept from the 'quick and dirty fix' originally envisaged, which by its very nature would almost automatically lead to the dispute being referred to arbitration or litigation, but it just shows that the adjudicators who have undertaken this complex, difficult, sometimes worrying but always rewarding task have provided a procedure that exceeds the expectations of many at the time that the process was introduced.

It is not just those with roots and experience in arbitration who are successful adjudicators, but it is clear that a major requirement for an adjudicator to be successful is an under-standing of the factors identified above.

It is a rather sad but interesting reflection that as adjudication has taken off, it has affected the approach of arbitrators and there is some evidence, albeit mainly apocryphal, that many arbitrators are now applying some of the procedures used by adjudicators to the benefit of the arbitration process. Some might say it is a pity that it took the advent of adjudication to encourage this – the powers are all there in the 1996 Arbitration Act – but better late than never. The climate has changed and parties are far more prepared to look at procedures in arbitration that would have been far too radical for them before the introduction of adjudication, even with the advances resulting from the 1996 Arbitration Act. This must be to the benefit of those parties who are unfortunate enough to be unable to resolve their dispute after an adjudication has taken place. At the time of writing, amendments to the Construction Industry Model Arbitration Rules are on the stocks to provide for arbitration with a 100-day limit.

Anyone who is selected as adjudicator should have at the very least received accredited training and to have done this they will also have been vetted on the necessary experience and understanding of the industry. Adjudicators are those who are the exception in their field rather than the rule. It is arguable that those who only meet the normal everyday standards are, albeit often innocently, the cause of the disputes rather than the solution. Many panels of adjudicators have been set up. The requirements placed upon those achieving a panel can be varied but it is unlikely that any member of a panel run by a recognised nominating body will not have undergone some form of training. The majority of nominating bodies have understood the importance of adjudication to the industry and how important it is that the adjudicators who are appointed are competent to undertake the task involved. They have been rigorous in their selection process. They have tested the back-ground and working experience in the industry of potential adjudicators. In addition, they have vetted particular dispute resolution experience and ability to understand and apply the law. There are adjudicators who have been found lacking but it is evident at the time of writing that many nominating bodies are starting to instigate procedures to review those already on their panels, to ensure that they are of a standard that satisfies the requirements of the industry.

In the same way that an arbitrator or a judge cannot have technical expertise in every field, nor can an adjudicator. It is important for an adjudicator to keep up to date in respect of current practices in his field. An adjudicator should be chosen for his track record and selection and training. What is needed from an adjudicator will necessarily include some managerial skills to find out the facts and the law and to control the parties. If particular knowledge is required the adjudicator can enlist experts in law or technical issues, either through the parties or himself. This should, however, be the exception rather than the rule; the time constraints of the adjudication process militate against the adjudicator who thinks of doing this.

Training of adjudicators

It is of great importance to a party seeking the appointment of an adjudicator, and of equal importance to the other party as well, that any adjudicator, whether selected by agreement or nominated by a third party, is competent to carry out the task. Competence covers not only his ability to deal with the procedural issues and make a decision, but also to understand the technicalities of the dispute.

An adjudicator's technical ability is not something that has formed any part of the training regimes that have been set up to cater for the demand for adjudicators since the passing of the Act. A certain level of assessment of technical ability goes on but the principal way in which those formulating lists of adjudicators have dealt with this requirement is to have required evidence of experience in the construction industry relevant to the type of dispute to which a particular individual is to be nominated. Where an agreed appointment is mooted, the customary way is for the parties to request the prospective adjudicator to submit his curriculum vitae setting out not only his experience and training as an adjudicator, but also his involvement in the industry at a practical level.

As far as training related to the procedural aspects of adjudication is concerned, all nominating bodies require some evidence of training in the specific area of adjudication and many provide it themselves.

A typical training programme would deal with the Act and the Scheme, the adjudication provisions in the commonly used forms of contracts and the various adjudication rules and procedures that are in common use. As far as the practicalities of the adjudication process are concerned, matters relating to the appointment, the procedure required to allow the parties to make their submissions and the adjudicator to make his decision within the short timescale allowed for the process, and such matters as decision-making and writing, are included in the training procedure. Two matters that have become of far greater relevance than some anticipated are a familiarity with the law of contract and matters relating to the jurisdiction of the adjudicator.

The one matter that is becoming evident is that the nature of the training offered needs to be reviewed. A three-day course originally put together as a conversion course for arbitrators who had already gone through an intensive and often lengthy training as arbitrators is clearly insufficient for those without such a background. A course culminating in a formal diploma in adjudication is being mooted at the time of writing and, in our view, this must be the way forward for the training of future adjudicators.

How can an adjudicator be appointed?

The appointment of an adjudicator can be made in the following ways:

- the parties can agree on a person to act as adjudicator
- an appointment can be made by prior agreement, almost invariably by naming an adjudicator in the contract
- an appointment can be made when the dispute occurs by agreement
- a party may appoint an adjudicator unilaterally by referring the dispute to an adjudicator nominated by an adjudicator nominating body (ANB).

Agreement on an adjudicator

There is always an advantage in consensus on who should be appointed. It saves time and acrimony to agree on a person to resolve the dispute. If somebody is selected who the parties know and can expect to do the job properly, there may be less of a problem with acceptance of the decision and therefore enforcement. An agreed person can be chosen with the particular dispute in mind. The particular specialism or area of expertise can be sought from a known source. The parties are then more likely to comply with any order as to timetable and more likely to co-operate in any investigation the adjudicator needs to undertake. If the parties have agreed, the selection process will have examined the competence of the adjudicator and therefore they will have confidence in the person and the process. ANBs have an excellent track record in finding the right adjudicator in the limited time they have, but there is always the possibility for whatever reason that an unsuitable person is nominated. An agreement between the parties will in all probability avoid this.

Naming an adjudicator by prior agreement

It is possible that the parties may suddenly come to the conclusion after a contract is made that they ought to agree an adjudicator, but the circumstances in which this may occur are difficult to envisage. It is far more likely that any such agreement is reached prior to the finalisation of the contract documents and that a name is inserted in them.

The earliest that an adjudicator's name can be put forward is in the tender documents. This may be because the party preparing the tender document is seeking to impose a particular adjudicator to act under the contract. Although there may be provision for naming the adjudicator at this stage, it may be that no name is provided until the tender is received. The intention may be to insert a name after discussion with the successful tenderer when the contract is signed. This often provides a formula for proceeding with the contract works with no named adjudicator at all. The problem that might arise in this instance, if there is no machinery in the contract for the nomination of an adjudicator by a third party, will be overcome by the provisions of the Scheme and the appointment procedures in them.

The particular benefit of having a named adjudicator is that there is no question as to who is going to adjudicate. The moment a dispute arises the named adjudicator should be immediately available to act. There is no delay in trying to agree an adjudicator or involving an ANB.

As far as the adjudicator himself is concerned, by reaching agreement the parties have the opportunity to choose someone with appropriate training and experience. The sort of questions that may require answering in such a case relate to such matters as: the nature of the training he has undertaken; his experience, both in respect of adjudication and the practical issues raised by the dispute; and his professional specialism. Almost certainly in these days of problems arising from the interpretation of what the agreement between the parties is, some familiarity with matters of contract law is essential.

There can be other benefits in naming an adjudicator from the outset, which may not apply if a dispute already exists. The contracting parties can select a person they know and trust. They can make the selection process as rigorous as they wish and interview unknown but eligible adjudicators. If there are a number of disputes throughout the currency of the works a named individual will acquire project knowledge, which should assist in eliminating learning curves in respect of each dispute. This should improve both the speed and the economics of resolving disputes. The decisions should also be consistent.

When the adjudicator is named in the contract the parties have made their selection of the appropriate person to resolve their disputes, but they have not necessarily secured an appointment and made a contract for the individual to resolve any particular dispute. In fact it is not uncommon for an adjudicator to be named without his knowledge. This is not always the best solution as the named adjudicator may well not be available or may have a conflict of interest when a dispute arises.

If the parties do want some certainty with a named adjudicator they can discuss the availability of the adjudicator should a dispute occur and, if they wish, provide for a retainer to ensure availability on the basis of a reasonable commitment. Parties to construction contracts are however notoriously unwilling to commit themselves to expenditure that may ultimately prove to have been unnecessary. Commitment fees for adjudicators have proved successful on larger projects and in the case of Dispute Review Boards that carry out a very similar function to an adjudicator, regular site visits to keep abreast with progress are not unknown. On small or medium-sized projects any form of retainer is however unlikely to be accepted by the parties.

Naming an individual from the outset can have the disadvantage that an appropriate specialist will not necessarily be chosen at the time to suit the dispute in question. An adjudicator could be named in the contract who is totally unsuited to the type of dispute that arises. A specialist electrical engineer could find himself involved with a structural engineering dispute. The parties might as well not have bothered to name him in the contract at all. There will be delay while an alternative adjudicator is found.

One way of overcoming the problem of an inappropriate or unavailable named adjudicator is the naming of more than one adjudicator so that disputes of various natures can be dealt with by an adjudicator with appropriate experience; this avoids the possibility that a single named adjudicator will either be unavailable or not have the competence to deal with the dispute. This can however lead to disagreement between the parties themselves if the nature of the dispute covers more than one discipline and is within the competence of more than one of the named adjudicators.

Naming an individual from the outset can have disadvantages relating to impartiality. If one of the parties to the contract seeks to impose an adjudicator on the other party, this could raise questions as to the impartiality of the individual named. If there is any question that the party naming the individual in the contract has sought to impose his 'own man' this must bring the whole question of his impartiality into doubt. This could inevitably lead to a challenge to the jurisdiction of the individual simply on the grounds he lacks the qualification on which to proceed, the qualification in this case being that he is impartial. There is also an argument that if an individual is named in a contract or a series of contracts he may develop a partiality. He could be named under a main contract and all the related sub-contracts. While he may be impartial at the outset he may become tainted and blasé, as the work proceeds, as he becomes familiar with the project and its disputes. He may develop, because of his involvement in earlier disputes, a lack of impartiality that was not there from the outset.

Early experience suggests there are obvious instances of naming an individual from the outset that flout the concept of impartiality. The naming of the project architect in a main contract or a director of the contractor in all the project sub-contracts are obvious examples. How can the architect rule as an impartial adjudicator on his own decisions made as architect under the contract, when those decisions are disputed? Equally, how can the director of the contractor make decisions that may affect the well-being of his firm and maybe his own remuneration, and yet remain impartial? Any suggestion that these arrangements are acceptable is untenable.

It should never be forgotten that the interpretation that is placed by the court on the impartiality of a tribunal is not that the tribunal will act impartially but that there can be no perception of bias in the eyes of an objective observer.

When naming an individual both parties should therefore carefully consider whom they name and that he satisfies the qualifying criteria. If the named adjudicator is not seen as being impartial, this will not prevent either party from pursuing adjudication. The contract will not have provided an impartial adjudicator and thus fails to provide the minimum for adjudication required by the Act. The right to refer disputes to adjudication will not be extinguished and the Scheme will apply.

The problem of unavailability of the named adjudicator cannot be entirely avoided. Simple matters such as the adjudicator planning a long holiday during the currency of the works can be established before the adjudicator is named in the contract. This will not avoid the possibility of illness or unexpected family commitments and the like, which cannot be predicted. It is unlikely that an individual adjudicator will accept being named in one contract and then say no to any further appointments. The ultimate scenario is that one adjudicator agrees to be named in a number of contracts and then, when demand for his service increases due to a number of disputes occurring at the same time, he is unable to cope with any one of them properly. One situation where there is a probability of such an occurrence is where an adjudicator's name is inserted in a contract without his knowledge. It is always advisable to ensure that any adjudicator so named is aware of that fact, albeit it is unlikely that a refusal will result; adjudicators are only human, and this problem may still occur.

A firm rather than an individual can be named in the contract, but caution needs to be exercised. This is a halfway solution between naming an individual and naming an appointing body. If a firm is to be named, the parties to the contract need to know all the individuals who are likely to be used as adjudicators should a dispute arise. This will avoid a 'pot luck' selection when the dispute occurs. It is more difficult to vet a firm than an individual. All the individuals within a firm need to be tested with the same rigour that a single named individual would be.

In the case of *Mowlem* v. *Hydra-Tight*, which has already been mentioned in this chapter in another context, Mowlem set out in the contract a 'list of approved adjudicators' which comprised the barrister members of Atkin Chambers, Gray's Inn, a perhaps surprising choice when made at the time of the letting of the contract as the probability is that any disputes arising will relate to payment and other technical matters. When it came to enforcement proceedings the court found that a defined pool of adjudicators is sufficient and a single adjudicator does not have to be named.

In naming a firm the parties could fall foul of the requirements in the contract, adjudication rules or the Scheme. The particular rules may specifically require an individual as adjudicator rather than a firm. The Scheme, for example, requires under paragraph 4 that 'Any person requested or selected to act as adjudicator . . . shall be a natural person acting in his personal capacity'. In this case the adjudicator who acts could not be the firm but would have to be a specific individual from the firm.

It was decided in *Mowlem* v. *Hydra-Tight* that it was not necessary to name the individuals within Atkin Chambers. The naming of that chambers was a valid way of deciding upon the adjudicator and an individual member of that chambers could be selected at the time the dispute arose, a procedure which did accord with paragraph 2.1(a) of the Scheme, 'the referring party shall request the person (if any) specified in the contract to act as adjudicator'. Common sense does however suggest that even in the instance of a barristers' chambers, if a grouping of individuals is used rather than a specific name, the individual competencies of those that might do the adjudication should be ascertained before such an inclusion is made in a contract.

Appointment when the dispute occurs

Waiting until a dispute arises overcomes the problems associated with expertise and availability. The individual appointed as adjudicator will have to indicate that he is available and that he has the correct expertise, otherwise he will not be appointed. The downside is the time that is likely to be needed to get the adjudicator appointed. The parties may not be able to agree on the name of someone to approach. An adjudicator of the agreed expertise may not be readily available. If the parties do agree on a particular individual, he may not be available due to other commitments. Experience does however show that where an adjudicator is agreed the process often takes longer than seven days, but it will be pretty clear to the party seeking the appointment of an adjudicator if the other party is not 'playing ball'. Given the ability of the majority of the ANBs who offer their services to the industry to ensure a speedy nomination if a reluctance on the part of the other party is recognised, the situation can be remedied by an application to an ANB for a nomination.

All the matters that have been considered above in respect of the appointment of an adjudicator at the outset of a contract apply if the parties set out to agree an adjudicator after a dispute has arisen, and this is probably the best way of ensuring that the right person is appointed. A large number of adjudicators are appointed in this fashion but in the majority of cases where the parties do not have a name in their contract an ANB is relied on.

The appointment by a third party is generally to be regarded as a fall back as the parties, once the application for an appointment is made, lose control over the individual to be appointed. The danger is always that the quality of the appointee will suffer due to the ANB having such a short time to find someone to do the job. It would appear however that this is not a general problem. There are apocryphal tales of inadequate adjudicators and all ANBs do from time to time receive complaints, but these are in the main from parties who have lost an adjudication. There is however no great outcry as to the inadequacy of the adjudicators that have been nominated by ANBs.

What are Adjudicator Nominating Bodies?

The Scheme went through a number of drafts and in the early stages of development it was always the intention that a list of approved bodies for nomination of adjudicators would be published and would form part of it. These would be known as Adjudicator Nominating Bodies (ANBs). The particular bodies who were to be named had been vetted and approved to be included in the list in the secondary legislation. However, in the final stages of the drafts for the Scheme the naming of ANBs in the Scheme was dropped. The grounds for this were that each time it was required to add to the list or to remove a name from the list this would require further parliamentary time and debate.

All was not lost for those who had invested in training, setting up panels and an appointing infrastructure. Some were to be appointers or nominators in the Standard Forms of Contract in any event. Others could offer their services as nominators under the definition in paragraph 2(3) of the Scheme: 'an "adjudicator nominating body" shall mean a body (not being a natural person and not being a party to the dispute) which holds itself out publicly as a body which will select an adjudicator when requested to do so by a referring party.' This wide definition gives almost anybody the opportunity to set up a firm to simply provide adjudicators at the request of a party, if they consider it a commercial proposition to do so, and some organisations have done just this.

There are commercial concerns who seek to sell either adjudication services or to appoint adjudicators or to act as ANBs. They may provide different permutations on appointment to those who were to be the original ANBs. Whether it is a commercial concern or an institution making the appointment, a fee will be charged for the service of providing an adjudicator.

The application for an appointment will in all probability require the referring party to provide details of both the dispute and the parties on an application form. The details requested on such application forms assist in the selection of an appropriate adjudicator and should eliminate problems with conflicts of interest. The application form may even require the inclusion of a suggested specialism for the adjudicator.

It would be invidious for the authors as individuals who are members of particular professional bodies to seek to draw comparisons between the various ANBs. We shall restrict our comments to the things that a party seeking a nomination should consider before approaching an ANB for a nomination.

A preferred ANB should be one that requires those individuals on its panel to have attained a senior level in a primary profession, to have experience of dispute resolution and adjudication and to have a sound knowledge of the law of contract. They should have been tested on those aspects before they are allowed on the list. There should also be a regime in place for ensuring that the panel members keep abreast of developments in adjudication and that they remain competent to carry out the functions of adjudicator.

When consulting an ANB, ask questions such as: how rigorous has the training process been? What is their appointing experience? What means of feedback and quality control are in place? How experienced are their adjudicators?

The question of how experienced the adjudicators are is of considerable relevance as adjudication has become more widely used. The fact that an organisation is making few appointments does not necessarily mean that their adjudicators have little or no experience as many adjudicators are on the panels of more than one ANB. It may also be the case that for geographical or other reasons an adjudicator who is on the panel of an ANB that makes lots of nominations has little experience.

The ability of an ANB to nominate sufficiently quickly so that the referral can be made in time, can be relevant. It is the general experience that most ANBs take this responsibility very seriously but there are some tales, possibly apocryphal, where an ANB has taken far longer than it should to make a nomination. The speed with which an ANB nominates is extremely important but this must be balanced with the quality of the adjudicator appointed and it may be that where an unusual specialism is required the time to nominate may be exceeded.

All ANBs generally appoint when asked, taking the line that the jurisdiction of the adjudicator is a matter between the parties and the adjudicator. There are times when it is obvious that the ANB has no jurisdiction to nominate and it will decline to act. Interestingly, in the case of *Mowlem* v. *Hydra-Tight*, mentioned earlier, the first ANB that was approached declined to nominate but another which was approached subsequently did nominate. The nominee of the latter was however found by the court to be without jurisdiction.

Each ANB has its own forms for an applicant to complete before a nomination is made. They will seek varying levels of information but this is generally nothing more than that they have jurisdiction to make the nomination and some idea of the nature of the dispute so that they can make an appropriate nomination. They will also require a fee for the service. Under the provisions of the Scheme, ANBs have five days to nominate and while they may well take note of the time between the application for a nomination and the date of the notice of adjudication, they will be quite entitled to wait until the information that they require is made available and the appropriate fee paid before making a nomination. It is not the role of a nominating body to ascertain the detailed nature of the dispute or to investigate such matters as jurisdiction. The referring party could say no more than that there is a dispute, pay the fee and ask the ANB to select an adjudicator, but a bit more than this is required to get the right person for the dispute. The proper and most effective way is to provide the ANB with a copy of the notice of adjudication that is required by section 108(2)(a) of the Act, preferably with a copy of the notice of adjudication.

The adjudicator and appointment

The individual who is asked if he is prepared to have his name entered into a contract as adjudicator must find out the nature of the work that is to be carried out under the contract that the parties are signing. An adjudicator will in all likelihood be making decisions on technical issues as well as applying the law to the particular facts of the dispute. If he does not have a clue concerning these technical issues he should not put himself forward. In the case of a party-agreed appointment in the contract, the adjudicator will at least know the nature of the contract and will as a result be able to identify his general suitableness for adjudicating upon disputes arising from that contract. There is a limit on the number of contracts an adjudicator should be named in at any one time. The adjudicator needs to keep a record of all those contracts he is named in and their duration. The potential disaster for an adjudicator is the call on his time to handle several disputes on different contracts all within the same 28-day timescale. There must come a point where an adjudicator has too much potential commitment and should refuse to be named in any further contracts. This must also be balanced against the need to earn a living in the adjudicator's primary profession, the commitment to that profession and the possibility of conflict of interest arising not only from his other commitments but possibly from allegations that he has been named too many times by a particular contractor or employer.

Where an enquiry comes as a result of a nomination by a third party, the problem for the prospective adjudicator is that, due to the timescales involved, he is on occasion unlikely to have sufficient details on the technical aspects or the magnitude of the dispute. There may be the briefing papers that the ANB has received from the applicant party but often these do not reveal the true nature or extent of the dispute. When considering whether to agree to be nominated, the adjudicator will at least be in full knowledge of his other commitments during the following 28 days and should at the very least ensure that he sees a copy of the notice of adjudication so that he has some idea of the nature of the dispute.

Section 108(2)(a) of the Act, which requires the giving of a notice of adjudication, is a requirement of the contract. The Scheme has more detailed requirements in paragraph 1 on the content of the notice of adjudication. The nominating bodies will not therefore be required to appoint on receiving a telephone call from site, at least until it is followed up by a copy of the notice. There may be specific requirements in the particular contract relating to the notice, as there are in the Scheme. These should, in principle, be satisfied before an ANB accepts any duty to appoint, but there is no doubt that the ANBs generally put the duty to appoint before any requirement that they ensure that the detailed requirements of the Scheme or any contract have been followed in the notice of adjudication. This may well result in a challenge to the nominated adjudicator's jurisdiction on the basis that the dispute is inadequately defined and thus does not exist. It is, however, in keeping with the generally held view that matters of jurisdiction are for the parties and the adjudicator to resolve and that ANBs should nominate within the requisite time limit when requested.

We have already mentioned the matter of dissatisfaction procedure which the ICE contracts and allied forms require to be followed prior to the commencement of an adjudication. We suggest that if a referring party is insistent on commencing an adjudication in the face of objections by the responding party, the adjudicator should proceed on the basis that to do otherwise is to interfere with the referring party's right to adjudication at any time. Any question as to the validity of the decision then becomes a matter of enforcement. This may, as noted elsewhere, be a non-existent problem if the ICE alters its provisions in this respect.

When the prospective adjudicator is approached with an enquiry as to whether he is prepared to take on the appointment, whether this be before or after the dispute arises, there are a number of matters that he must consider. After deciding whether he has the time and the desire to take the adjudication on, the principal matter for the adjudicator, in view of the wording of section 108(2)(e) of the Act, is can he act impartially? This impartiality has to be seen in the context of the common law. It is not just that he sees himself as being able to act impartially, it is that he can do this in the eyes of the parties and if necessary in the eyes of the court if his impartiality is challenged. In common parlance it is not actual bias that we are concerned with – the fact that no adjudicator will knowingly be biased has to be taken as read – it is the perception of the parties regarding any impression of bias that might be given by the adjudicator's actions or personal or business connections.

Ideally the prospective adjudicator should also satisfy himself in respect of the following matters prior to accepting appointment:

- Is there a dispute? Has it crystallised?
- Does the dispute relate to a construction contract within the meaning of the HGCRA? See sections 104 and 105. Also be aware of section 106 (contracts with residential occupiers). Adjudication of other disputes is always possible by agreement between the parties. This may even be written into their contract.
- Does the dispute come within the terms of section 108(1) or extend those terms?
- Has the person who has referred the dispute to adjudication the authority to do so?
- Is the ANB which has approached him empowered to do so?

Where the appointment of the adjudicator follows a nomination by an ANB, the timescales in the Act generally preclude any detailed prior investigation of these matters before the referral notice is received, and general experience appears to be that the notice of adjudication is issued, the adjudicator nominated and the dispute referred to him in such quick time that any matters of this nature come up for the first time as jurisdictional objections from the responding party.

It was confirmed in *Christiani & Neilsen* v. *The Lowry Centre*[4] that there is no absolute obligation upon an adjudicator to investigate his jurisdiction even if it is challenged. In any event, as decided in *The Project Consultancy* v. *The Trustees of the Gray Trust*[5], he has no power to decide his own jurisdiction unless the parties give him that power. It is however obvious that if there is a clear problem at the time that the adjudicator is approached, he ought to identify it even if the ANB has not done so and decline the nomination, thus saving a lot of wasted time.

The adjudicator will need to respond to any jurisdictional objection that may be made by

---

[4] *Christiani & Neilsen Limited* v. *The Lowry Centre Development Company Limited* (16 June 2000).
[5] *The Project Consultancy Group* v. *The Trustees of the Gray Trust* [1999] BLR 377.

the responding party but must not allow any concern regarding these matters to delay him from acting; his job is to get on with it and make his decision within 28 days of the dispute being referred to him. The court has during 2002 and 2003 considered a number of questions relating to the existence of a dispute and we consider these in some detail in Chapter 9.

The adjudicator will be making decisions that relate to the rights and duties of the parties. These will principally be in connection with the contractual agreement between the parties. He must therefore be confident that he has undertaken the necessary training and has the necessary knowledge to deal with the technical and contractual/legal aspects of the dispute before agreeing to be nominated or accepting an appointment by agreement.

If the adjudicator does not understand the law of contract and does not have the knowledge to interpret the nature of the agreement between the parties, how can he ascertain these rights and duties? More than just a basic understanding of the law is necessary to deal with the matters that are likely to be faced by an adjudicator, and an individual lacking this ability should not take on the task.

Section 108(1) of the Act states that only disputes arising under the contract are within the adjudicator's jurisdiction. If an adjudicator is unfamiliar with legal concepts he may deal with a dispute that relates to negligence, something that does not arise under the contract. The party required to pay by the decision may well succeed in resisting an action to enforce it. The adjudicator may not get taken to task as a result but if nothing else there has been a considerable waste of time and money.

These considerations are very important if the individual is asked if he is prepared to have his name entered into a contract as adjudicator. When that enquiry comes after the dispute is in being, they are vital.

The adjudicator's terms and conditions

It is extremely unlikely that any adjudicator will agree to adjudicate without some record of the terms that he will apply to his appointment.

If the appointment is by agreement with the parties, most adjudicators will act only after terms are agreed with the parties.

Some contract drafting bodies publish adjudication agreements in a standard format, of which more later, and require their use by the terms of the contract. This will be the case whether the appointment is by party agreement or after a nomination. Where a standard form contract does not apply it is the usual practice of adjudicators to prepare terms and send them to the parties.

Some adjudicators seek the agreement of the parties to their terms, others just send their terms with a covering letter to the effect that 'these are the terms that apply to my appointment'. While the former method has the benefit of some certainty for the adjudicator in that the parties will have agreed his hourly rate, he is in fact only entitled to be paid his reasonable fees and expenses and, however low his hourly rate may be, if his charges relate to what the parties consider to be an inordinate number of hours it may be that his fee will be challenged on that basis.

It is clear that an adjudicator cannot insist on agreement of his terms in the face of a reluctant party before he starts the adjudication as he would not then be complying with his obligations to the ANB which nominates him, as the nomination will be on the basis that the adjudicator is ready and willing to do the adjudication in the requisite 28 days.

The setting of the terms on the basis that 'they will apply' is fine as long as the terms suggested are not contentious. In general the adjudicator completes the adjudication, charges his fee and that is the end of it. Where the hourly rate appears high to the parties there may be a conceptual problem but at the end of the day it is the overall amount charged that is important to the parties, not the hourly rate. An adjudicator who charges a high hourly rate may charge an overall fee that is considerably lower than an adjudicator with a lower hourly rate and there is some comment from unofficial surveys of the adjudication process that this may be the case.

Adjudicators who are registered for VAT or who issue their fee notes through their firms who charge VAT need to make the parties aware of that fact in their terms.

It does not hurt to record the adjudicator's charge per mile for using his car, and a confirmation of the class of travel to be utilised on train journeys will not go amiss.

Some adjudicators request an upfront payment to give them security for all or part of their fees and expenses and others seek to impose some form of lien by which they will not release their decision until they are paid. We shall discuss liens when considering the finalisation of the decision in Chapter 11. A request for an upfront fee of possibly some thousand pounds immediately after the nomination is notified can cause some consternation, particularly on the part of the referring party who will no doubt have to pay it. There is no real objection to such a procedure provided that the adjudicator does not refuse to act within the requisite 28 days if he does not receive it, and provided that it represents a payment upfront for fees yet to be earned. Any connotation of an appointment fee that is non-returnable is to be deprecated. Members of the RICS should of course be aware that any payment for work not yet done must be placed in a client's account as it is a breach of the RICS's regulations to do otherwise and may become a disciplinary matter.

Some adjudicators seek an indemnity from the parties in respect of possible action by third parties. Such a provision may, however, be difficult to enforce without specific agreement from the parties.

There are various other pet items that adjudicators include in their terms but all the common ones are noted above.

The standard forms of contract and appointment

All the standard forms of contract recognise that it is generally preferable that the parties agree the adjudicator but they allow for the fact that it is often the case that such agreement will be impossible and that a fall-back procedure will be necessary.

These contracts also recognise that adjudication is a right that can be invoked unilaterally and the procedures work on this premise.

The common way in which this is dealt with in the standard forms of contract is to allow for the appointment of an adjudicator by agreement, failing which a person or body is named as nominator or appointer. There is often a list of persons or bodies from which the parties can choose one by deleting those not wanted, with the fall-back position of one of those named being identified as appointer/nominator if no choice is made. In this way adjudication can be commenced unilaterally. Even where the parties have failed to select an appointer or nominator in their contract, the lists will include a body by default to make good the failure to complete the contract correctly.

At the time that the statutory right to adjudication came into force, the contract-writing bodies did not have their contract amendments in place in sufficient time for the commencement date of 1 May 1998. Ideally, had the contracts been in place about six weeks before that date, enquiries and bids could have recognised and incorporated the new contract regimes for contracts to be entered into after 1 May 1998. Because the contracts were not available this far in advance, there were a number of contracts entered into after 1 May 1998 that did not have the new adjudication and payment provisions or had the pre-existing provisions that did not comply with the Act. On larger projects such contracts may still be current and the Scheme applies to these contracts unless the parties agree to use the subsequent amendments to the standard form to include compliant adjudication provisions. Rather surprisingly, even five years down the line, there are occasional instances where previous editions of the standard contracts are still being used and the parties will need to use the Scheme to invoke their right to adjudication.

Some organisations in the construction industry ignore standard forms, and many hybrid or non-standard forms are produced by such organisations to suit their individual purposes. There are no signs that this practice will stop. There are novel attempts to comply with the legislation including, in the early days, clauses drafted in an attempt to skirt round or modify the legislation, but these have been nipped in the bud by the support of the courts for the adjudication process.

Joint Contracts Tribunal main contracts

The Joint Contracts Tribunal (JCT) published its 1998 editions during that year but not until September 1998. These include the Standard Form of Building Contract which is published in the Private Editions (With and Without Quantities), the equivalent Local Authority editions and the With Contractor's Design edition. There are also the Intermediate and Minor Works Forms. These are the most commonly used versions but there are several others published to fill various special requirements.

In order to bridge the gap between 1 May and September 1998, Amendment 18 to the 1980 edition was published. There may well be contracts to which Amendment 18 applies which have not yet reached finalisation and adjudication may still be required. That said, the adjudication provisions in all the various JCT 1998 editions and those in Amendment 18 are, other than in two respects, identical. We deal with the whole of the provisions in JCT 98 first and then look at these relatively minor differences afterwards.

All the JCT forms of contract are set up to provide the right to adjudication in one of the 'Articles' which are included in the early pages of the forms concerned. It is in Article 5 in the Private Editions, stating simply 'If any dispute or difference arises under this Contract either Party may refer it to adjudication in accordance with clause 41A'. Clause 41A includes detailed adjudication procedures. In some versions the Article number changes and the number of the clause detailing the adjudication procedures also varies. The wording remains effectively identical. From now on we shall consider the Private Editions and use the numbering of these.

Two definitions relating to adjudication are included in clause 1.3 of the Private Editions and the With Contractor's Design Form (clause 8.3 of the Intermediate Form). There are no definitions in the Minor Works Form. The words so defined are an 'Adjudication Agreement' – cross referred to clause 41A.2.1 in the Private Editions which refers to 'the JCT Adjudication Agreement' and an 'Adjudicator' defined as 'any individual appointed pursuant to clause 41A as the Adjudicator'.

The JCT Adjudication Agreement is published separately from the contracts themselves. For a discussion on this see Chapter 8, Adjudicators' Agreements. The adjudicator must be an individual and not a firm. This follows the requirements of the Scheme in paragraph 4.

The JCT has taken a relatively simple approach to their standard route for appointment. They have sought to comply with the Act and also some of the representations they had from industry through their constituent bodies. The identity (appointment) of the adjudicator is dealt with in clause 41A.2 as follows.

Identity of adjudicator

41A.2   The adjudicator to decide the dispute or difference shall be either an individual agreed by the Parties or, on the application of either Party, an individual to be nominated as the adjudicator by the person named in the Appendix ('the nominator'). Provided that

    41A.2.1   no adjudicator shall be agreed or nominated under clause 41A.2 or clause 41A.3 who will not execute the Standard Agreement for the appointment of an adjudicator issued by the Joint Contracts Tribunal (the 'JCT Adjudication Agreement') with the Parties, and

    41A.2.2   where either Party has given notice of his intention to refer a dispute to adjudication then

      –   any agreement by the Parties on the appointment of an adjudicator must be reached with the object of securing the appointment of, and the referral of the dispute or difference to, the adjudicator within 7 days of the date of the notice of intention to refer (see clause 41A.4.1);

      –   any application to the nominator must be made with the object of securing the appointment of, and the referral of the dispute or difference to, the adjudicator within 7 days of the date of the notice of intention to refer;

Upon agreement by the Parties on the appointment of the adjudicator or upon receipt by the Parties from the nominator of the name of the nominated adjudicator the Parties shall thereupon execute with the adjudicator the JCT Adjudication Agreement.

Footnote: [vv] The nominators named in the Appendix have agreed with the JCT that they will comply with the requirements of clause 41A on the nomination of an Adjudicator including the requirement in clause 41A.2.2 for the nomination to be made with the object of securing the appointment of, and the referral of the dispute or difference to, the Adjudicator within 7 days of the date of the notice of intention to refer; and will only nominate Adjudicators who will enter into the 'JCT Adjudication Agreement'.

The JCT deal with the eventuality that the adjudicator might die or be otherwise unable to adjudicate as follows.

Death of adjudicator – inability to adjudicate

41A.3   If the adjudicator dies or becomes ill or is unavailable for some other cause and is thus unable to adjudicate on a dispute or difference referred to him, the Parties may either agree upon an individual to replace the adjudicator or either party may apply to the nominator for the nomination of an adjudicator to adjudicate that dispute or difference; and the Parties shall execute the JCT Adjudication Agreement with the agreed or nominated adjudicator.

The standard route for appointment is based on the parties agreeing on an adjudicator at the time their dispute occurs. Where no such agreement can be reached, one or both of the parties can approach the nominator identified in the appendix to the contract for an adjudicator to be nominated. The nominators are the 'President or a Vice-President/ Chairman or a Vice-Chairman: Royal Institute of British Architects, Royal Institution of Chartered Surveyors, Construction Confederation and National Specialist Contractors Council'. If the parties have failed to complete their contract correctly by striking out all but one of the names of the nominators in the contract, by default the nominator is the President or Vice-President of the Royal Institute of British Architects. The concept of presidential or chairman nominations might well have been dropped in view of the liberal definition of ANBs in the Scheme. However, this is the chosen method and the organisations named do generally manage to overcome the difficulties created by the personal nature of the nomination and still meet the timetable required in the contract. Some of the institutions operate under delegated powers so that an individual other than the President or Vice-President can deal with the appointment.

There is an important qualification for the appointment of an adjudicator. In clause 41A.2.1 no adjudicator should be agreed or appointed who will not execute a JCT Adjudication Agreement with the parties. This must be considered by the parties when reaching agreement. They should notify their potential adjudicator that one of the qualifications on appointment will be to execute the JCT Adjudication Agreement. Equally, any nominator must not nominate any person who is not aware of, or who will not agree to the execution of, the JCT Adjudication Agreement. We understand at the time of writing that the JCT is considering doing away with the JCT Adjudication Agreement and if this does happen this will no longer be a matter of concern to the parties or an ANB.

The JCT seeks to comply with the Act in terms of securing the appointment and referring the dispute to adjudication by using the words of the Act itself. In clause 41A.2.2 there is a notice of intention to refer a dispute to adjudication. This is the notice generally described as the notice of adjudication. This follows section 108(2)(a) of the Act. What follows is the stated intention from the Act, the object of securing the appointment and the referral to the adjudicator within seven days of the notice of intention to refer. This complies fully with the requirements of the Act and it is of no consequence therefore if an adjudicator is not in place within seven days of the notice of intention to refer. There is no effect other than the delay to the ultimate decision, and the failure to comply with the appointment timetable is not fatal to the process.

The final provision of clause 41A.2 requires that, as soon as the parties have agreed on the appointment or an adjudicator has been nominated, all three execute the JCT Adjudication Agreement. If the parties fail to execute this agreement this is not fatal to the process. Clause 41A.5.6 makes it clear that failure by one or both of the parties will not invalidate the decision. In practice there will be many instances where one or both parties fail to sign the agreement. This leads to the question as to why the agreement is needed at all and as noted above, the JCT is reviewing the Agreement at the time of writing.

The two differences in Amendment 18 are firstly an alternative procedure that was offered for appointment of the adjudicator in that the parties could by agreement name their adjudicator in the contract. The JCT decided that there were fewer benefits than advantages offered by this facility and, perhaps bearing in mind that general experience seems to be that the space for naming the adjudicator was left blank, it was discarded.

The second difference was a provision that the referral could be delayed until such time as the adjudicator signs the JCT Adjudication Agreement. This of course does not comply with the requirement that the contract is drafted with the intention of the referral being made within seven days of the notice of adjudication, which is why it was dropped in the 1998 edition. If faced with a contract that is based upon Amendment 18, the adjudicator will have to consider whether to proceed in accordance with the contract provisions or use the Scheme. He will possibly decide to use the Scheme but the parties may be quite content for him to use the contractual provisions and if they are there would appear to be no reason for the adjudicator not to do so.

Joint Contracts Tribunal sub-contracts

The JCT has published sub-contracts for use under the Standard Form and the Intermediate Form for nominated sub-contractors and named sub-contractors. These are the Nominated Sub-Contract Conditions NSC/C (for use with nominated sub-contractors under the Standard Form) and the Sub-Contract Conditions NAM/SC (for use where sub-contractors are named under the Intermediate Form).

The respective adjudication provisions are included in section 9 of NSC/C (related to Article 4 of the Form of Agreement NSC/A which is the basic provision relating to the right to adjudicate) and section 35 of NAM/SC (similarly related to Article 4 of Tender and Agreement NAM/T). These sections include provisions for adjudication, arbitration and legal proceedings but it is only the adjudication aspect that we are concerned with here. Other than the numbering the clauses are identical.

Apart from the clause numbering and references to form NSC/T the provisions are identical.

Since 2002 the JCT has produced its own domestic sub-contract DSC/A (Agreement) and DSC/C (Conditions) which include a list of appointing bodies with the RIBA as the default. The wording of the adjudication provisions follows that of the Standard Form Contracts.

Construction Confederation sub-contracts

The field of sub-contracting is probably where the adjudication provisions and also the payment provisions are of greatest importance. The standard forms of sub-contract are the benchmark documents for the industry. Despite this, there are a great number of sub-contracts let on hybrid in-house forms of sub-contract. As far as the appointment of the adjudicator is concerned, the situation seems to be very little different from what happens generally. Some contractors seem to like to name one or a list of adjudicators, others to rely on an ANB. There are, of course, those contractors who do not have formal procedures for appointing sub-contractors and these will in general have to rely on the Scheme provisions for appointment of the adjudicator.

Notwithstanding the emergence of the JCT DSC forms and the consequent obsolescence of the DOM/1 and DOM/2 forms, the habit of the industry to cling to forms with which it is familiar suggests that these forms will remain in use for a considerable time to come.

Not surprisingly, the DOM/1 and DOM/2 Standard Forms of Sub-Contract follow the approach adopted in JCT 98 and apart from the clause numbering the terms are identical. As with JCT 98 there is no optional amendment to permit the naming of the adjudicator in the contract. This differs from previous editions of this form – where adjudication only dealt with set-off – which were based on the adjudicator being named in the contract as the standard approach.

The nominators are set out in the Appendix and are the same as in JCT 98 but in this case the default nominator is 'the President or Vice-President of the Royal Institution of Chartered Surveyors'.

As with JCT 98 this contract requires that the adjudicator shall enter into the JCT Adjudication Agreement with the parties. A party cannot disrupt the adjudication by failing to do this and it is confirmed by clause 38A.5.6 that failure by one or both of the parties to complete the agreement does not invalidate the decision.

Institution of Civil Engineers' main contracts

The ICE has adopted an entirely different approach to the JCT. The adjudication procedure is contained in a separate document to the contract. They have also sought to redefine the point at which a dispute occurs. Before the adjudicator can be appointed the contract requires that there is another reference on a matter of dissatisfaction before the dispute comes into being. The new clause 66 works on the premise of avoiding disputes.

We have briefly considered this provision earlier but it is worth further consideration here. The object of this provision appears to be to exhaust any contractual mechanism before a formal means of dispute resolution is adopted. The old provisions on engineers' decisions have been replaced by matters of dissatisfaction. A dispute does not exist until the engineer has made his decision on the matter of dissatisfaction. The question is, even though the parties have contracted on this basis, does this prevent a party from taking a dispute to adjudication at any time? This is important from the viewpoint of whether or not an adjudicator can be appointed until the matter of dissatisfaction procedure has been exhausted. Is a matter of dissatisfaction a lesser being than a dispute or can a dispute exist as well as a matter of dissatisfaction? If the dispute can exist independently of a matter of dissatisfaction the parties can proceed to adjudication at any time. There is also a question as to whether the matter of dissatisfaction itself could be construed as a dispute for the purposes of the Act. Under the drafting of this particular clause the parties agree that a dispute does not exist until the matters of dissatisfaction provisions have been exhausted and a notice of dispute has been served.

Having contracted on the basis that in the civil engineering forms a dispute cannot come into being before the matter of dissatisfaction has been dealt with, it is arguable that in the absence of a dispute no adjudication can take place. The matter of dissatisfaction is dealt with either by an engineer's decision or by the lapse of one month of the reference to the engineer of the matter. There must be scope for a dispute on whether the engineer has actually dealt with the matter in his decision or indeed whether the matter was properly referred to him. If the parties take no issue with this procedure, it is, as we have already noted, an excellent method of ensuring that a dispute has properly crystallised before the adjudication commences. In the case of one party wanting adjudication in face of objection by the other party, it is probably correct for the adjudicator to proceed and leave the unresolved questions as to the validity of the matter of dissatisfaction procedure for any enforcement proceedings that may occur after he has completed his decision.

Clause 66(1) introduces the contractual procedures to facilitate the clear definition and early resolution (whether by agreement or otherwise) of disputes.

Clause 66(2) deals with matters of dissatisfaction by requiring that any party who is at any time dissatisfied as a result of any matter arising under or in connection with the contract or the carrying out of the works refers such a matter to the engineer who shall notify his written decision to the employer and the contractor within one month of the reference to him.

Clause 66(3) is to the effect that the employer and the contractor agree that no matter is a dispute unless and until the time for the giving of a decision by the engineer on a matter of dissatisfaction has expired or the decision given is unacceptable or has not been implemented and in consequence the employer or the contractor has served on the other and on the engineer a notice in writing, called the notice of dispute, or an adjudicator has given a decision on a dispute and the employer or the contractor is not giving effect to the decision, and in consequence the other has served on him and the engineer a notice of dispute. The dispute is then defined as that stated in the notice of dispute. The clause continues to confirm that for the purposes of all matters arising under or in connection with the contract or the carrying out of the works, the word 'dispute' shall be construed accordingly and shall include any difference.

Clause 66(4) requires that the parties continue to perform their obligations under the contract notwithstanding the existence of a dispute following the service of a notice under clause 66(3) unless the contract has already been determined or abandoned.

Clause 66(4) also requires that the parties give effect forthwith to every decision of the engineer on a matter of dissatisfaction given under clause 66(2) and of the adjudicator on a dispute given under clause 66(6) unless and until that decision is revised by agreement of the employer and contractor or pursuant to clause 66.

There is further confusion when the matter of dissatisfaction procedure has been exhausted. A party may at any time refer a dispute to conciliation under the ICE Conciliation Procedure.

This provision in Clause 5 can be summarised as follows: The parties may at any time before commencing arbitration proceedings seek the agreement of the other for the dispute to be considered under the Institution of Civil Engineers' Conciliation Procedure (1994) and if the other party agrees to this procedure any recommendation of the conciliator shall be deemed to have been accepted as finally determining the dispute by agreement unless a Notice of Adjudication or a Notice to Refer to arbitration has been served in respect of that dispute not later than 1 month after receipt of the recommendation by the dissenting party.

This simply gives another option to the parties to choose the way in which to resolve the dispute. This procedure cannot be instigated once a notice of arbitration has been issued but there is nothing that prevents a conciliation after an adjudication, or even during the adjudication if the parties so desire.

Clause 66(6) in the ICE contracts, which deals with adjudication, is simple because it refers to the separate adjudication procedure. The principles of this clause are as follows:

The parties each have the right to refer any dispute as to a matter under the contract for adjudication and either party may give notice in writing to the other at any time of his intention so to do. The adjudication shall be conducted under the Institution of Civil Engineers' Adjudication Procedure (1997). Unless the adjudicator has already been appointed he is to be appointed by a timetable with the object of securing his appointment and referral of the dispute to him within seven days of such notice. The adjudicator shall reach a decision within 28 days of referral or such longer period as is agreed by the parties after the dispute has been referred.

The Institution of Civil Engineers' Adjudication Procedure

Unfortunately Thomas Telford, the publishers of the Procedure, have made it financially impossible for us to reproduce the Procedure in full and the reader will note its absence from our appendices.

Paragraph 3 of the Procedure deals with the appointment and it can be summarised as follows.

Paragraph 3.1 provides that where an adjudicator is named in the contract or has been agreed by the parties prior to the issue of the notice of adjudication, the referring party issues the notice of adjudication to the responding party and to the adjudicator, together with a request for confirmation from the adjudicator within four days of the date of issue of the notice of adjudication that he is able and willing to act.

Paragraph 3.2 provides that where an adjudicator has not been so named or agreed, the referring party may include with the notice the names and addresses of one or more persons who have agreed to act for selection by the responding party who shall select and notify the referring party and the selected adjudicator within four days of the date of issue of the notice of adjudication.

Paragraph 3.3 provides that if an adjudicator is not selected in the aforesaid manner then either party may within a further three days request the person or body named in the contract, or if none is so named The Institution of Civil Engineers, to appoint the adjudicator.

Paragraph 3.4 provides that the adjudicator shall be appointed on the terms and conditions set out in the ICE Adjudicator's Agreement and Schedule and shall be entitled to be paid a reasonable fee together with his expenses. The parties shall sign the agreement within seven days of being requested to do so.

Paragraph 3.5 provides that if for any reason whatsoever the adjudicator is unable to act, either party may require the appointment of a replacement adjudicator in accordance with the procedure in paragraph 3.3.

The provisions on appointment deal with all possible options. The adjudicator can be appointed in the contract, agreed by the parties prior to the issue of the notice of adjudication, or where the adjudicator cannot be agreed the Institution of Civil Engineers will appoint. There is also the option to propose names of potential adjudicators at the time the notice of adjudication is issued. There is a standard form of Adjudicator's Agreement and Schedule. This covers the terms and conditions on which the adjudicator is appointed.

There is also a standard form on which to apply to the ICE for the selection or appointment of an adjudicator.

Civil Engineering Contractors Association sub-contracts

The Civil Engineering Contractors Association (CECA) is the successor body to the FCEC. The FCEC were responsible for publication of the old FCEC Blue Forms of Sub-contract for use with the ICE main contract forms. The main theme of those forms was to seek to make as far as possible the sub-contract forms 'back to back' with the main contracts. A similar approach has been adopted on those amendments brought about by the Act.

CECA produces a wide range of standard forms which relate to the following forms:

- ICE Conditions of Contract 5th Edition
- ICE Conditions of Contract 6th Edition
- ICE Conditions of Contract for Design and Construct
- GC & PC/Works/1 with quantities (including the without quantities supplement)
- GC & PC/Works/Single Stage Design and Build.

The provisions of the CECA forms of sub-contract relating to the ICE 6th edition are as follows:

18(1) If any dispute or difference shall arise between the Contractor and the Sub-Contractor or in connection with or arising out of the Sub-Contract, or the carrying out of the Sub-Contract Works (excluding a dispute concerning VAT but including a dispute as to any act or omission of the Engineer) whether arising during the progress of the Sub-Contract Works or after their completion it shall be settled in accordance with the following provisions.

(2) (a) Where the Sub-Contractor seeks to make a submission that payment is due of any amount exceeding the amount determined by the Contractor as due to the Sub-Contractor or that any act, decision, opinion, instruction or direction of the Contractor or any other matter arising under the Sub-Contract is unsatisfactory, the Sub-Contractor

shall so notify the Contractor in writing, stating the grounds for such submission in sufficient detail for the Contractor to understand and consider the Sub-Contractor's submission

(b)   Where in the opinion of the Contractor such a submission gives rise to a matter of dissatisfaction under the Main Contract, the Contractor shall so notify the Sub-Contractor in writing as soon as possible. In that event, the Contractor shall pursue the matter of dissatisfaction under the Main Contract promptly and shall keep the Sub-Contractor fully informed in writing of progress. The Sub-Contractor shall promptly provide such information and attend such meetings in connection with the matter of dissatisfaction as the Contractor may request. The Contractor and the Sub-Contractor agree that no such submission shall constitute nor be said to give rise to a dispute under the Sub-Contract unless and until the Contractor has had the time and opportunity to refer the matter of dissatisfaction to the Engineer under the Main Contract and either the Engineer has given his decision or the time for the giving of a decision by the Engineer has expired.

The first part of this clause gives the link between the main and the sub-contracts. There are no actual provisions concerning matters of dissatisfaction themselves within the sub-contract but the link is to matters of dissatisfaction arising under the main contract. This enables the main contractor to postpone any adjudication with the sub-contractor where in the main contractor's opinion the matter in dispute gives rise to a matter of dissatisfaction under the main contract. This may involve the sub-contractor with a wait of more than one month before the engineer under the main contract has given his decision or the time for giving that decision has expired.

The sub-contractor and main contractor have an agreement similar to that in the main contract that a dispute does not arise until the matter of dissatisfaction procedure under the main contract has been completed. All the same arguments as we have considered in respect of the main contract apply here also and we have no doubt that the adjudicator should proceed with an adjudication when so requested by a sub-contractor even if the main contractor objects.

The provisions on conciliation are similar to those under the main contract:

(3)   (a)   The Contractor or the Sub-Contractor may at any time before service of a Notice to Refer to arbitration under sub-clause 18(7) by notice in writing seek the agreement of the other for the dispute to be considered under the Institution of Civil Engineers' Conciliation Procedure (1994) or any amendment or modification thereof being in force at the date of such notice

(b)   If the other party agrees to this procedure any recommendation of the conciliator shall be deemed to have been accepted as finally determining the dispute by agreement so that the matter is no longer in dispute unless a Notice of Adjudication under sub-clause 18(4) or a Notice to Refer to arbitration under sub-clause 18(7) is served within 28 days of receipt by the dissenting party of the conciliator's recommendation.

Note that again conciliation cannot be commenced where there is a notice to refer to arbitration but this restriction does not apply to adjudication.

The adjudication provisions rely on the ICE Adjudication Procedure:

(4)   (a)   The Contractor and the Sub-Contractor each has the right to refer any dispute under the Sub-Contract for adjudication and either party may at any time give notice in writing (hereinafter called the Notice of Adjudication) to the other at any time of his intention to refer the dispute to adjudication. The Notice of Adjudication and the appointment of the adjudicator shall, save as provided under sub-clause 18(10)(b), be as provided at para-

graphs 2 and 3 of the Institution of Civil Engineers' Adjudication Procedure (1997). Any dispute referred to adjudication shall be conducted in accordance with the Institution of Civil Engineers' Adjudication Procedure (1997) or any amendment or modification thereof being in force at the time of the appointment of the adjudicator.

(b) Unless the adjudicator has already been appointed he is to be appointed by a timetable with the object of securing his appointment and referral of the dispute to him within 7 days of such notice.

...

(10) (b) If a Main Contract Dispute has been referred to conciliation or adjudication under the Main Contract and the Contractor is of the opinion that the Main Contract Dispute has any connection with a dispute which is to be (but has not yet been) referred for conciliation or adjudication under this Sub-Contract (hereinafter called a Connected Dispute), the Contractor may by notice in writing require that the Connected Dispute be referred to the conciliator or adjudicator to whom the Main Contract Dispute has been referred.

Clause 66(10)(b) places a bar on any sub-contract dispute being referred to a separate adjudicator if the main contractor so requires. This does place an obligation on the main contractor to tell the sub-contractor on receipt of a notice of adjudication from the sub-contractor that it is a connected dispute. If this is not done and the sub-contractor obtains the appointment of a different adjudicator there may be some arguments put forward by the contractor but the use of the same adjudicator is not a contractual obligation, being merely at the discretion of the main contractor, and it does seem that the adjudicator procured by the sub-contractor should get on with the adjudication as referred by the sub-contractor in such circumstances. If the adjudicator already appointed under the main contract dispute were to decline to take the connected dispute for any reason, this could not in any way prevent another adjudicator from taking it.

Government contracts

Adjudication existed before 1998 under government contracts; the predecessor (GC Wks/1 Edition 3) to the new forms of contract had an adjudication provision. There are now four new forms, which all have adjudication provisions to comply with the Act. The premise is that the adjudicator is appointed from the outset and that he should be named in the Abstract of Particulars.

The adjudication provisions on appointment are as follows.

Condition 59 Adjudication

(1) The Employer or the Contractor may at any time notify the other of intention to refer a dispute difference or question arising under, out of, or relating to, the Contract to adjudication. Within 7 Days of such notice, the dispute may by further notice be referred to the adjudicator specified in the Abstract of Particulars.

(2) The notice of referral shall set out the principal facts and arguments relating to the dispute. Copies of all relevant documents in the possession of the party giving the notice of referral shall be enclosed with the notice. A copy of the notice and enclosures shall at the same time be sent by the party giving the notice to the PM, the QS and the other party.

(3) (a) If the person named as the adjudicator in the Abstract of Particulars is unable to act, or ceases to be independent of the Employer, the Contractor, the PM and the QS, he shall be substituted as provided in the Abstract of Particulars.

(b)　It shall be a condition precedent to the appointment of an adjudicator that he shall notify both parties that he will comply with this Condition and its time limits.

(c)　The adjudicator, unless already appointed, shall be appointed within 7 Days of the giving of a notice of intention to refer a dispute to adjudication under paragraph (1). The Employer and the Contractor shall jointly proceed to use all reasonable endeavours to complete the appointment of the adjudicator and named substitute adjudicator. If either or both such joint appointments has not been completed within 28 Days of the acceptance of the tender, either the Employer or the Contractor alone may proceed to complete such appointments. If it becomes necessary to substitute as adjudicator a person not named as adjudicator or substitute adjudicator in the Abstract of Particulars, the Employer and Contractor shall jointly proceed to use all reasonable endeavours to appoint the substitute adjudicator. If such joint appointment has not been made within 28 Days of the selection of the substitute adjudicator, either the Employer or Contractor alone may proceed to make such appointment. For all such appointments, the form of adjudicator's appointment prescribed by the Contract shall be used, so far as is reasonably practicable. A copy of each such appointment shall be supplied too [sic] each party. No such appointment shall be amended or replaced without the consent of both parties.

One of the qualifications for appointment as adjudicator under this form is given in paragraph (3)(b). Any adjudicator appointed must comply with the requirements of condition 59. This should not cause any delay to an adjudication even if an adjudicator has not been named in the contract as this undertaking can be obtained very quickly by exchange of fax at the same time as the initial approach is made to the adjudicator.

NEC Engineering and Construction Contract

The New Engineering Contract was the form that Latham sought to be used as the model for a new universal construction contract. The recommendations of the Latham report have not borne fruit. The industry is not adopting this as its model form. What was the New Engineering Contract became the NEC Engineering and Construction Contract. The original New Engineering Contract contained its own provisions on adjudication. These provisions did not comply with the requirements of the Act. It was therefore necessary to make new adjudication provisions that do comply with the Act. The printed forms of contract have not been amended to reflect the Act but there is an addendum booklet, ADDENDUM Y(UK)2 (Ref NEC/ECC/Y(UK)2/April 1998, which accompanies all forms of the NEC contract.

Clause 90 Avoidance and settlement of disputes

Clause 90.1 requires that the parties and the project manager follow this procedure for the avoidance and settlement of disputes.

Clause 90.2 allows the contractor if dissatisfied with an action or a failure to take action by the project manager to notify that dissatisfaction to the project manager no later than four weeks after he became aware of the action or after he became aware that the action had not been taken. Within two weeks of such notification of dissatisfaction, the contractor and the project manager attend a meeting to discuss and seek to resolve the matter.

Clause 90.3 provides that if either party is dissatisfied with any other matter, he notifies his dissatisfaction to the project manager and to the other party no later than four weeks after he became aware of the matter. Within two weeks of such notification of dissatisfaction, the parties and the project manager attend a meeting to discuss and seek to resolve the matter.

Clause 90.4 provides that the parties agree that no matter shall be a dispute unless a notice of dissatisfaction has been given and the matter has not been resolved within four weeks.

Clause 90.5 provides that either party may give notice to the other party at any time of his intention to refer a dispute to adjudication. The notifying party refers the dispute to the adjudicator within seven days of the notice.

Document NEC/ECC/Y(UK) 2 April 1995 also includes an addendum clause 90 relating to sub-contracts which is drafted in the same terms as clause 90 of the main contract addendum.

Not surprisingly with their engineering background, both these contracts adopt similar provisions to the ICE contracts for avoidance and settlement of disputes. Matters of dissatisfaction form the preliminary procedure before the dispute can come into being. When the matter of dissatisfaction procedure is complete, either party can simply then refer the dispute to adjudication.

The Institution of Chemical Engineers' Form of Contract 2001

Clause 46 of the Red Book, Lump Sum Contracts, includes the following clauses leading up to the appointment of an adjudicator:

46.1 This Clause 46 shall only apply to disputes under a construction contract as defined in the Housing Grants, Construction and Regeneration Act 1996, or any amendment or re-enactment thereof.

46.2 Notwithstanding any provision in these General Conditions for a dispute to be referred to an *Expert* in accordance with Clause 47 (Reference to an Expert) or to Arbitration in accordance with Clause 48 (Arbitration), either party shall have the right to refer any dispute or difference (including any matter not referred to the *Project Manager* in accordance with Sub-clause 45.3) as to a matter under or in connection with the *Contract* to adjudication and either party may, at any time, give notice in writing to the other of his intention to do so (hereinafter called a 'Notice of Adjudication'). The ensuing adjudication shall be conducted in accordance with the edition of the 'Adjudication Rules' (the 'Rules') published by IChemE current at the time of service of the Notice of Adjudication.

46.3 Unless the adjudicator has already been appointed, he is to be appointed to a timetable with the object of securing his appointment within seven days of the service of the Notice of Adjudication. The appointment of the adjudicator shall be effected in accordance with the Rules.

Unfortunately the IChemE, while allowing reproduction of excerpts of its contract, has not given permission to reproduce its adjudication rules. We therefore précis the nomination procedure as follows.

Clause 3: Procedure for nomination

Clause 3.1 provides the right required by the Act to refer any dispute to adjudication at any time. The extent of the matters that can be referred includes matters arising in connection with the contract.

Clause 3.2 is in identical words to clause 3.1 of the ICE Adjudication Procedure. Where an adjudicator is named in the contract or has been agreed by the parties prior to the issue of the Notice of Adjudication, the referring party issues the Notice of Adjudication to the responding party and to the adjudicator, together with a request for confirmation from the adjudicator within four days of the date of issue of the notice of adjudication that he is able and willing to act.

Clause 3.3 is again in identical terms to clause 3.2 of the ICE Adjudication Procedure. This provides that if an adjudicator is not selected in the aforesaid manner then either party may within a further three days request the person or body named in the contract or if none is so named The Institution of Civil Engineers to appoint the adjudicator.

Clause 3.4 is framed in similar terms, other than the nominator, to clause 3.3 of the ICE Adjudication Procedure. If an adjudicator is not selected in the aforesaid manner then either party may within a further three days request the President for the time being (or a Past President) of the IChemE to appoint an adjudicator. This application is required to be in a specific form which is included as Annex C to the Rules.

Clause 3.5 follows clause 3.5 of the ICE Procedure and provides that if for any reason whatsoever the adjudicator is unable to act, either party may require the appointment of a replacement adjudicator in accordance with the procedure in paragraph 3.3.

These clauses are very clear as to the procedure to be adopted and cater for every eventuality including the inability of an appointed adjudicator to act. The inclusion of an application form to the IChemE as the ANB is not something that is generally done elsewhere because where there are alternative ANBs named in a contract, or none at all, the inclusion of the application form for one ANB would not be proper and the inclusion of a form to cater for the individual requirements of every ANB that might be used is clearly impossible.

The requirements of the notice (of adjudication) under this form are that 'details of the matter in dispute' are given. This is far less prescriptive than the Scheme but providing that the details are sufficient for the other party, the IChemE as ANB and the adjudicator to identify the extent of the dispute referred and the adjudicator's jurisdiction, it will be perfectly adequate.

Clauses 4 and 5

The form then continues in clauses 4 and 5 to deal with the acceptance of nomination and confirmation of appointment.

Clause 4 provides that in the event that the selected nominee is willing to act as the adjudicator, he is to inform the parties in writing. He is also required to complete the form of agreement as given in Annex B and issue it to the parties for signature by them within seven days.

Clause 5 provides that signing the form of agreement and its return to the adjudicator by either of the parties shall constitute confirmation of the adjudicator's appointment, and thereafter the requirements of clauses 6 to 8 (dealing with the procedure after the adjudicator's appointment) shall apply.

The Association of Consulting Engineers

The ACE issued the second edition of its conditions of engagement in 1998. This includes a very simple adjudication provision as follows:

> Either party may refer any dispute arising under this agreement to adjudication in accordance with the Construction Industry Council Model Adjudication Procedure.

The appointment of the adjudicator will therefore follow the procedures set out in the CIC Procedure.

The Construction Industry Council Model Adjudication Procedure

The appointment provisions in the CIC Procedure 3rd Edition are as follows:

10. If an adjudicator is named in the Contract he shall within 2 days of receiving the Notice confirm his availability to act. If no adjudicator is named, or if the named adjudicator does not so confirm, the referring Party shall request the body named in the Contract, if any, or if none, the Construction Industry Council, to nominate an Adjudicator within 5 days of receipt of the request. The request shall be in writing, accompanied by a copy of the Notice and the appropriate fee. Alternatively the Parties may, within 2 days of the giving of the Notice appoint the Adjudicator by agreement.

The CIC Procedure provides for the eventuality of the nominated adjudicator being unable to act by allowing for a further nomination. It also has a standard Adjudicator's Agreement.

Clause 12 of the CIC Procedure provides for objections to a particular adjudicator in the following way:

12. If a Party objects to the appointment of a particular person as adjudicator, that objection shall not invalidate the Adjudicator's appointment or any decision he may reach.

Centre for Effective Dispute Resolution

The CEDR Adjudication Rules require that a copy of the notice of adjudication is sent by the referring party to the adjudicator if he is named in the contract and that the adjudicator is required to notify the parties in writing that he is available to act.

The CEDR Rules also deal with the situation where no adjudicator is named in the contract, as follows:

2. If no Adjudicator is named in the contract or if the named Adjudicator does not confirm his or her availability to act then the Referring Party shall immediately apply to the Centre for Effective Dispute Resolution ('CEDR Solve') using CEDR Solve's application form to nominate an Adjudicator. CEDR Solve shall nominate an Adjudicator and communicate the nomination to all the parties within 5 days of receipt of:
    *   The completed application form
    *   A copy of the Notice of Adjudication
    *   CEDR Solve's nomination fee.
3. The Adjudicator shall, within 24 hours of receipt of the nomination, confirm in writing to the Parties that he or she is available to act, whether in response to receiving the Notice or to a nomination by CEDR Solve. The Adjudicator shall provide to them, at the same time, a copy of the terms on which he or she is prepared to act including information regarding fees and expenses.

As we have already discussed, it will be usual for any nominated adjudicator to confirm his terms to the parties on receipt of the nomination. These rules merely make than obligatory.

The Technology and Construction Solicitors' Association

The provisions relating to the appointment of the adjudicator in Version 2.0 of the TeCSA Adjudication Rules are very similar to those in the CEDR Rules and they too require that the adjudicator confirms availability.

The relevant provisions are as follows:

4. Where the Parties have agreed upon the identity of an adjudicator who confirms his readiness and willingness to embark upon the Adjudication within 7 days of receipt of the notice requiring adjudication, then that person shall be the Adjudicator.

5. Where the Parties have not so agreed upon an adjudicator or where such person has not so confirmed his willingness to act, then any Party shall apply to the Chairman of TeCSA for a nomination. The following procedure shall apply:
    (i) The application shall be in writing accompanied by a copy of the Contract, a copy of the written notice requiring adjudication and TeCSA's appointment fee of £100.
    (ii) The Chairman of TeCSA shall endeavour to secure the appointment of an Adjudicator within 7 days from the notice requiring adjudication.
    (iii) Any person so appointed, and not any person named in the contract whose readiness and willingness is in question, shall be the Adjudicator.

6. Within 7 days from the date of the [notice requiring adjudication]:
    (i) provided that he is willing and able to act, any agreed Adjudicator under Rule 4 or nominated Adjudicator under Rule 5(ii) shall give written notice of his acceptance of appointment to all parties, and;
    (ii) the referring party shall serve the Referral Notice on the Adjudicator and the Responding Party.

Final comments

There are, as described above, numerous variants in terms of the detailed procedure that enables a party desiring adjudication to get an adjudicator appointed. The ideal way is to agree an adjudicator who is available and suitable to deal with the dispute. Failing this the Act requires the contract to include a procedure to allow for nomination by an ANB within the required seven-day period. Either by means of the procedure in the contract or, should there be none, by the appointment provisions in the Scheme, an adjudicator will be in place and in all normal circumstances be ready to receive the referral by the seventh day.

Chapter 8
Adjudicators' Agreements

General introduction

No-one with any sense who enters into any form of contract does so without having something in writing to record what has been agreed. They need a framework to govern the rights and obligations that they receive and undertake by entering into the contract. Memories can be very short when troubles arise and a written agreement will give them something to refer to if the nature or detail of the agreement is questioned and needs to be confirmed.

The parties to a construction contract who are involved in adjudication and the adjudicator himself are in a contract. The adjudicator agrees with the parties that he will adjudicate. The terms of that contract will ideally be recorded formally in order for the rights and obligations of the parties themselves and of the adjudicator to be readily ascertainable.

The best solution is for the parties to the construction contract and the adjudicator to enter into a formal adjudication agreement, which will define with precision what each party to that agreement has to do and can expect the other to do. The commonly used standard forms of contract provide such an agreement; the Scheme for Construction Contracts does not.

There are two types of agreement: one that envisages the agreement being entered into with a named adjudicator at the time that the construction contract is originally formed, and one that is entered into after the dispute has arisen. The Joint Contracts Tribunal (JCT) when it produced Amendment 18 in 1998 proposed two separate, albeit very similar, forms for these two situations, but when the 1998 editions of the various JCT contracts were published shortly after, they did not include the option to name the adjudicator in the contract. The other bodies that have produced formal agreements have only produced one and this can be used or adapted for either situation. The Institution of Civil Engineers (ICE) has two separate families of contract, the ICE Forms of Contract and the NEC Engineering and Construction Contract. In the NEC there are also two separate forms but the difference between these does not relate to the time of appointment of the adjudicator (see later in this chapter).

The realities of construction contracts are, however, that there will be a significant number of occasions when a formal adjudicator's agreement is not completed. The construction contract may not be in a standard form or the standard form of agreement that accompanies a standard form contract may not be completed. The appointment may result from a nomination under the Scheme that has no accompanying adjudicator's agreement. In these instances the parties and the adjudicator have to fall back upon what can be interpreted as being the contract between them. In the main these will be terms that are either implied by law or expressly applicable because of the way that the adjudicator's appointment has been procured.

This being the case, in this chapter we look first at what the parties and the adjudicator should be looking for in respect of the terms that will apply to the agreement (contract) between them. We then look at the terms that govern such an agreement where there is nothing that sets out in any detail the agreement between the adjudicator and the parties to the construction contract. Finally, we consider the terms that apply to the agreement where the formal adjudication agreements published by contract writing bodies and others are utilised.

The terms of an adjudicator's agreement

As with most agreements there is a tension between the parties, each wanting to obtain terms that they see as being most favourable to themselves while accepting that there will have to be some element of compromise if agreement is to be reached. This is precisely the situation where the parties to a construction contract have decided that it is in their mutual interest to agree an adjudicator. They will see the agreement from their perspective, the adjudicator from his.

There are two principal matters that are most likely to be in the forefront of the minds of the parties to the construction contract. The first and most important will be 'can he do the job?' This will have a supplementary element, 'will he perform in the 28 days allowed?' The second question will be 'how much will it all cost?' and this will probably have two main supplementary points, 'what is the adjudicator's fee?' and 'what other matters will the adjudicator propose for inclusion in the terms of his agreement?' The parties may also wonder if a ceiling can be set to the amount that they will have to pay to the adjudicator.

The adjudicator will probably have much the same thoughts in his mind: 'can I do the job?', 'have I the time available?' and 'what terms should I seek?'.

In respect of the third of these points the adjudicator will have an ideal set of terms that he will provide to the parties to the construction contract on receiving their enquiry. The parties will often seek to negotiate these terms and this then becomes the normal situation of a pre-contract negotiation. The principal difference from what might be described as 'normal' contractual negotiations is that there is a pressure, possibly from both parties and certainly from the party which sees the result of the adjudication as being the receipt of further payment, to get on with it and it is incumbent on all involved to reach agreement pretty quickly otherwise that party is likely to get fed up with the delay and go off to an ANB for a nomination. It will then probably be the case that the nominated adjudicator states his terms and proceeds with the adjudication without agreeing them. The parties then lose their opportunity to negotiate and their only fall back is to challenge the reasonableness of the adjudicator's charges if they consider them to be excessive.

The principal matter that the adjudicator will want to agree is his fee. Except in the case of adjudications where the issues are limited and clearly identifiable at this stage, the adjudicator is unlikely to be willing to accept a lump sum arrangement as this will put him totally at the mercy of the way that the parties conduct themselves during the course of the adjudication.

If the adjudicator is registered for VAT he will want to be sure that he can recover this on top of his basic fee.

He will also wish to see as many as possible of the following matters covered by his agreement with the parties:

- that he can recover any expenses (and probably the nature of those expenses such as a mileage rate for the use of his car for site visits for example) that he incurs while carrying out the adjudication, including the cost of any legal or other advice that he might take;
- that there is no dispute, if he has to travel to a meeting or a site, as to whether he can charge for his time spent travelling;

- whether he can, if he so wishes, apply a lien to his decision and refuse to release it until he gets paid;
- the nature of the liability of the parties to him in respect of his fee (is it joint and several?);
- the nature of his liability to the parties for his actions as adjudicator (his immunity);
- whether the parties are prepared to offer an indemnity against third party action in connection with his activities as adjudicator; and
- what obligations will be placed upon him by implication due to his acceptance of the position of adjudicator under the particular construction contract between the parties.

Both the parties and the adjudicator will, if they think about it, also probably want to know what happens if the adjudicator fails to perform.

The agreement between the adjudicator and the parties

The agreement between the adjudicator and the parties will preferably be in writing. It may be, however, where there is no reference to any published adjudicator's agreement that the time pressures are such that the adjudicator has to proceed without anything in writing from the parties other than the nomination and the referral notice. In this situation the provisions of the contract between the parties will govern the adjudicator's own agreement with the parties either as implied or express terms. If the contract does not comply with the Act, the Scheme will apply and the requirements of the Scheme will govern the agreement between the adjudicator and the parties. The contract may, however, have been drafted to do no more than comply with the Act. In that event the requirements placed on the adjudicator are those in the Act, which are set out in the compliant construction contract.

There is a certain difficulty here in that the construction contract will, either specifically or statutorily, include certain provisions that relate to the adjudicator but to which the adjudicator is not a party. The construction contract may also include other terms that seek to place further requirements upon the appointed adjudicator. How do the parties ensure that the adjudicator complies? The answer is of course that they should enter into a written agreement.

But this will not always be the case. No adjudicator should, however, ever agree to adjudicate without a sight of the construction contract, if for no other reason than to identify whether he is adjudicating under the terms of the contract or the Scheme. Having been made aware of the terms of the construction contract it is our view that such provisions must, at the very least, be implied into the adjudicator's agreement with the parties. If it is the Scheme that applies, the provisions of the Scheme will equally, as a minimum, be implied.

It is also quite possible that the terms will in fact be express terms on the basis that the adjudicator will have seen the contract or be aware that his appointment as adjudicator arises under the Scheme and he will be deemed to have accepted the task of adjudicator in the full knowledge of those requirements.

The most likely situation where there will be no standard adjudicator's agreement will be in the case of a construction contract that does not comply with the Act. It may be, in that situation, that the parties have agreed to go to adjudication, have thrashed out a form of adjudicator's agreement between themselves and possibly with an agreed adjudicator, but it is more than likely that nothing of the sort will have happened.

It is far more likely that the adjudicator will have been put in place by a unilateral application by a party to an adjudicator nominating body (ANB) and his subsequent nomination by that ANB. In that case the terms of the agreement between the adjudicator and the parties are likely to be those implied as a result of the nomination of the adjudicator by the ANB and those that are either impliedly or expressly incorporated from the contract between the parties or, as in all likelihood where the Scheme will apply, from the Scheme itself.

Before we look at the terms that the adjudicator's agreement will include as a result of the Scheme, it is worth looking at the rights and duties of the parties and the adjudicator in the situation where the contract complies with the Act but includes nothing more concerning adjudication.

On the principles discussed above, the requirements of the Act will govern the adjudicator's agreement. These terms will be:

- The parties have a right to a decision within 28 days.
- The adjudicator has a duty to reach that decision within 28 days.
- The parties have a right to extend that period by agreement (after the dispute has been referred).
- The adjudicator has a right to extend the period by up to 14 days with the consent of the referring party.
- The adjudicator has a duty to act impartially.
- The parties cannot hold the adjudicator liable for anything done or omitted in the discharge or purported discharge of his functions as adjudicator save in the case of bad faith. Any employee or agent of the adjudicator is similarly protected.

The above list relates only to the rights and duties set out in the agreement between the adjudicator and the parties, not to the contract between the parties themselves.

It must be noted that the Act is silent on fees and the adjudicator's right to payment will arise in common law as an implied term that he has a right to a reasonable fee.

For the purpose of comparison, the subject of a reasonable fee for arbitrators is discussed in *Commercial Arbitration*[1] (Mustill & Boyd). The parameters set out there are as follows:

- The arbitrator starts out by setting a daily or hourly rate that takes into account his skill, qualifications and status.
- He then multiplies this rate by the length of time involved.
- The arbitrator then considers if the resulting figure needs adjustment in respect of any of the following factors:
  - the complexity of the dispute and the difficulty or novelty of the questions involved;
  - the skill, specialised knowledge and responsibility required of him;
  - the number and importance of the documents studied;
  - the importance of the dispute to the parties; and
  - the value of the property involved or the amount of the sum in issue.

There is no reason why similar parameters should not apply to adjudicators but, and it is a big but, the adjudicator should always bear in mind that he will generally be utilising the skills that he uses in his general day-to-day practice. We can see no reason for any adjudicator to seek to take advantage of the system. We are strongly of the view that the rates for adjudicating should not be set at a level that is significantly different from the usual hourly rates attained by an individual of the adjudicator's position and status within his profession.

One general point on adjudicators' fees relates to the liability of the parties for that fee. Many adjudication agreements, contractual adjudication clauses, adjudication rules and the

---

[1] *Commercial Arbitration*, 2nd Edition, Mustill & Boyd, Butterworths, 1989, at p. 237.

Scheme, generally clarify this position by stating that the liability for the adjudicator's fee is joint and several. This means that each party is liable for the whole fee in the event that the other party fails to pay part or the whole fee. It is worth noting that even where the procedure for the adjudication requires that the adjudicator sets out in his decision who is responsible for paying his fee, until that fee is paid there is still a joint and several liability to the adjudicator until the fee is paid in full.

Where there is nothing in the adjudication agreement, contract or rules concerning the liability of the parties for the adjudicator's fees and expenses, the situation can only be governed by the common law (that is the law created by the courts). The problem is that there is as yet little common law relating to this aspect of adjudication.

As far as the liability for fees is concerned it is our view that the courts may well draw a parallel with arbitration before the 1996 Arbitration Act was brought into force. The 1950 and 1979 Arbitration Acts were silent on the nature of the parties' liability to the arbitrator and the courts had found that they were jointly liable. Mustill & Boyd cite *Crampton and Holt* v. *Ridley*[2] and *Brown* v. *Llandovery*[3], in support of joint liability. They go on to suggest that this liability is not only joint but joint and several.

There is some indication that the courts will approach liability for adjudicators' fees in accordance with the views set out in *Commercial Arbitration* even where there is no formal written agreement between the parties. It is the authors' experience that parties generally pay adjudicators' fees and the occasions when recourse to the courts has been necessary are thankfully few in proportion to the number of adjudications that have taken place, albeit almost every practising adjudicator will have his own story of difficulty in recovering fees. One of the authors has himself had experience of an action in the court to recover fees from a recalcitrant responding party where the losing referring party had gone bust and had failed to pay the adjudicator. In this case the court was prepared to accept the adjudicator's terms, which contained a statement that liability was joint and several, even though neither party had indicated acceptance of those terms.

Two matters relating to the fees of adjudicators have resulted in judgments: *Paul Jensen* v. *Staveley Industries*[4] and *Stubbs Rich* v. *Tolley*[5]. In the former, the adjudicator found that he did not have jurisdiction and had to seek the court's aid in recovering his fee for the time he spent in reaching the conclusion that he had no jurisdiction. The second again related to the recovery of fees and the principal issue related to the amount that the adjudicator had charged. In this case the court considered an allegation that the hours charged were unreasonably excessive. Initially, when the adjudicator's fee was challenged on this basis, the adjudicator was required to pay them back. On appeal it was however found that the adjudicator's fees were something that did fall within the provision that 'The adjudicator shall not be liable for anything done or omitted in the discharge or purported discharge of his function as adjudicator unless the act or omission is in bad faith'. On this basis unless it can be shown that the adjudicator had acted in bad faith his fees are payable in full.

In most normal circumstances however the parties honour their commitment to the adjudicator and pay his fees within a reasonable time of the decision.

[2] *Crampton and Holt* v. *Ridley & Co* (1887) 20 QBD 48.
[3] *Brown* v. *Llandovery Terra Cotta etc. Co Ltd* (1909) 25 TLR 625.
[4] *Paul Jensen Ltd* v. *Stavely Industries Plc* (27 September 2001).
[5] *Stubbs Rich Architects* v. *W.H. Tolley & Son Limited* (8 August 2001).

Terms where the Scheme applies

There is no adjudicator's agreement produced to accompany the Scheme. The terms of the agreement between the parties and the adjudicator must, as discussed above, be those that are required of him by the Scheme. The rights and duties of the parties and of the adjudicator in respect of such an agreement are as follows:

- The adjudicator, if he is named in the contract, has the duty to indicate whether or not he is willing to act within two days of receiving a request to act. (If he is nominated by an ANB this duty will be a requirement placed upon him by the ANB in consideration of his being on their list. This arises from the implied undertaking of the ANB, when appointing under the Scheme, to only nominate adjudicators who will comply with the Scheme.)
- The adjudicator has the duty to operate as a natural person in his personal capacity.
- The adjudicator has the duty not be an employee of any of the parties to the dispute.
- The adjudicator has the duty to declare any interest, financial or otherwise, in any matter relating to the dispute.
- The referring party has the duty to refer the dispute in writing (the 'referral notice') to the adjudicator within seven days of the date of the notice of adjudication and to send copies to every other party to the dispute.
- The adjudicator may, by agreement, adjudicate at the same time on one or more disputes under the same contract.
- The adjudicator may, by agreement, adjudicate at the same time on related disputes on different contracts.
- The adjudicator may resign at any time on notice to the parties.
- The adjudicator has the duty to resign when the dispute to which he is appointed is the same or substantially the same as one that has been previously referred to adjudication and a decision has been taken in that adjudication.
- The adjudicator may resign where he is not competent to decide the dispute because it varies significantly from the dispute that was referred to him. (Note that this relates only to the relationship between the actual dispute and the referral notice. If the dispute in the referral notice is significantly different from that described in the notice of adjudication there is no jurisdiction save in respect of that part of the referral that clearly relates to the notice of adjudication.) It is a matter of interpretation of individual circumstances as to what the term 'varies significantly' means.
- Where the adjudicator resigns because the dispute has already been adjudicated or it varies significantly from that described in the referral notice, he is entitled to reasonable fees and expenses.
- The parties have the right to agree to revoke the appointment of the adjudicator.
- Where the adjudicator's appointment is revoked, he is entitled to reasonable fees and expenses save where such revocation is due to the adjudicator's default or misconduct.
- The adjudicator has the duty to act impartially.
- The adjudicator has the duty to act in accordance with any relevant terms of the contract between the parties.
- The adjudicator has the duty to reach his decision in accordance with the applicable law in relation to the contract.
- The adjudicator has the duty to avoid incurring unnecessary expense.
- The parties have the duty to comply with any request or direction of the adjudicator in relation to the adjudication.
- The adjudicator has the duty to reach his decision not later than 28 days after the date of

the referral notice or 42 days after the date of the referral notice if the referring party so consents or in such longer period as both parties may agree.

- The adjudicator has the duty to decide the matters in dispute. He may take into account any other matters that the parties to the dispute agree should be within the scope of the adjudication.
- The adjudicator has the duty to give reasons if one party so requests.
- The adjudicator is entitled to the payment of such reasonable amount as he may determine by way of fees and expenses reasonably incurred by him.
- The parties are jointly and severally liable for any sum which remains outstanding following the making of any determination on how the payment shall be apportioned.
- The adjudicator shall not be liable for anything done or omitted in the discharge or purported discharge of his functions as adjudicator, unless the act or omission is in bad faith and any employee or agent of the adjudicator is similarly protected.

It can be seen from the above that a very comprehensive and detailed set of terms and conditions govern the agreement between the parties and the adjudicator when the Scheme applies.

We have excluded procedural matters here although these may quite easily be considered to be matters that could form a part of the adjudicator's agreement with the parties.

Even with this lengthy list of matters governing the adjudicator, he will probably still want to ensure that all the points that we have discussed earlier, particularly his hourly rate, are covered somewhere.

Published adjudicators' agreements

We review the following commonly used adjudicators' agreements:

- The Joint Contracts Tribunal (JCT)
- The Construction Industry Council (CIC)
- The Institution of Civil Engineers (ICE) (also the Adjudicator's Contract for use with the NEC Engineering and Construction Contract (NEC))
- GC/Works/1
- The Institution of Chemical Engineers (IChemE)
- The Centre for Dispute Resolution (CEDR).

The Technology and Construction Solicitor's Association (TeCSA) does not publish a form of agreement.

The JCT adjudication agreement

The JCT has two forms of adjudicator's agreement, one for where the adjudicator is named in the contract, notwithstanding the fact that the contract itself does not include the amendments to allow for this that were included in Amendment 18, and one for where the adjudicator is agreed or nominated at the time that the dispute arises.

These forms of agreement apply to the JCT main and sub-contract forms and to the Construction Confederation forms DOM/1 and DOM/2.

The provisions of these two forms are essentially the same. The contracting parties are named, as is the adjudicator. There is no provision for more than two parties other than the adjudicator. Next comes the description of the works and details of the main contract. This last has to include all amendments. Only three lines are allowed. Given the extensive amendments that are commonly seen in contracts, there is almost certainly going to be rather more than just the title of the form itself if the full extent of the amendments are to be shown.

There then follow five clauses and a Schedule, as follows.

Appointment and acceptance

The appointment is by both parties. They are both required to execute the JCT Adjudication Agreement with the adjudicator by clause 41A.2.3 of the contract. The adjudicator is also required to execute the agreement by clause 41A.2.3. A party who does not want the adjudication for any reason is therefore in breach of contract if it does not sign the agreement. The JCT obviously anticipates that a party may not sign the agreement by including in clause 41A.5.6 the provision that failure to enter into the JCT Adjudication Agreement does not invalidate the decision of the adjudicator. (See also below in relation to the adjudicator's fees.)

Adjudication provisions

The adjudication provisions of the main contract are expressly incorporated into the adjudicator's agreement by this clause. There is as a result no need for any debate as to whether they are express or implied terms.

For the avoidance of needless repetition we do not set out the full detail of the terms expressly incorporated into the Adjudication Agreement here (see Chapter 10, which deals with adjudication clauses). It is worth noting, however, that the JCT has decided that the adjudicator's immunity remains in the adjudication clause in the main contract and this does not appear in the Agreement. The timescales are also only mentioned in the adjudication clause as is the impartiality requirement. There is also a specific requirement that the adjudicator should allocate responsibility for his fee as between the parties and if he should fail to do this they bear it in equal shares.

It is worth mentioning specifically that the question of a lien for the adjudicator on his decision is not dealt with and also that there is no requirement that reasons be given.

Adjudicator's fee and reasonable expenses

The parties are jointly and severally liable for the adjudicator's fee and for all expenses reasonably incurred. Note that there is no provision for the fee to be reasonable. The level of fee is set in the Schedule appended to the Agreement as a lump sum or an hourly rate. There is, however, no requirement for the number of hours that he charges to be reasonable. We would suggest, however, that there must be an implied term of reasonableness but note the decision in *Stubbs Rich* v. *Tolley* relating unreasonableness to bad faith, discussed above.

A possible problem arises as a result of these provisions and that is where there has been a nomination by a named body and the parties and the adjudicator cannot agree a fee. The parties are, as noted above, required to execute the adjudication agreement by the contract and the adjudicator has to undertake to execute it as a condition of being named in the nominating body's list. It is to be hoped that all concerned will be reasonable in such circumstances but it is always possible that a party who has not instigated the adjudication will be unenthusiastic about the whole idea and just refuse to agree a fee for the sake of it. This is clearly a breach of contract and can ultimately be dealt with by the provisions of that contract. The adjudicator will, in these circumstances, have to proceed in the absence of any agreement as to his fees.

Unavailability of the adjudicator to act on the referral

The adjudicator is required to give immediate notice to the parties if he is unable to complete the adjudication if he becomes ill or unavailable. Interestingly, in the form of agreement for an adjudicator named in the contract, there is no requirement to notify the parties in the event that this occurs before he adjudicates or between two adjudications. Such a notification might possibly avoid delay in the commencement of an adjudication as the parties could then go immediately to a substitute adjudicator who was already in place at the time that the dispute arises.

Termination

The parties can terminate the adjudicator's agreement or an adjudication at any time on written notice to him. This cannot be done unilaterally. The adjudicator is entitled to his fee in these circumstances save where the termination arises because of his failure to give his decision within the required timescales or at all.

The Schedule

This includes no more than a blank into which must be inserted the adjudicator's lump sum fee or hourly rate. It is not unusual for an adjudicator to seek to include additional matters relating to his terms in the Schedule.

The agreement has provision for the parties and the adjudicator to append their signatures and a witness is required to each signature.

The CIC adjudicator's agreement

The basic premise of this agreement is that the dispute has been referred to adjudication under the CIC Model Adjudication Procedure.

As with the JCT agreement discussed above, this agreement is short and there are matters expressly incorporated into the agreement by the reference to the adjudication procedure.

The agreement itself is laid out in similar fashion to the JCT agreement, with the parties and the adjudicator being named and a provision being included for the date of the contract between the disputing parties to be inserted. The agreement also provides a space following the words 'in connection with' and it is to be presumed that the intention here is to insert not only the nature of the contract but also the details of the dispute.

There then follow six clauses and a Schedule as follows:

1. The rights and obligations of the Adjudicator and the Parties shall be as set out in this Agreement.
2. The Adjudicator agrees to adjudicate the dispute in accordance with the Procedure.
3. The Parties agree jointly and severally to pay the Adjudicator's fees and expenses as set out in the attached schedule and in accordance with the Procedure.
4. The Adjudicator and the Parties shall keep the adjudication confidential, except so far as is necessary to enable a party to implement or enforce the Adjudicator's decision.

5. The Parties acknowledge that the Adjudicator shall not be liable for anything done or omitted in the discharge or purported discharge of his functions as Adjudicator (whether in negligence or otherwise) unless the act or omission is in bad faith, and any employee or agent of the Adjudicator shall be similarly protected from liability.
6. This Agreement shall be interpreted in accordance with the law of England and Wales.

The second clause, as with the JCT agreement, is an express inclusion of the provisions of the Procedure into the Adjudicator's Agreement. Again, we shall not set out these terms in detail save that it is worth mentioning that such things as impartiality and timescales are dealt with in the Procedure rather than in the Adjudicator's Agreement. This Procedure deals with the allocation of the adjudicator's fees in a similar fashion to the JCT. This Procedure used to specifically provide the adjudicator with a lien on his decision. This is now deleted in the Third Edition. The original First Edition of this Procedure did not require the adjudicator to give reasons but this is changed in the Second Edition 'unless the parties agree at any time that he shall not be required to give reasons'.

The CIC Procedure confirms that the adjudicator has the right to charge for any legal or technical adviser. The adjudicator may direct a party to pay all or part of his fees. If no such direction is made the parties shall pay them in equal shares. The party requesting adjudication is liable for the adjudicator's fees and expenses if the adjudication does not proceed.

The Third Edition includes a further paragraph numbered 4 as follows:

> The Adjudicator may destroy all documents received during the course of the adjudication six months after the delivery of his decision provided that he shall give the parties 14 days notice of his intention to do so and that he shall return the documents to the parties if they so request.

An adjudicator will generally have to keep all the documents for at least six years. Adjudicators accumulate a vast number of files after dealing with a number of adjudications and storage can become a tremendous problem. Many adjudicators instigate a system whereby, once the adjudication is complete, the parties are allowed the opportunity to collect all documents submitted within a limited time, failing which he will destroy them. This is generally not recognised formally other than by the ICE and now the CIC. Where an adjudication is a one-off, many adjudicators allow a very limited time between completion of the decision and informing the parties that the documents provided will be destroyed unless collected. In the case of an adjudicator named in the contract this would, of course, not be until the expiry of all possibility of his being asked to act in respect of further disputes. Correspondence files should be kept.

The provision for the destruction of documents in the ICE Adjudication Procedure does not set a time limit for retention. The similar provision in the NEC Adjudicator's Contract requires a date for the termination of the adjudicator's appointment to be inserted and the adjudicator is not required to retain the documents after that date.

The CIC now requires retention of documents for six months. This seems to be an unnecessarily long time where the adjudication is not going to be one of a series. Once the adjudicator has produced his decision his duties are complete. He has nothing to do with any enforcement proceedings or settlement negotiations on a personal basis. If either or both parties think that the documents will be needed in the future they can just as easily request their retention or return after four weeks as six months.

These provisions are followed by a Schedule which covers the adjudicator's hourly rate (in respect of all time spent on the adjudication including travelling time) and sets a maximum charge per day. It also sets out reasonable expenses and allows for 'other extraordinary expenses necessarily incurred'. The Schedule also deals with VAT.

There is then provision for the agreement to be signed by the parties and the adjudicator. There is no requirement for these signatures to be witnessed.

The ICE adjudication agreement

This form of agreement is very similar to that produced by the CIC. There are, however, a few differences.

This agreement is set up to cover both the appointment of the adjudicator at an early stage and after a dispute has arisen. The confidentiality clause is similar to that of the CIC Agreement save that there is provision for this to be waived with the consent of the other parties, which consent shall not be unreasonably refused. The immunity clause is included in the adjudication procedure but it is worth mentioning here that, as distinct from the other agreements, the NEC Adjudicator's Contract excepted, this immunity is extended to an indemnity against the actions of third parties.

The adjudicator is, by this agreement, required to inform the parties if he intends to destroy documents.

This agreement includes an extensive schedule dealing with the adjudicator's terms. It covers his hourly rate and that he is entitled to be paid at this rate for his travelling time. It gives a non-exclusive list of disbursements that may be charged. These are specifically cited as examples and there is no restriction on other expenses being charged. It includes provision for an appointment fee which is to be paid in equal amounts by each party within 14 days of the adjudicator's appointment. If the appointment fee exceeds the final amount payable to the adjudicator, the balance is returnable. It deals with VAT. One further inclusion is a provision for interest at 5% per annum above the Bank of England base rate on fees and expenses that remain unpaid after the expiry of seven days from receipt of invoice. The period between sending out the invoice and receipt is not defined. As a general rule the adjudicator will send his invoices by first class post and provided that this is done, if there is any problem in this respect, their receipt would as a general principle be deemed to have occurred two days after the date of posting, Saturdays and Sundays excluded.

The Adjudicator's Contract for the NEC Engineering and Construction Contract

This book is not the place to indulge in a consideration of the approach adopted as to the form of words used in the NEC. Suffice it to say that the use of the present tense throughout can, initially, be disconcerting.

The approach adopted by the ICE committee that was charged with the drafting of this adjudicator's contract is quite different from that adopted by the other contract-writing bodies whose forms of agreement are considered in this book. This difference is just as marked when comparison is made with the ICE Adjudication Procedure, designed for use with all the ICE forms of contract other than the NEC.

All the other forms of adjudicator's agreement run to no more than three pages in total. The NEC Adjudicator's Contract has two alternative forms of agreement which are followed by two and a half pages of conditions and a further half page of 'contract data'. With the introductory pages and index the document runs to a total of 12 pages. This is not all. The Adjudicator's Contract itself is accompanied by a separate document entitled *Guidance Notes and Flow Charts*, which runs to a total of 19 pages.

As noted above there are two forms of agreement, one of which is stated to be inappropriate where a 'United Kingdom Scheme for Construction Contracts applies', and the other is to be used only where such a Scheme does apply. The guidance notes make it clear that this wording is used to cover both the England and Wales and the Scotland Schemes.

These two agreements are very simple in form. They both name the parties and the adjudicator and have provision for their signatures. They each then have clauses which do no more than state that the adjudicator is appointed by the parties and that the adjudicator accepts the appointment and undertakes to carry out the adjudicator's duties as described in the conditions of contract. The form that applies to the Scheme adds the words 'and the Scheme for Construction Contracts' in certain places. The Scheme version also places a requirement on the adjudicator to apportion his fee and expenses between the parties.

These simple forms of agreement are followed by detailed conditions of contract, which are worth considering in some detail as they include a number of provisions that do not appear in other forms of adjudicator's agreement.

The provisions worthy of note for this reason are as follows:

- There is an obligation on the adjudicator to notify the parties as soon as he becomes aware of a conflict of interest or that he is unable to act.
- The conditions identify in considerable detail the expenses that an adjudicator can charge. The list is not, as in other forms of agreement, non-exhaustive. If the adjudicator is involved in an expense not covered in the list we would suggest that he cannot insist upon it being reimbursed without having obtained prior agreement.
- There is a priority clause between the contract between the parties and the adjudicator's contract, the latter prevailing in the case of conflict – a pragmatic move by the authors of the adjudicator's contract in the anticipation of this contract being used in cases where the NEC itself is not being used.
- There is a detailed communications clause. It specifies that communications required by the contract must be in a form that can be 'read, copied and recorded'. It also states that a communication has effect when received at the last address notified by the recipient or, if none is notified, at the address of the recipient stated in the Form of Agreement.
- The adjudicator under this contract does not receive 'advice'; he may 'obtain from others help that he considers necessary'.
- There is a confidentiality clause.
- The adjudicator is required to keep the documents provided to him until 'termination' (see below).
- The adjudicator's fees are payable in equal shares unless otherwise agreed. He is allowed to claim for time spent travelling and the adjudicator is required to invoice each party its share of his fee and expenses after he has communicated his decision to the parties. There is no lien on his decision.
- The parties are required to pay the adjudicator within three weeks of invoice unless another period is agreed. If a payment is late, interest is payable. Interest is compounded annually.
- If one party fails to pay the adjudicator, the other has to pay him and this payment has to include any interest due to the adjudicator because of the late payment. That party then has to seek recovery of both the original fee and the interest from the other party.
- This contract has the required immunity clause but also includes an indemnity in respect of claims, compensation and costs arising out of the adjudicator's decision.
- The parties may agree to terminate the adjudicator's agreement at any time.
- The adjudicator himself can terminate his appointment because of a conflict or if he is

unable to act. More interestingly, he can terminate if he is not paid within five weeks of a payment becoming due, albeit this will, of course, only apply where the adjudicator is appointed for the duration of the contract and is still in post five weeks after he has made a decision.

● There is also provision for the insertion of a termination date in the contract.

The schedule in this form of contract is called the 'Contract Data'. In addition to the adjudicator's fee and certain items required by the NEC format, such as the nature of the contract between the parties and the law, language and currency of that contract, it contains provision for the interest rate to be set at a certain percentage above the rate of a chosen bank. It also includes the termination date of the adjudicator's appointment.

The detail of this Adjudicator's Contract means that the contract adjudication clause is very simple, covering little more than the requirements of the Act (see Chapter 6). A point of interest is that there is no cross reference to the Adjudicator's Contract in the NEC itself.

GC/Works/1 Model Form 8 Adjudicator's Appointment (Condition 59)

This form is drafted on the presumption that it will be completed at the time the contract for the works is let. There is as a result no provision for a description of the dispute in the preamble to the form. The preamble, in common with other adjudicators' agreements, names the parties and the adjudicator. It also provides for the situation if the adjudicator is a 'named substitute adjudicator'. This is a useful provision, not seen in other standard forms of adjudicator's agreement, to cover the situation where the original adjudicator is not available for any reason. Under this contract it is therefore possible to have the principal adjudicator and one or more substitutes in place at the outset of the contract works. This is most useful in overcoming the delays that are likely to occur where the adjudicator who is named in the contract proves not to be available and there is no provision for a substitute.

This agreement has eight clauses and a schedule that are sufficiently different from the other agreements to be worth setting out in detail. They are as follows:

1. The adjudicator shall, as and when required, act as [adjudicator] OR [named substitute adjudicator] in accordance with the Contract, except when unable so to act because of facts or circumstances beyond his reasonable control.

This makes it quite clear that in taking up the position of named adjudicator, the person so named must make himself available at all reasonable times to adjudicate.

2. The Adjudicator confirms that he is independent of the Employer, the Contractor, and the Project Manager and Quantity Surveyor under the Contract, and undertakes to use reasonable endeavours to remain so, and that he shall exercise his task in an impartial manner. He shall promptly inform the Employer and the Contractor of any facts or circumstances which may cause him to cease to be so independent.

This provision introduces the concept of independence. It is not something that is spelt out in other forms of adjudicator's agreement or in the contracts upon which they are based. It may be that it has been introduced because there have historically been forms of adjudication in the public sector in which the adjudicator has specifically been an officer of the authority. It is a useful provision and will prevent the abuse of the system that may occur in commercial contracts where the architect or engineer is appointed as adjudicator in order to 'avoid his power being eroded'. This is to be deprecated. It is almost inevitable in such a situation that the contract administrator, however well meaning, will be seen by the contractor to be lacking in impartiality if, as adjudicator, he confirms a decision that he has made as contract administrator. The end result will be a continuance rather than a resolution of the dispute.

3. The Adjudicator hereby notifies the Employer and the Contractor that he will comply with Condition 59 (Adjudication) of the Contract, and its time limits.

This is an important provision as Condition 59(8) prevents the adjudicator from varying or overruling 'any decision previously made under the Contract by the Employer, the PM or the QS' in respect of certain matters where the remedy is limited to financial compensation.

4. The Adjudicator shall be entitled to take independent legal and other professional advice as reasonably necessary in connection with the performance of his duties as adjudicator. The reasonable net cost to the Adjudicator of such advice shall constitute expenses recoverable by the Adjudicator under this Agreement.

In all the other contracts, save for the NEC, this provision is covered as a procedural matter in the adjudication clause rather than in the adjudicator's agreement.

5. The Adjudicator shall comply, and shall take all reasonable steps to ensure that any persons advising or aiding him shall comply, with the Official Secrets Act 1989 and, where appropriate, with the provisions of Section 11 of the Atomic Energy Act 1946. Any information concerning the Contract obtained either by the Adjudicator or any person advising or aiding him is confidential, and shall not be used or disclosed by the Adjudicator or any such person except for the purposes of this Agreement.

The matters set out in the first sentence of this provision are not covered by any other form of contract and relate to the nature of the works likely to be carried out under the GC/Works form of contract.

6. The Employer and the Contractor shall pay the Adjudicator fees, expenses and other sums (if any) in accordance with the Contract and the Schedule, plus applicable Value Added Tax.
7. The Adjudicator is not liable for anything done or omitted in the discharge or purported discharge of his functions as adjudicator, unless the act or omission is in bad faith. Any employee or agent of the Adjudicator is similarly protected from liability.

This agreement follows the normal trend in including this provision in the adjudicator's agreement.

8. The proper law of this Agreement shall be the same as that of the Contract. Where the proper law of this Agreement is Scots law, the parties prorogate the non-exclusive jurisdiction of the Scottish courts.

The schedule that follows is left completely blank save for the provision that it is to cover the adjudicator's 'Fees, Expenses, etc.'. It is therefore up to the adjudicator and the parties to agree what goes in here.

The IChemE adjudicator's agreement

The adjudicator's agreement in this contract is an interesting departure from all other forms in that it is set out in the form of a draft letter to the parties from the adjudicator.

It starts with a statement that the adjudicator has been nominated by (agreement between the parties/the IChemE) as adjudicator in accordance with the contract and that he requires the parties' agreement to matters that follow.

The use of the words 'require your agreement' seems a bit draconian. As has been discussed elsewhere, either or both the parties may be reluctant to agree to the adjudicator's proposals and a more sensible approach might be to say 'seek your agreement'. Then at least there is no question as to the adjudicator not proceeding as soon as the referral arrives on his desk.

There then follow eleven paragraphs, as follows.

- The matter to be determined.
- Appointment shall be made and investigation shall be carried out in accordance with the IChemE's Adjudication Rules.

The use of the word 'investigation' is interesting. At first glance there seems to be a presumption and possibly even a requirement that the adjudicator will investigate the facts and the law but this is modified by paragraphs 5 and 6.

- Particulars of the contract.
- The representatives of the parties to be contacted in connection with the adjudicator's investigation are as follows: ... (if known...).

This appears to presume that the adjudicator and the parties will have been in contact before the nomination is received but this is modified by the words 'if known'. It is of course possible that this will have happened but in the case of a unilateral application for the nomination of an adjudicator direct to the ANB, this is unlikely.

- The adjudicator will decide the method of adjudication.
- This adjudication will not be an arbitration and the adjudicator may reach a decision by:
  - using his personal knowledge and expertise and that of others, if required;
  - requesting information from the parties and considering the responses;
  - conducting an investigation which may include the examination of documents, inspection of the relevant parts of any premises, site or engineering works and listening to oral submissions by the parties.
- The adjudicator will be allowed free and timely access to all records, information, persons and premises under the parties' control that the adjudicator may deem necessary to determine the matters in dispute.

These three items state pretty clearly what is expected of the parties and the adjudicator. The sub-paragraphs make it clear that the adjudicator may use his own expertise and conduct his own investigations, but this should be tempered by the requirement that has now been made clear by the court that the outcome of the use of his own expertise and the results of any investigation should be put to the parties before the adjudicator makes his decision[6].

The requirements on the parties are pretty clear with regard to the provision of access to documents and the like for the adjudicator. This may create some difficulty as an adjudicator should not use information that he obtains without the other party being made aware of it, and some might say that the third point in particular might result in the release of commercially sensitive information that might be disclosable in arbitration or court proceedings but that a party would not normally be prepared to disclose in an adjudication.

---

[6] *RSL (South West) Ltd* v. *Stansell Ltd* (16 June 2003).

- The adjudicator shall give reasons for his decision.
- The adjudicator's fees and expenses shall be in accordance with Annex D.

Annex D requires the adjudicator to provide appropriate details for the calculation of his fees, as explained below.

- Payment of fees and expenses will be due not later than 15 days after the date of the relevant invoice and the adjudicator reserves the right to charge interest at the daily rate of …% per day on unpaid amounts with interest on late payments chargeable from the due date for payment of the relevant invoice.
- The adjudicator, his firm and IChemE shall not be liable for anything done or omitted in the discharge or purported discharge of the adjudicator's functions unless that act or omission is in bad faith and any employee or agent thereof shall similarly be protected from liability. The parties shall save harmless and indemnify the adjudicator, his firm and IChemE and any employee or agent thereof against all claims by third parties in respect of the adjudication and in respect of this the parties shall be jointly and severally liable.

It is worthy of note that the IChemE takes the view that joint and several liability is important in respect of the matters covered by this last point, but that there is no mention of the joint and several liability which most adjudicators would expect to apply to their fees and expenses in order that the adjudicator can seek the whole of his fee from one party where the other party, who may have been found liable, fails to pay. Any adjudicator sending a letter on the basis of the IChemE's draft would be well advised to consider this point and make a suitable amendment.

The letter then finishes with a request that an enclosed copy be signed and returned within seven days.

There is then an addendum entitled 'Agreement to these terms' which only provides for signature by one party. If each party does this on their copy the adjudicator will presumably have to take a copy of each copy himself and send it to the parties in order to complete the agreement.

Annex D, which deals with the method of calculation of the adjudicator's fees, follows the format of the rest of the IChemE contract in that it is very detailed. It is headed 'a suitable format might be as follows'. A summary of the provisions is as follows:

- A detailed, non exclusive, note of what would be included as 'time spent in connection with the adjudication'.
- Confirmation that the adjudicator's hourly rate includes overhead costs.
- A provision for increasing the hourly rate on 1 January on the basis of the Retail Price Index.
- A 'disruption fee' in the event of cancellation of the adjudication.
- A statement that a day is 8 hours.
- Travelling time is regarded as working time up to a limit of 8 hours per 24 hour period for any one journey.
- Expenses reimbursed at cost.
- Invoices payable within 15 days.
- Fee payable in equal part by each party.

Centre for Effective Dispute Resolution

Prior to 2002 CEDR had a very complicated arrangement which required a four-party agreement that included CEDR as well. This has now been swept aside and all that CEDR produces is a set of 'Rules for Adjudication' which do not include any form of agreement.

Technology and Construction Solicitors' Association

TeCSA does not include an adjudicator's agreement with its Rules.

Conclusion

There is a wide selection of adjudicators' agreements available. All of them cover the basic requirements of a set of terms and conditions under which the adjudicator will operate either as an express term in the agreement itself or by the implication of terms from the adjudication clause in the construction contract into the agreement. In most instances the adjudicator and the parties will be faced with a nomination at the time that the dispute arises and they will have to use the standard agreement that applies to their particular form of contract.

Where the Scheme applies or there is a compliant contract without an agreement incorporated, there is nothing to stop the parties using any of the standard forms of agreement but reality suggests that the urgency of the situation will often mean no formal agreement being signed. In those instances the adjudicator will put in writing to the parties the terms that he intends to apply as adjudicator at the earliest time possible after he gets involved. In all normal circumstances these will include the level of his fee and the expenses that he will charge. His agreement with the parties will include those matters that are implied by the HGCRA and/or the Scheme but he may consider it prudent to set them out in writing at this stage. There will no doubt be other matters that it would be appropriate to include to cover the specific circumstances of the case.

At the end of the day, provided that the fee he charges is not seen by the parties to be unreasonable, it will be paid. Even if he has to recourse to an action in the court to recover his fees, there is precedent in the case of *Stubbs Rich* v. *Tolley* where the court found that the measure of unreasonableness was equated with bad faith.

Where the appointment occurs at the outset of the contract, the adjudicator and the parties should consider whether it is appropriate to enter into a standard form agreement or whether to formulate an ad hoc agreement to cover the particular contract involved. In the latter event there are a wide range of situations already covered by the various forms of agreement and the adjudicator and the parties should have no difficulty in finding appropriate wording already prepared.

Chapter 9
Jurisdiction, Powers and Duties

The three subjects of this chapter are closely linked. They are analogous to an egg. The shell is jurisdiction, the white or albumen is the powers and the yolk the duties. The white is always bounded by the shell. The yolk is bounded within the white and also the shell. Jurisdiction delimits the thing that the adjudicator is there to do. Powers deal with the way in which he can do it. The adjudicator's duties relate to both his jurisdiction and his powers. The parties also have duties placed upon them. We shall touch on these as we develop this chapter.

What are jurisdiction, powers and duties in essence?

The word jurisdiction is used to describe the nature and extent of the adjudicator's task. Jurisdiction is the authority granted to a man so that he can exercise justice in respect of matters brought before him. It will arise from the agreement between the parties that disputes will be referred to adjudication. It will, however, be limited by the character of the questions to be answered as are properly referred to him based on the notice of adjudication.

The powers of the adjudicator are what he is permitted to do in carrying out his task. They can arise in one of two ways. They can be set out within the parties' contract, either in the contract conditions themselves or within adjudication rules adopted by the parties, or they can come from the Scheme in the absence of a regime in the parties' contract. In either case the adjudicator has extremely wide powers as the very nature of the adjudication process, particularly the time limitation, means that he must be able to require the parties to assist him in his task and also be able to take steps to overcome any failure on the part of one or both parties in this respect.

The adjudicator's principal duty is to reach a decision within the time limit. He has one other overriding specific duty and that is to be impartial. There may be other duties imposed upon him, for example the requirements in paragraph 12 of the Scheme to the effect that he must reach his decision in accordance with the applicable law in relation to the contract and that he avoids incurring unnecessary expense.

Jurisdiction

The basic principles relating to the jurisdiction of an adjudicator are very simple. There must be a dispute and it must arise under a construction contract as defined in the Act. There are certain aspects in which this basic concept is widened, for example the parties can agree that adjudication applies to contracts that are not within the definition of construction contracts or that it can apply to matters wider than those arising under the contract.

The courts have now considered the adjudicator's jurisdiction in several cases. Many of these relate to the question of whether or not there was a dispute at the time that the notice of adjudication was given. This point has been considered in two recent cases, *Beck Peppiatt* v. *Norwest Holst*[1] and *Orange* v. *ABB*[2].

In both these cases it was confirmed that the proper way to define whether or not a dispute exists is that in *Halki* v. *Sopex*[3]. That definition is that 'there is a dispute once money is claimed unless and until the defendants admit that the sum is due and payable'.

In the former, Forbes J, the judge in charge of the Technology and Construction Court, said that the law is satisfactorily stated in *Sindall* v. *Solland*[4] as follows:

> '4. . . . For there to be a dispute for the purposes of exercising the statutory right to adjudication it must be clear that a point has emerged from the process of discussion or negotiation that has ended and that there is something which needs to be decided.
>
> 5. In my view, Judge Lloyd's definition is simple and easily applied. It accords with the ordinary meaning of the English word "dispute" and has much to commend it. It is not in conflict with Halki. . .'

In *Cowling* v. *CFW*[5] it was confirmed that for there to be a dispute capable of being referred to adjudication, the referring party must be careful to ensure that the responding party has had sufficient prior opportunity to consider the matters which are intended to be raised.

In *Pegram* v. *Tally Wiejl*[6] the Court of Appeal considered questions relating to whether an adjudicator was properly appointed under the Scheme. This judgment deals with a specific situation relating to the adjudicator's appointment but paragraphs 8 to 12 inclusive are well worth reading for the Court of Appeal's appraisal of adjudication.

The Court of Appeal has considered a number of points regarding adjudicators' jurisdiction that relate to specific aspects of interpretation of the Act and we deal with these elsewhere.

Can the adjudicator determine his own jurisdiction?

The general rule is that the adjudicator has no jurisdiction to decide his own jurisdiction. This comes from arbitration before the Arbitration Act 1996. The two cases which confirm this point are *Smith* v. *Martin*[7] and the *Christopher Brown* v. *Oesterreichischer* case.[8] The principle established in this case is that 'Arbitrators whose jurisdiction is challenged are entitled to make their own inquiries into the question whether or not they have jurisdiction in order to determine their own course of action, although the result of their inquiry can have no effect on the rights of the parties. Their award is in no way affected by the fact that it expressly or impliedly refers to a finding by the arbitrators as to their jurisdiction.'

In *Project Consultancy* v. *The Gray Trust*[9], an adjudication case, the *Christopher Brown* case was used in argument as to whether or not there was an ad hoc submission to allow the adjudicator to determine his own jurisdiction. The argument failed; the complaining party

---

[1] *Beck Peppiatt Ltd* v. *Norwest Holst Construction Ltd* [2003] BLR 316.
[2] *Orange EBS Ltd* v. *ABB Ltd* [2003] BLR 323.
[3] *Halki Shipping Corporation* v. *Sopex Oils Ltd* [1998] 1 WLR 726.
[4] *Sindall Ltd* v. *Solland and Others* (15 June 2001).
[5] *Cowling Construction Ltd* v. *CFW Architects* (15 November 2002).
[6] *Pegram Shopfitters Ltd* v. *Tally Weijl (UK) Ltd* [2003] EWCA Civ 1750, CA.
[7] *Smith* v. *Martin* [1925] 1 KB 745.
[8] *Christopher Brown* v. *Oesterreichischer Waldesbesitzer* [1954] 1 QB 8.
[9] *The Project Consultancy Group* v. *The Trustees of The Gray Trust* (16 July 1999).

had successfully reserved its position on jurisdiction. The distinction made in this case with the *Christopher Brown* case was as follows:

'7. Ms Rawley draws an analogy between the position of an adjudicator and that of an arbitrator as it was at common law before section 30 of the Arbitration Act 1996 came into force. At common law, an arbitrator was able to inquire into his jurisdiction in order to determine what course of action to follow, but the result of his inquiry could have no effect on the rights of the parties. She draws my attention to *Christopher Brown Ltd* v. *Genossenschaft Oesterreichischer* [1954] 1 QB 8, 12-13. I do not find this analogy helpful. The question in the present case is one of statutory interpretation: what does "decision" in section 108(3) mean? I do not see how the common law position of arbitrators in relation to their own jurisdiction can shed any light on that. In any event, it is to be noted that Devlin J said that the result of an arbitrator's inquiry as to his own jurisdiction "has no effect whatsoever upon the rights of the parties". A decision by an adjudicator does have an effect on the rights of the parties in the sense that, if an adjudicator decides to make an award, the paying party is obliged to pay up at once, since the decision is binding until the dispute is finally resolved by one means or another.'

In our view the best guidance that currently exists on whether or not an adjudicator can decide his own jurisdiction is found in *Fastrack* v. *Morrison*[10]:

'31. If a party challenges the entire jurisdiction of the adjudicator, as Morrison does, it has four options. Firstly, it can agree to widen the jurisdiction of the adjudicator so as to refer the dispute as to the adjudicator's jurisdiction to the same adjudicator. If the referring party agrees to that course, and the appointed adjudicator accepts the reference to him of this second dispute, the jurisdiction of the adjudicator could then be resolved as part of the reference. The challenging party could, secondly, refer the dispute as to jurisdiction to a second adjudicator. This would not put a halt to the first adjudication, if that had already led to an appointment, since the adjudicator has a statutory duty, unless both parties agree otherwise, to decide the reference in a very short timescale. The challenging party could, thirdly, seek a declaration from the court that the proposed adjudication lacked jurisdiction. This option is of little utility unless the adjudicator has yet to be appointed or the parties agree to put the adjudication into abeyance pending the relatively speedy determination of the jurisdiction question by the court. The Technology and Construction Court can, for example, resolve questions of that kind within days of them being referred to it. Fourthly, the challenging party could reserve its position, participate in the adjudication and then challenge any attempt to enforce the adjudicator's decision on jurisdictional grounds. That is the course adopted by Morrison.

32. The adjudicator can, of course, investigate any partial or entire jurisdictional challenge. He could, if he was satisfied it was a good one, decline to adjudicate on the part of the reference he regarded as lacking jurisdiction. Alternatively, he could decide that the challenge was a bad one and proceed with the substance of the adjudication. That is what happened in this adjudication. However, unless the parties have vested the jurisdictional dispute in the hands of the adjudicator in addition to the underlying dispute, the adjudicator cannot determine his own jurisdiction and the challenging party may seek to avoid enforcement proceedings by showing that the sum claimed was decided upon without jurisdiction. The court would give appropriate weight to any findings of fact relevant to that jurisdictional challenge but would not be bound by them and would either have to bear out the challenge with evidence or, if that was not necessary, determine the challenge and either enforce or decline to enforce the whole or part of the adjudicator's decision depending on the decision reached as to jurisdiction. The role of the court in a jurisdictional challenge summarised here is supported by recent decisions in the Technology and Construction Court, particularly the decision of my own in *Sherwood & Casson Ltd* v. *Mackenzie*, unreported, 30 November 1999; *The Project Consultancy Group* v.

---

[10] *Fastrack Contractors Limited* v. *(1) Morrison Construction Limited (2) Imreglio UK Limited* (4 January 2000).

The Trustees of the Gray Trust, unreported, 16 July 1999, Dyson J; and dicta in *Macob Civil Engineering Ltd* v. *Morrison Construction Ltd* [1999] Building Law Reports 93, Dyson J.'

There are interesting points in *Christiani & Neilsen* v. *The Lowry Centre*[11] concerning ad hoc agreements to decide jurisdiction and jurisdiction to decide a particular dispute referred:

> **'The First Issue – Did the parties agree that the adjudicator could determine his own jurisdiction?**
> It is trite law that the adjudicator had no jurisdiction to decide whether he had jurisdiction to act as an adjudicator under the scheme provided for by the HGCRA. This limitation was one which the adjudicator clearly accepted that he was subject to. However, the parties to an adjudication can always agree to vest in the adjudicator ad hoc jurisdiction to determine his own jurisdiction. Thus, the parties could have agreed to vest the adjudicator with the power to decide whether or not the relevant contract under which the dispute arose was entered into before 1 May 1998. What the status of such a decision would have been, and whether or not it could be challenged on the ground that it disclosed an error of law can only be decided following a consideration of the express and implied terms of the agreement to confer such ad hoc jurisdiction. I must therefore first determine whether such an agreement was entered into by the parties.
> It has to be borne in mind when considering whether the parties did reach such an agreement that an adjudicator, faced with a challenge to his own jurisdiction, has a choice as to how to proceed. The adjudicator has three options:
>
> 1. He can ignore the challenge and proceed as if he had jurisdiction, leaving it to the court to determine that question if and when his decision is the subject of enforcement proceedings.
> 2. Alternatively, the adjudicator can investigate the question of his own jurisdiction and can reach his own conclusion as to it. If he was to conclude that he had jurisdiction, he could then proceed to decide the dispute that had been referred to him. That decision on the merits could then be challengeable by the aggrieved party on the grounds that it was made without jurisdiction if the adjudicator's decision on the merits was the subject of enforcement proceedings.
> 3. Having investigated the question, the adjudicator might conclude that he had no jurisdiction. The adjudicator would then decline to act further and the disappointed party could test that conclusion by seeking from the court a speedy trial to determine its right to an adjudication and the validity of the appointment of the adjudicator.
>
> It is clearly prudent, indeed desirable, for an adjudicator faced with a jurisdictional challenge which is not a frivolous one to investigate his own jurisdiction and to reach his own non-binding conclusion as to that challenge. An adjudicator would find it hard to comply with the statutory duty of impartiality if he or she ignored such a challenge. Thus, given that the adjudicator in this case was clearly conscious of, and conscientiously seeking to comply with, his duty to act impartially, I have to consider whether the procedure adopted by him prior to his reaching his conclusion as to his own jurisdiction was one that followed an agreement to confer on him ad hoc jurisdiction to determine his own jurisdiction or was one which followed his adoption of an impartial procedure to assist him in reaching his own non-binding conclusion as to that jurisdiction.'

The important point here is that the adjudicator can make his own non-binding conclusion on jurisdiction. There is little point in framing such a conclusion as 'forming a view'; it is a decision which does not bind the parties and can be dealt with at enforcement. This decision does not give the adjudicator jurisdiction per se; it merely allows him to proceed or resign.

Where the parties give the adjudicator ad hoc jurisdiction to decide his own jurisdiction they will be bound by the decision he makes (*Whiteways* v. *Impresa Castelli*[12]). The jurisdiction was given here in the form of a letter and written submissions by the parties on jurisdiction.

---

[11] *Christiani & Neilsen Limited* v. *The Lowry Centre Development Centre Limited* (16 June 2000).
[12] *Whiteways Contractors (Sussex) Limited* v. *Impresa Castelli Construction UK Limited* (9 August 2000).

There is one area where there is express authority under the Scheme for the adjudicator to make a binding decision on jurisdiction. This is contained in paragraph 9(2) where the adjudicator can decide to resign when the dispute is substantially the same as one which has been previously decided (see *Sherwood & Casson* v. *Mackenzie*[13]).

Independence

An adjudicator may be acting beyond his jurisdiction if he is not independent of the parties or the subject matter of the dispute, so it is worth examining this concept here.

There is no general requirement that the adjudicator is independent. Independence in this sense means having no connection with either party or the subject matter of the dispute. The Act does not require the adjudicator to be independent. The Scheme, in paragraph 4, has a requirement that goes some way towards a requirement of independence. The adjudicator has to be a natural person acting in his personal capacity. Under the Scheme the adjudicator cannot be an employee of any of the parties to the dispute. If he has any interest, financial or otherwise, in any matter relating to the dispute he is not eligible to act as adjudicator. This would bar a firm of consultants who are acting for the employer either on the project in question or on another project.

It may be a requirement of the contract that the adjudicator is independent. Forms GC/Works/1, 2, 3 and 4 have this requirement. Such a term, although it goes beyond the requirements of the Act, is nevertheless enforceable and would go to the adjudicator's jurisdiction to act.

If there is no requirement for independence, as we have discussed in Chapter 7, there is nothing to prevent the employer naming his architect as adjudicator, or the main contractor appointing a commercial director as adjudicator in a sub-contract. Such a prospective appointee would have to consider the rules of conduct of his professional institution which may have a bearing on such a matter. The real problem with lack of independence is that it may encroach on impartiality as judged by the innocent bystander. If an adjudicator is perceived not to be impartial in this way he will not be complying with a basic requirement of the Act and thus will fail in this aspect of the jurisdictional requirements of an adjudicator.

Impartiality, natural justice, fairness and bias

The adjudicator has a duty to act impartially. This is founded in the Act. All of the contracts contain this duty. The Scheme in paragraph 12 requires that adjudicators shall act impartially. To act with impartiality or to be impartial is no more than to treat the parties equally. This does not mean that each is entitled to an equal amount of time. It is simply a matter of balance and the quality rather than some distribution of a timetable. In the way in which he conducts his inquiry and concludes with a decision the adjudicator must constantly remain impartial. Lack of impartiality manifests itself in bias. The adjudicator must be unbiased both in terms of showing no actual bias and there being no impugned bias. Impugned bias is to be seen to be biased whether you are actually biased or not. While impartiality may seem simple to define, it is difficult to exercise in practice. If a party does not feel the sense of being treated equally it will probably seek to attack the adjudicator or the decision on grounds of bias.

---

[13] *Sherwood & Casson Limited* v. *Mackenzie* (30 November 1999).

Does the duty to act impartially include the rules of natural justice or fairness? The drafting of the Act took certain provisions from the Arbitration Act 1996. Section 33(1)(a) of the Arbitration Act 1996 states: 'act fairly and impartially as between the parties, giving each party a reasonable opportunity to appoint his case and dealing with that of his opponent'. Fairly and impartially are not synonymous. There are those who would argue that a procedure to resolve disputes which is limited to a 28-day duration is inherently unfair. It is also argued that the failure to repeat the word 'fairly' from the phrase in the Arbitration Act 1996 means that the adjudicator need not be fair. It follows that if the adjudicator can act in an unfair way this supports the premise that the whole process is unfair. It may be reasonable to seek to imply a term of fairness, but it is not necessary to do so. The system of adjudication can work without such an implied term. To act fairly is to be just and unbiased and to operate in accordance with the rules. There is a link here with impartiality in that the adjudicator must behave in an unbiased way. It must follow that an adjudicator who acts impartially will also act fairly. The system with the restraint of a 28-day process, may give an impression that this will lead to unfairness. The adjudicator should always act in an even-handed manner insofar as the time and the conduct of the parties permit.

Do the rules of natural justice apply to adjudication? Adjudication is a judicial process. The adjudicator is applying the facts and the law to reach a decision on the rights of the parties to the contract. It also permits an inquisitorial approach rather than the traditional adversarial approach to resolving disputes. The term 'procedural fairness' has been coined to describe the way in which an adjudicator should act and this can be summarised as follows:

> the adjudicator must be, and must be seen to be, disinterested and unbiased; every party to the dispute must be given a reasonable opportunity to present its case and to answer the case of its opponent.

It would be difficult to operate outside of these rules and comply with the express duty to act impartially. However, in a procedure of such short duration, the rules of natural justice may prove to be particularly difficult to comply with. They will also be read having regard to the constraints of the timetable. Giving each party a fair opportunity to present its case and to deal with that of its opponent does not mean that inordinate time should be required or allowed. Providing the adjudicator maintains balance between the parties and tailors his inquiry to that properly necessary to reach a decision, he should not fall foul of the rules of natural justice.

There have been a number of cases where the courts have examined allegations of lack of impartiality by adjudicators or procedural unfairness during the course of the adjudication.

The basic principle was set in *Macob* v. *Morrison*, which we have already mentioned on a number of occasions. An adjudicator's decision is enforceable summarily in the courts whether or not there has been procedural irregularity, error or breach of natural justice.

This concept has been considerably eroded by subsequent decisions of the court, a number of which we set out briefly below.

In *Glencot* v. *Ben Barrett*[14] it was confirmed that the adjudicator has to conduct the proceedings in accordance with the rules of natural justice or as fairly as the limitations imposed by Parliament permit.

An adjudicator may, due to the time restraints, speak to a party separately but he should

---

[14] *Glencot Development & Design* v. *Ben Barrett & Son (Contractors) Ltd* [2001] BLR 207.

inform the other party himself of the matters discussed. It is not appropriate to rely on the party he has spoken to informing the other party[15].

The adjudicator may not take evidence from third parties and not give the parties the opportunity to comment on it, and it is inappropriate for an adjudicator to give a witness statement for use in enforcement proceedings as this casts doubts on his impartiality[16].

In circumstances where the adjudicator was common in a number of adjudications between different parties on the same project, there where undoubtedly problems of obtaining information in one adjudication and its use in another which would give rise to concerns in regard to breaches of natural justice. In such circumstances and in the face of an objection by one of the parties, an adjudicator should withdraw from the appointment[17].

Where an adjudicator reaches his decision in part on reliance upon documents which the parties have not seen, there is breach of natural justice. Natural justice requires that the parties have an opportunity to know the case against them and an adjudicator should give them opportunity to comment on any material from whatever source including from his own knowledge and experience, if such information is to be given any significance in his decision[18].

An adjudicator has been found to be acting properly in ascertaining points of law and deciding them without reference to the parties even if the point has not been raised by either of them. While coming to this conclusion the court expresses some concern at the adjudicator's action[19].

The provisions of the Human Rights Act (HRA) have a bearing on questions of fairness. This Act has only been raised in two cases that we are aware of, *Elenay* v. *The Vestry*[20] and *Austin Hall* v. *Buckland Securities*[21]. In the former it was decided that the HRA did not apply to adjudication as it is not finally determinative of the parties' rights. This conclusion was accepted in the latter case in which it was also decided that an adjudicator is not a public authority and that adjudication does not contravene sections 6 or 7 (the right to a public hearing) of the HRA.

The Act

While the origin of all three aspects is to be found in the Act, the Act itself does not actually grant any jurisdiction, powers or duties at all. What it does provide, however, are the basic parameters and the things that have to be in the construction contract as a minimum for compliance with the Act. Adjudication and the adjudicator are not particularly well defined under the Act.

The right given in section 108(1) is the 'right to refer a dispute arising under the contract for adjudication under a procedure complying with this section'. This sets the boundaries for jurisdiction. It only deals with disputes arising under the contract (unless the parties agree in their contract to widen this). The procedure must comply with section 108. It is only parties

---

[15] *Discain Project Services Ltd* v. *Opecprime Developments Ltd* [2000] 8 BLR 402.
[16] *Woods Hardwick Limited* v. *Chiltern Air Conditioning Limited* [2001] BLR 23.
[17] *Pring & St Hill Limited* v. *C.J. Hafner T/A Southern Erectors*, TCC, 31 July 2002.
[18] *RSL (South West) Limited* v. *Stansell Limited* (16 June 2003).
[19] *Karl Construction (Scotland) Limited* v. *Sweeney Civil Engineering (Scotland) Limited* (21 December 2000) – Judicial Review 2002.
[20] *Elanay Contracts Ltd* v. *The Vestry* [2000] BLR 33.
[21] *Austin Hall Building Limited* v. *Buckland Securities Limited* [2000] BLR 272.

to a construction contract who have this right given to them by statute. If it is not a construction contract there is no statutory right, but there may be a contractual one and will be if the parties have included an adjudication provision in their contract. The definition of a construction contract contained in the provisions of sections 104 to 107 inclusive prescribes the boundaries of what constitutes a construction contract. These boundaries are further limited by the Exclusion Order. We have examined these provisions in earlier chapters. There is nothing to prevent the parties to the contract making provisions for adjudication that apply to work that goes beyond the definition of a construction contract in the Act. Nevertheless, the definition of construction contracts in the Act is complex and the provisions in the Exclusion Order further complicate this. This has already proved to be a ripe source of jurisdictional arguments for reluctant parties, as is evidenced by the nature of the arguments put up in court proceedings where enforcement has been resisted. If the contract that the parties enter into provides that all disputes that arise under the contract are subject to the right to adjudication (it may go further and include disputes that are in connection with the contract), the adjudicator's jurisdiction would not be fettered by the definition of construction contracts in the Act and it would be extended by the agreement of the parties.

As far as the adjudicator is concerned, he is only there to reach a decision on a dispute that has been referred to him under the contract. The Act does not give the adjudicator jurisdiction to deal with disputes that do not arise *under* the contract. He would not have jurisdiction to deal with disputes that do not fall within the various definitions in the Act of a construction contract or a contract in writing. Nor does he have the jurisdiction to operate under a procedure that is outside of the requirements of the Act. This in plain terms gives some boundaries that help to define jurisdiction.

Jurisdiction, powers and duties which give the adjudicator greater scope than the Act and do not conflict with it, are enforceable under the law of contract. An adjudication clause that permits the adjudicator to decide disputes which arise both under the contract and in connection with it, is a compliant clause. It provides the minimum required by the Act and goes beyond it. Parties cannot rely on the Act to restrict the provisions they have made in their contracts. Its provisions will bind them once the contract is made between the parties. It cannot be argued afterwards that the adjudication clause should be taken back to the minimum required by the Act. Such attempts to avoid a decision should not receive the support of the courts. It is incumbent upon the adjudicator to comply with the parties' agreement and he would not be doing his job properly if he were to ignore procedures that have been agreed between the parties.

An area of possible difficulty arises when the parties have failed to include the requisite provisions in their contract to comply with the Act, but they have included additional provisions that do not conflict with the Act. An example would be an agreement setting out a provision for a response by the responding party 14 days after the referral within a contract that does not comply with the Act. As the contract is non-compliant, the Scheme applies to the adjudication. Paragraph 13 of the Scheme however gives the adjudicator the right to decide the procedure for the adjudication. Does the adjudicator have jurisdiction to set his own procedure and ignore the agreement by allowing a more sensible period of seven days for the response or is he bound by the agreement of the parties to a 14-day period? Some might suggest that the adjudicator would be within his rights if he were to apply Paragraph 13 of the Scheme and set his own procedure. Others might say that the agreement of the parties is paramount, it does not conflict with the Act, albeit that it creates difficulties for the adjudicator in allowing the referring party a response and holding a meeting if one proves necessary within the remaining 14 days, and the adjudicator is bound by the parties' agreement to 14 days for the response.

Something that would perhaps go directly to the adjudicator's jurisdiction in respect of the matters referred would be an agreement that disputes arising in connection with the contract can be adjudicated in an otherwise non-compliant contract. Does the adjudicator deal with any issues that are clearly in connection with rather than under the contract? Whatever the adjudicator chooses to do, it is absolutely vital that he makes it quite clear in his directions at the earliest possible time what he intends. Difficulties are likely to result if the adjudicator decides to set seven days for the response and does not inform the responding party of that immediately on receipt of the referral, and the responding party is working towards the delivery of his response 14 days after the referral. Similarly, difficulties can occur if the adjuducator intends to deal with matters arising in connection with the contract, and the responding party makes no submissions in respect of them.

Section 106 of the Act further restricts jurisdiction. The statutory right to adjudication is excluded from contracts that concern works for residential occupiers. The case of a development which was only partly for the occupation on a residential basis by one of the parties was considered in *Thomas* v. *Bick & Bick*[22] where it was found that the proportion that was for the occupation of one of the parties, even though it exceeded 50%, was insufficient to come within the ambit of section 106, and the adjudicator's decision was enforced.

One point of note is that if a residential occupier enters into a contract with a contractor under an unamended JCT Minor Works Form, this contains an adjudication clause. The provisions of this clause will bind the parties and either party will have the right to have any dispute arising under the contract dealt with by adjudication.

This matter was explored fully in *Lovell* v. *Legg and Carver*[23]. This was a JCT Minor Works Form between a residential occupier and a builder. The residential occupier sought to argue in the court that the provisions on adjudication were not applicable under the Unfair Terms in Consumer Contracts Regulations 1999. In this case, rather than argue that the residential occupier was exempt from the adjudication provisions in the contract, entire reliance was placed on the Unfair Terms in Consumer Contracts Regulations 1999. The judge rejected this argument on the basis that nothing in the contract offended the legislation:

> 'To be unfair the terms must cause a significant imbalance in the parties' rights and obligations under the contract to the detriment of the consumer. Any imbalance will not do: it must be a significant imbalance. Moreover that significant imbalance must be caused by the adjudication provisions contrary to the requirement of good faith. In my view neither requirement is satisfied in the present case.'

The fact that the parties are bound by a contract that contains adjudication provisions was also explored in *Mohammed* v. *Bowles*[24]. This contract contained adjudication provisions that complied with the Act. Here the residential occupier took adjudication proceedings against the builder. He then sought to enforce the decision by use of statutory demand. The builder argued that contracts with residential occupiers were exempt and therefore the adjudicator had no jurisdiction. The judge rejected this argument:

> '29. On the basis of the evidence before me I am satisfied that the parties entered into a contract for residential construction works that included a form of dispute resolution which adopted the framework of the dispute resolution procedure contained in the 1996 Act. For present purposes I do not have to consider whether the exchange of letters in October 2001 created a new contract, one which

---

[22] *Samuel Thomas Construction* v. *Bick & Bick (aka J&B Developments)* (28 January 2000).
[23] *Lovell Projects Ltd* v. *Legg and Carver* (July 2003).
[24] *Jamil Mohammed* v. *Dr Michael Bowles* (11 March 2003).

replaced the existing contractual arrangements, although I comment that it seems doubtful to me that they did. The adjudicator has already determined this issue and it is not for this court to look behind the adjudicator's decision. If the applicant is unhappy with the adjudicator's determination upon the question of jurisdiction then his remedy is to apply to the court to have that decision set aside on the basis that he disputes the adjudicator's jurisdiction and/or to seek a declaration on the question of jurisdiction. I note that to date he has not chosen to adopt that course.'

There is one other case where a contract with a residential occupier was explored by the court; this was in *Picardi* v. *Cuniberti*[25]. In this case, the adjudicator's decision was not enforced. The distinction between this contract and *Lovell* v. *Legg and Carver* was that in the latter the contract had been insisted upon by the defendants in the enforcement proceedings, who were the employers under the contract and had been legally advised at the time the contract was drawn up. In *Picardi* v. *Cuniberti* it was found that specific clauses of the contract, including adjudication, had not been drawn to the residential occupiers' attention and that the adjudication, and other, provisions were not applicable.

Section 107 deals with agreements that are in writing. The definition of writing in section 107 is necessarily wider than would normally constitute writing. The definition is taken from the Arbitration Act 1996, which was designed to avoid the interminable arguments as to whether or not the arbitration clause was incorporated in the contract. What this does is restrict the instances in which it would be arguable that the agreement is not in writing. Where an agreement is not in writing under the definition in section 107, an adjudicator would not have jurisdiction to deal with any dispute.

The requirements of section 107 were examined by the Court of Appeal in *RJT* v. *DM Engineering*[26] where it was found that all the terms, and not merely the existence of a construction contract, had to be evidenced in writing if the contract was to be capable of being referred to adjudication. Leave to appeal to the House of Lords was refused on 10 October 2002.

In the context of a 'back of fag packet' agreement, which is not all that unusual in the construction industry, there is some difficulty caused by this decision. Even the briefest of agreements will generally have some record in writing of the amount that one party is to pay the other or, in the case of a daywork contract, how it is to be calculated. If there are no payment provisions the critical payment terms are imported from the Scheme. Other than a time for completion it is difficult to see the problem, but there it is. The result is that a large sector of the construction industry, which is one that perhaps needs adjudication more than most, is thus probably prevented from using it. A case perhaps for adjudicators to be bold in deciding whether there is a contract in writing?

There was a further examination of contracts in writing by the Court of Appeal in *Thomas-Fredric's* v. *Wilson*[27] but in this case the issue related to the name of one of the contracting parties.

Section 108 of the Act also has a bearing on the adjudicator's jurisdiction. In terms of the timetable, once the adjudicator is appointed, he is required to reach a decision within 28 days of the referral. He will have neither the jurisdiction nor the powers to operate outside that 28-day period without the agreement of the parties. However, the adjudicator's jurisdiction can be extended by the contract allowing the period of 28 days to be increased by up to 14 days with the consent of the referring party at the time. The adjudicator has a duty

---

[25] *Picardi (t/a Picardi Architects)* v. *Mr & Mrs Cuniberti* (19 December 2002).
[26] *RJT Consulting Engineers Ltd* v. *DM Engineering (Northern Ireland) Ltd*, CA (8 March 2002).
[27] *Thomas-Fredric's (Construction) Ltd* v. *Wilson* [2003] EWCA Civ 1367, CA.

to act impartially. This is both a duty and concerns jurisdiction. It is a duty because it is spelled out as a duty in section 108. It concerns jurisdiction because the adjudicator who does not act impartially is acting not in accordance with his jurisdiction and this may lead to any decision that he makes being void. In addition, the adjudicator may take the initiative in ascertaining the facts and the law. This is clearly a power that the adjudicator has from the Act but there is no compulsion that he exercises it.

While the Act may help with the understanding of the jurisdiction, powers and duties of the adjudicator, it is not of itself the real basis of these matters. The Act requires that all of the points in section 108 are incorporated in the construction contract. It will therefore be the construction contract itself that provides the basis of the adjudicator's jurisdiction, powers and duties.

One other aspect of jurisdiction that is important relates to what has been described by the court in *Carter* v. *Nuttall*[28], an action for an injunction to prevent an adjudication from proceeding, as 'threshold' and 'internal' jurisdiction. The court identified that threshold jurisdictional questions relate to the ability to set in train an adjudication process at all, whereas internal jurisdictional matters relate to the extent of the matters with which a properly appointed adjudicator can deal. For setting the adjudication in 'train' see the commentary in Chapter 3. Internal jurisdiction deals with the tasks the adjudicator has to carry out to reach a decision on the dispute or difference identified in the notice of adjudication.

Subsidiary questions

Making the decision can be characterised as answering a question. This is clear from the statement in *Nikko* v. *MEPC*[29]:

'If he has answered the right question in the wrong way, his decision will be binding. If he has answered the wrong question, his decision will be a nullity.'

By its nature a subsidiary question must go to the main point or main question. The subsidiary question must be something which is essential to answer in order to answer the main question. It must be an integral part of the decision to be made.

This was first explored in a Scottish case where a claim for payment was inextricably linked to the terms of the contract on when the payment was due. This was the case of *Karl* v. *Sweeney*[30]:

'The adjudicator rightly or wrongly concludes that the sub-contract does not provide an adequate mechanism for deciding when instalment monthly payments become due. In the circumstances the adjudicator's position seems to be that she could not answer the central issue in the referral which was when in terms of the sub-contract provisions the relevant instalment becomes due and payable. She could not give effect to one or other of the parties' respective contentions because the sub-contract did not permit her to do this. Accordingly she would require in her decision to make a finding explaining why she was not going along one or other of the paths suggested to her by the parties. Such a finding would be an integral part of her decision on the reference. If the adjudicator rejects a suggestion that the sub-contract contains provisions about when payments become due,

---

[28] *R.G. Carter Limited* v. *Edmund Nuttall Limited* (21 June 2000).
[29] *Nikko Hotels (UK) Ltd* v. *MEPC Plc* [1991] 2 EGLR 103, at p. 108B.
[30] *Karl Construction (Scotland) Limited* v. *Sweeney Civil Engineering (Scotland) Limited* (21 December 2000).

then because of the implications of this on her resolution of the referral it is difficult to say that the question of whether or not the sub-contract contains an adequate provision for governing when payment of instalments is due is not part of the dispute which has been referred and that the matter constitutes a separate and independent dispute requiring a different referral. Sweeney's claim for payment in terms of the Application for Payment No 5 is inextricably linked to the construction of the Sub-Contract. If the construction and application of the Sub-Contract payment provisions required (even in part) separate adjudication, it is easy to see what a tangle could result if a different adjudicator reached a different conclusion. We should find the very protracted wrangling which the summary remedy for provisional decisions by an adjudicator is designed to avoid. As it happens I do not find the reasons for the adjudicator's finding about the inadequacy of the payment provisions in the contract to be clearly set out.'

In *Joinery Plus* v. *Laing*[31] the courts looked at a necessary step along the route:

'61. This decision is, in effect, that the construction of contractual terms, however erroneous, gives rise to a question of law within jurisdiction if that issue of construction arises as a necessary step along the route that the adjudicator must travel in order to determine the question that has been referred to adjudication. The clause in question formed part of the contract but its dismembered state gave rise to a question of construction as to what if any effect should be given to it. Thus, the decision as to that issue was one arising out of the contract and the underlying dispute that had been referred to adjudication. It must be said that this decision, which of course I must give effect to, goes near to the limits of errors of law within jurisdiction.'

The Scheme

If the construction contract does not meet the requirements of section 108(1) to (4) of the Act then the Scheme applies. We consider the machinery of how this is done elsewhere. In contrast to the Act, which has very minimal criteria, the Scheme is orientated towards providing 'nuts and bolts' procedures from appointment to the decision and the effect of the decision. In this event the Scheme, Part 1 of which deals with adjudication, becomes one of the bases of the adjudicator's jurisdiction, powers and duties.

The first likely question as to the adjudicator's jurisdiction is the validity of his appointment. The timetables and procedures for securing appointment are both complex and written in strict language. This is partly because there are a number of permutations from which appointments can occur. The Scheme deals with these in paragraphs 2 to 6 inclusive. The language concerning the number of days involved in the appointment process is strict. The words 'must' and 'shall' are used frequently. This could be interpreted that if the strict timetable is not met, the appointment may not be valid. Fortunately this does not appear to be creating difficulties in practice and the absence of any sanction in connection with the use of the word 'shall' in paragraph 7(2) coupled with the wording of the Act that there need only be an intention that appointment is achieved within seven days of the notice of adjudication, reinforces this. It is a fact that many referral notices are issued after the expiry of the seven-day period and also that adjudicators proceed to make their decisions in such circumstances, and it is relevant that, to the authors' knowledge, in the 170 or so enforcement proceedings in respect of which the judgments have been made available, not one has been refused enforcement because of a late referral.

Paragraph 1 of Part 1 of the Scheme sets out the requirements for a notice of adjudication

[31] *Joinery Plus Limited (in administration)* v. *Laing Limited* (15 January 2003).

and thus how, under the Scheme, the adjudicator's jurisdiction is delimited. Paragraph 1(1) follows the Act in terms of disputes arising under the contract. The notice required (the notice of adjudication) narrows the jurisdiction. This was dealt with in *Whiteways* v. *Impresa Castelli*[32] where it was held that the adjudicator should not look outside the four corners of the claim when dealing with abatement unless it has been mentioned in a notice of with-holding. The jurisdiction is therefore to deal with the disputes in the notice of adjudication. The notice must comply with the requirements of paragraph 1(2) and (3)(a) to (d) inclusive otherwise it will not constitute a valid notice. The identity of the dispute, (3)(b), and the redress sought, (3)(c), narrows the matters the adjudicator can deal with.

Paragraph 7 deals with the referral notice. Nowhere does this paragraph give the referring party the right to refer a different dispute in the 'referral notice' to that in the 'notice of adjudication'. This matter is not assisted by paragraph 9(4), which permits the adjudicator's resignation, without loss of fees, where he finds he is no longer competent to deal with the dispute due to the dispute varying significantly from that referred to him. It is difficult to see where an adjudicator would have jurisdiction to decide a dispute other than the one referred to him in the absence of consent of the parties (*The Massalia*[33] and *The Kostas Melas*[34]).

An adjudicator will not have jurisdiction to make a decision where the dispute is the same or substantially the same as one which has previously been referred to adjudication, and a decision has been taken in that adjudication (paragraph 9(2) of the Scheme).

An adjudicator will not have jurisdiction to continue where both parties have revoked the appointment under paragraph 11. He will still have jurisdiction to determine his fees.

Paragraphs 12 to 19 inclusive are headed 'Powers of the adjudicator'. This is actually a mixture of jurisdictional points, powers and duties. Paragraph 12 covers the duties of acting impartially and avoiding unnecessary expense. The most important point in paragraph 13 is that the 'adjudicator may take the initiative in ascertaining the facts and the law necessary to determine the dispute'. This gives powers to conduct an inquiry rather than adopting an adversarial approach. The remainder of the paragraph amounts to no more than a shopping list. Paragraph 14 imposes a duty on the parties to comply with any request or direction of the adjudicator. Paragraph 15 gives the powers of sanction against parties who do not comply with any request or direction of the adjudicator.

Paragraph 20 gives the option to increase the scope of the adjudication to take into account any other matters that the parties agree should be within the scope of the adjudication and even allows the adjudicator, of his own volition, to take into account matters under the contract which he considers are necessarily connected with the dispute. The adjudicator should always be mindful of *Balfour Beatty* v. *Lambeth*[35] and *RSL* v. *Stansell*[36] and ensure that he acquaints the parties of his intentions in such circumstances. He is, by paragraph 20(a), given the particular power to open up, revise and review any decision taken or any certi-ficate given by any person referred to in the contract. If the decision or certificate is stated to be final and conclusive, the adjudicator will have no powers to examine the contents at all.

The remaining paragraphs of Part I deal with the effects of the decision. The parties have a basic duty to comply with the decision and there may be some assistance if enforcement is necessary by making the decision peremptory, but this device is generally not seen as being necessary since the decision in *Macob*.

---

[32] *Whiteways Contractors (Sussex) Limited* v. *Impresa Castelli Construction UK Limited* (9 August 2000).

[33] *Societe Franco-Tunisienne D'armement-Tunis* v. *The Government Of Ceylon* [1959] 2 Lloyd's Rep 1.

[34] *SL Sethia Liners Ltd* v. *Naviagro Maritime Corporation* [1981] 1 Lloyd's Rep 18.

[35] *Balfour Beatty Construction Ltd* v. *The Mayor & Burgesses of the London Borough of Lambeth* (12 April 2002).

[36] *RSL (South West) Ltd* v. *Stansell Ltd* (16 June 2003).

The Standard Forms of Contract

Joint Contracts Tribunal Main Contract Form

The JCT has adopted the compliance points of the Act as the basis of its adjudication clause. Nevertheless, this is a lengthy clause, also adopting some parts of the Scheme in respect of the adjudicator's powers. We hear a rumour at the time of writing that the JCT may be considering reverting to the Scheme as the 'book of rules' for adjudication, but that is for the future. The recently published Major Projects Form adopts the Scheme as its adjudication provisions.

   Clause 41A deals with settlement of disputes using adjudication. The first part of the clause deals with appointment, which we have looked at in Chapter 7. The section examined here, with relevant parts set out in the following extracts, deals with the referral of the dispute through to the decision.

Dispute or difference – notice of intention to refer to adjudication – referral

41A .4 .1 When pursuant to article 5 a Party requires a dispute or difference to be referred to adjudication then that Party shall give notice to the other Party of his intention to refer the dispute or difference, briefly identified in the notice, to adjudication. If an Adjudicator is agreed or appointed within 7 days of the notice then the Party giving the notice shall refer the dispute or difference to the Adjudicator ('the referral') within 7 days of the notice. If an Adjudicator is not agreed or appointed within 7 days of the notice the referral shall be made immediately on such agreement or appointment. The said Party shall include with that referral particulars of the dispute or difference together with a summary of the contentions on which he relies, a statement of the relief or remedy which is sought and any material he wishes the Adjudicator to consider. The referral and its accompanying documentation shall be copied simultaneously to the other Party.

41A .4 .2 The referral by a Party with its accompanying documentation to the Adjudicator and the copies thereof to be provided to the other Party shall be given by actual delivery or by FAX or by special delivery or recorded delivery. If given by FAX then, for record purposes, the referral and its accompanying documentation must forthwith be sent by first class post or given by actual delivery. If sent by special delivery or recorded delivery the referral and its accompanying documentation shall, subject to proof to the contrary, be deemed to have been received 48 hours after the date of posting subject to the exclusion of Sundays and any Public Holiday.

Conduct of the adjudication

41A .5 .1 The Adjudicator shall immediately upon receipt of the referral and its accompanying documentation confirm the date of that receipt to the Parties.

The referral of the dispute basically follows the Act. The notice of adjudication itself need only briefly identify the dispute. Notices of adjudication under this form can therefore be drafted on a general basis giving the adjudicator wide jurisdiction. There are one or two points in the referral that may give rise to argument concerning jurisdiction. The essentials of the referral for it to be valid are particulars of the dispute or difference, together with a summary of the contentions on which the referring party relies, a statement of the relief or remedy sought and any material the party wishes the adjudicator to consider. A copy of these materials must be sent to the other party.

Where contracts stipulate procedural matters, this usually gives rise to problems. If one of the parties does not follow the procedures stipulated in the contract, is there a breach? This may be merely a technical matter and have no other consequences, but it may be sufficient to suggest that there is no jurisdiction. In this case, if the notice of referral does not follow the requirements of the contract, it can be argued that there is no referral at all and therefore there is no jurisdiction. It is also important that the adjudicator confirms receipt of the documents from the referring party. Again, there is an argument here that if he does not do so there may be no jurisdiction.

> 41A .5 .2 The Party not making the referral may, by the same means stated in clause 41A.4.2, send to the Adjudicator within 7 days of the date of the referral, with a copy to the other Party, a written statement of the contentions on which he relies and any material he wishes the Adjudicator to consider.
>
> 41A .5 .3 The Adjudicator shall within 28 days of the referral under clause 41A.4.1 and acting as an Adjudicator for the purposes of S. 108 of the Housing Grants, Construction and Regeneration Act 1996 and not as an expert or an arbitrator reach his decision and forthwith send that decision in writing to the Parties. Provided that the Party who has made the referral may consent to allowing the Adjudicator to extend the period of 28 days by up to 14 days; and that by agreement between the Parties after the referral has been made a longer period than 28 days may be notified jointly by the Parties to the Adjudicator within which to reach his decision.

The responding party 'may' within seven days issue a written statement of the contentions on which he seeks to rely. It can be argued if the responding party submits these matters at a later date, that the written statement or any material on which the responding party wishes to rely is not admissible and therefore the adjudicator would not have the jurisdiction to deal with or consider it. The word used is however 'may', not 'shall' and while such arguments have been tried by referring parties, the adjudicator should treat them for the tactical ploys that they are and allow the submissions. In any event it is usual for a responding party, who is unacceptably restricted by this provision, to seek an extension of time before the expiry of the seven days. The adjudicator is master of procedure in any event under clause 41A.5.5 and can amend any timetable unless it is written in mandatory terms. As long as the referring party is not prejudiced as a result, or the adjudicator's ability to deal with the dispute, a short extension of time should be granted.

Probably the most important provision in the whole of this adjudication procedure is contained in clause 41A.5.3. This requires the adjudicator to act as an adjudicator for the purposes of section 108 of the Act and not as an expert or an arbitrator. One of the problems here is that the Act does not actually define what an adjudicator is or what acting as an adjudicator might be. There is also a further problem that arises from this point. Nowhere in this clause is there any restriction that requires that adjudication only applies to construction contracts as defined in the Act. On this basis in any contract entered into on the JCT form there will be the facility for taking disputes to adjudication regardless of the restrictions of the Act. This was the JCT's intention. Whether they have succeeded in achieving a provision that goes beyond the limitation of the definition in construction contracts is debatable. Section 108(1) provides that 'A party to a construction contract has the right to refer a dispute arising under the contract for adjudication under a procedure complying with this section'. Therefore if the adjudicator under clause 41A.5.3 is to act as an adjudicator for the purposes of section 108, how can the limitation of construction contract in section 108(1) be avoided?

This point has not however to the authors' knowledge, been raised in an adjudication or in enforcement proceedings. This is an instance of a contract that complies with the requirements of section 108 and then goes beyond them. It is both an Act compliant and contractual scheme for adjudication. As a matter of contract it should be enforceable.

> 41A .5 .4 The Adjudicator shall not be obliged to give reasons for his decision.

The provision in clause 41A.5.4 leaves it for the adjudicator to decide whether or not he will give reasons for his decision. This differs from the Scheme, where a request from either party does oblige the adjudicator to give reasons. Current thinking is that reasons should be given for a decision. Clause 41A.5.5 requires the adjudicator to act impartially and he may use his initiative in ascertaining the facts and the law. This follows both the Act and the Scheme. The important power is that the adjudicator sets his own procedure. The adjudicator is very much master of the proceedings and needs to be having regard to the timetable.

> 41A .5 .5 In reaching his decision the Adjudicator shall act impartially and set his own proce-
> dure; and at his absolute discretion may take the initiative in ascertaining the facts and
> the law as he considers necessary in respect of the referral which may include the
> following:
>
> .5 .1 using his own knowledge and/or experience;
> .5 .2 opening up, reviewing and revising any certificate, opinion, decision, requirement
> or notice issued, given or made under this Contract as if no such certificate, opi-
> nion, decision, requirement or notice had been issued, given or made;
> .5 .3 requiring from the Parties further information than that contained in the notice of
> referral and its accompanying documentation or in any written statement pro-
> vided by the Parties including the results of any tests that have been made or of
> any opening up;
> .5 .4 requiring the Parties to carry out tests or additional tests or to open up work or
> further open up work;
> .5 .5 visiting the site of the Works or any workshop where work is being or has been
> prepared for this Contract;
> .5 .6 obtaining such information as he considers necessary from any employee or
> representative of the Parties provided that before obtaining information from an
> employee of a Party he has given prior notice to that Party;
> .5 .7 obtaining from others such information and advice as he considers necessary on
> technical and on legal matters subject to giving prior notice to the Parties together
> with a statement or estimate of the cost involved;
> .5 .8 having regard to any term of this Contract relating to the payment of interest,
> deciding the circumstances in which or the period for which a simple rate of
> interest shall be paid.

The list of powers in clauses 41A.5.5.1 to 41A.5.5.8 inclusive is similar to that in paragraph 13 in the Scheme. The list is not exhaustive. Although it does not say so in clause 41A.5.5.7 any advice that the adjudicator receives should be revealed to the parties before the decision is made (*RSL* v. *Stansell*). Clause 41A.4.1 requires that the adjudicator and the parties execute the JCT Adjudication Agreement. This could be a source for challenging jurisdiction, where one of the parties fails to execute the agreement, if it were not for the provision in clause 41A.5.6. The adjudicator must execute the JCT Adjudication Agreement but if one or both of the parties fails to execute the Agreement this will have no effect on the decision. If the JCT Adjudication Agreement is to form the contract between the parties and the adjudicator, all this provision does is protect the decision. It does nothing to protect the relationship between the parties and the adjudicator and particularly the fees of the adjudicator. This clause also states 'or to comply with any requirement of the adjudicator under clause 41A.5.5'. This was explored by the courts in *Costain* v. *Wescol*[37] where it was held that these provisions were non-mandatory as far as the decision was concerned.

> 41A .5 .6 Any failure by either Party to enter into the JCT Adjudication Agreement or to comply with any requirement of the Adjudicator under clause 41A.5.5 or with any provision in or requirement under clause 41A shall not invalidate the decision of the Adjudicator.
>
> 41A .5 .7 The Parties shall meet their own costs of the adjudication except that the Adjudicator may direct as to who should pay the cost of any test or opening up if required pursuant to clause 41A.5.5.4.

The parties costs are dealt with in clause 41A.5.7. The parties shall bear their own costs, although there is nothing to prevent both parties agreeing that the adjudicator can deal with party costs and the adjudicator will have to do this under his obligation, once having accepted appointment as adjudicator, to adjudicate in accordance with the parties' agreement. The adjudicator may direct who bears the costs of any opening up or testing. The presumption here is that one of the parties will already have paid for this. There is no express mention of how any fees for obtaining technical or legal advice (covered by clause 41A.5.5.7) are to be paid. The JCT Adjudication Agreement is silent on this point but given that there must be a statement or estimate of the costs involved before the advice is obtained, the implication is that the parties will be liable for these costs.

> 41A .5 .8 Where any dispute or difference arises under clause 8.4.4 as to whether an instruction issued thereunder is reasonable in all the circumstances the following provisions shall apply:
>
> .8 .1 The Adjudicator to decide such dispute or difference shall (where practicable) be an individual with appropriate expertise and experience in the specialist area or discipline relevant to the instruction or issue in dispute.
>
> .8 .2 Where the Adjudicator does not have the appropriate expertise and experience referred to in clause 41A.5.8.1 above the Adjudicator shall appoint an independent expert with such relevant expertise and experience to advise and report in writing on whether or not any instruction issued under clause 8.4.4 is reasonable in all the circumstances.
>
> .8 .3 Where an expert has been appointed by the Adjudicator pursuant to clause 41A.5.8.2 above the Parties shall be jointly and severally responsible for the expert's fees and expenses but, in his decision, the Adjudicator shall direct as to who should pay the fees and expenses of such expert or the proportion in which such fees and expenses shall be shared between the Parties.
>
> .8 .4 Notwithstanding the provisions of clause 41A.5.4 above, where an independent expert has been appointed by the Adjudicator pursuant to clause 41A.5.8.2 above, copies of the Adjudicator's instructions to the expert and any written advice or reports received shall be supplied to the parties as soon as practicable.

Clause 41A.5.8 was not in the early amendment to the contract. This specifically deals with the provisions of clause 8.4.4 of the contract and whether or not any instruction issued under that clause is reasonable. This covers the powers of the architect when the work is not in accordance with the contract. The points to note are that these provisions require an adjudicator with expertise and experience in the specialist area or discipline relevant to the instruction.

---

[37] *Costain Ltd* v. *Wescol Steel Ltd* (24 January 2003).

Where this is not possible or practicable, the adjudicator is compelled to obtain specialist advice via an independent expert and he must issue his instructions to the expert and the report to the parties.

The fees of the expert are to be met by the parties who are jointly and severally responsible save that the adjudicator in his decision shall direct who pays those fees or the proportional liability for those fees.

Adjudicator's fee and reasonable expenses – payment

41A .6 .1   The Adjudicator in his decision shall state how payment of his fee and reasonable expenses is to be apportioned as between the Parties. In default of such statement the Parties shall bear the cost of the Adjudicator's fee and reasonable expenses in equal proportions.

.2   The Parties shall be jointly and severally liable to the Adjudicator for his fee and for all expenses reasonably incurred by the Adjudicator pursuant to the adjudication.

Under clause 41A.6.1 the adjudicator should state how his fees should be apportioned between the parties. Where he fails to make such a statement the fees are borne equally. Some adjudicators have confused this statement in the contract to mean that the parties should always bear the fees equally. This is incorrect. Fees should always be allocated on a 'costs follow the event' basis. This provision is purely about liability and has nothing to do with the joint and several liability the parties have for the fees and expenses under clause 41A.6.2.

Effect of Adjudicator's decision

41A .7 .1   The decision of the Adjudicator shall be binding on the Parties until the dispute or difference is finally determined by arbitration or by legal proceedings or by an agreement in writing between the Parties made after the decision of the Adjudicator has been given.

41A .7 .2   The Parties shall, without prejudice to their other rights under this Contract, comply with the decision of the Adjudicator; and the Employer and the Contractor shall ensure that the decision of the Adjudicator is given effect.

41A .7 .3   If either Party does not comply with the decision of the Adjudicator the other Party shall be entitled to take legal proceedings to secure such compliance pending any final determination of the referred dispute or difference pursuant to clause 41A.7.1.

The decision is binding on the parties until it is finally determined in arbitration or litigation or by an agreement in writing. Both parties are under the duty to comply with it. There is also an additional provision in clause 41A.7.2 requiring the parties to give effect to the adjudicator's decision. Failure to give effect to the decision will be a breach of contract. This provision was no doubt inserted to assist when it comes to enforcing the decision but in light of the approach of the courts to the enforcement of adjudicators' decisions it is probably unnecessary.

Immunity

41A .8   The Adjudicator shall not be liable for anything done or omitted in the discharge or purported discharge of his functions as Adjudicator unless the act or omission is in bad faith and this protection from liability shall similarly extend to any employee or agent of the Adjudicator.

The alternative provisions, which permit naming the adjudicator in the contract rather than appointing when the dispute occurs, only affect one point in respect of jurisdiction. The adjudicator must be an individual and not be an employee or otherwise engaged by either party.

Joint Contracts Tribunal Sub-Contract forms

At the time of writing the JCT has published three sub-contracts for use with the JCT main contract forms. They are the Nominated Sub-Contract documents and Domestic Sub-contract DSC/C (replacement for DOM/1) for use with JCT 80 and the Named Sub-Contract documents for use with the Intermediate Form of Contract. The provisions which cover adjudication are so close to the JCT 80 provisions described in the above section that they do not warrant further repetition here. There is also at the time of writing a further suite of Domestic Sub-contract documents to be produced which will replace the DOM suite of documents.

Construction Confederation

The Construction Confederation has published the revised DOM/1, DOM/2 and IN/SC forms. Save for a difference in numbering, these forms follow the adjudication provisions in the JCT main forms of contract. These will shortly be withdrawn from publication.

The predecessor to the new DOM/1 form was also called DOM/1 and contained its own version of adjudication provisions to deal with set-off. There are no longer any special provisions to deal with set-off. Set-off features in the payment provisions sections of the Act and is subject to the notices required in dealing with payment. Any dispute concerning set-off will therefore be dealt with in the same way as any other. This was the cause of some consternation by the sub-contract organisations in the consultation process on the revised DOM/1. They felt that a special procedure for dealing with set-off was still warranted. The details of the discussions are not known but the new DOM/1 was not endorsed by some of the sub-contract organisations.

Institution of Civil Engineers' main contracts

One of the most common challenges to jurisdiction is whether or not a dispute exists. The Institution of Civil Engineers has chosen the 'matter of dissatisfaction' route as discussed in Chapter 6. This, we suggest, even though it seems that it does not comply with the requirement that a party can take a dispute to adjudication at any time, is actually one of the best ways of ensuring that any matters referred to adjudication are actually in dispute between the parties and the issues are clearly identified for the adjudicator. As we have already noted, we understand that a modified procedure is under discussion at the time of writing.

The Institution of Civil Engineers has retained their conciliation procedure in clause 66(5). In clause 66, which is headed 'Avoidance and Settlement of Disputes', four ways are set out in which to resolve disputes, three of which depend on the matter of dissatisfaction procedure occurring first. There is then a further procedure for enforcing the adjudicator's decisions.

We set out a précis below.

Clause 66(6) has six sub-clauses, each of which reproduces one of the six compliance points in section 108(2) of the Act. There is an additional provision to the effect that the adjudication shall be conducted under the Institution of Civil Engineers' Adjudication Procedure (1997) or any amendment or modification of it.

Clause 66(7) provides that the decision of the adjudicator shall be binding in the precise terms of section 108(3) of the Act, and that the adjudicator and any employee or agent of the adjudicator has contractual immunity as required by section 108(4) of the Act.

Clause 66(7) also provides that all disputes arising under or in connection with the contract or the carrying out of the works shall be finally determined by reference to arbitration, and when an adjudicator has given a decision under clause 66(6) in respect of the particular dispute the notice to refer must be served within three months of the giving of the decision otherwise it shall be final as well as binding.

Failure to give effect to a decision of an adjudicator is excluded from the arbitration provision.

This adjudication clause states that disputes can be referred at any time but it must be read with the definition of a dispute set out in clause 66(2) which purports to prevent a party from doing this until the matter of dissatisfaction procedure has been gone through. As noted above, this may well be a non-compliant provision which would mean that the whole adjudication procedure has to be discarded and the Scheme comes into play. That is obviously a matter for the adjudicator to resolve at the time that a party, who will almost certainly be the contractor or sub-contractor referring party, seeks to have a matter adjudicated.

As for whether or not the adjudicator should proceed with the adjudication if the prior matter of dissatisfaction procedure has not been followed beforehand, the authors' view is that he should. The referring party has the right to adjudication and wants it. It then becomes a matter of enforcement as it did in *Mowlem* v. *Hydra-Tight*[38] where it was held that the matter of dissatisfaction procedure under the NEC contract does not comply with section 108 of the Act. This decision was however not subject to legal argument as the parties had agreed that it did not comply and did not ask the judge to consider the point.

The procedural matters in the contract are straightforward. The adjudication shall be conducted under the Institution of Civil Engineers Adjudication Procedure (1997). Clause 66(6), (7) and (8) contain the essentials required by the Act. The duties concerning timetable, impartiality and the power to take the initiative in ascertaining the facts and the law are included in these clauses. The relationship with the arbitration clause is also covered. Three months are allowed in which to serve any notice of arbitration following the decision of an adjudicator, otherwise that decision shall be final and binding. This may give rise to jurisdictional problems if a party seeks to move the decision on to arbitration outside the stipulated period of three months.

The provisions of the ICE Adjudication Procedure in respect of the conduct of the adjudication are as follows.

Clause 5.1 provides the standard time scales of 28 days, such longer period as is agreed by the parties after the dispute has been referred and the 14-day extension with the consent of the referring party.

Clause 5.2 provides that the adjudicator shall determine the matters set out in the notice of adjudication, together with any other matters which the parties and the adjudicator agree should be within the scope of the adjudication.

---

[38] *John Mowlem & Company Plc* v. *Hydra-Tight Ltd (t/a Hevilifts)* (6 June 2000).

In any event he does not have jurisdiction to do other than this. There is a provision to extend the jurisdiction by agreement with the parties to include any other matter which it has been agreed should be within the scope of the adjudication. There would be nothing to prevent the parties and the adjudicator making such an agreement even if this provision were not there.

Clause 5.3 provides that the adjudicator may open up, review and revise any decision (other than that of an adjudicator unless agreed by the parties), opinion, instruction, direction, certificate or valuation made under or in connection with the contract and which is relevant to the dispute. He may order the payment of a sum of money or other redress, but no decision of the adjudicator shall affect the freedom of the parties to vary the terms of the contract, or the engineer or other authorised person to vary the works in accordance with the contract.

This sits uneasily with the provision for adjudication in clause 66(6)(a) which refers to matters 'under the Contract' without any extension. The provision here seems to widen that referral to matters in connection with as well as arising under the contract. It is limited to those things listed such as opinions, instructions, directions, certificate or valuations. Presumably if an independent opinion were sought on a matter in connection with the contract, the adjudicator under this provision could revise that opinion.

Nothing the adjudicator does can affect the freedom of the parties to vary the terms of the contract, or the engineer or other authorised person to vary the works in accordance with the contract. The adjudicator himself cannot vary the terms of the contract. This is strictly a matter between the parties. They can choose to vary the contract at any time by agreement, whether in response to an adjudicator's decision or prior to that decision. Adjudicators simply do not have any jurisdiction to vary the contract itself and impose new terms on the parties. There would be jurisdiction to deal with any new terms the parties agreed upon in deciding any future dispute. This is giving effect to the terms rather than varying the contract. An adjudicator does not have any express authority to vary the works. He cannot restrict the authority of others under the contract in this respect. He can reach decisions which state that there is an entitlement to an instruction or a variation under the contract. This action does not constitute the issuing of the instruction or variation.

Clause 5.4 gives the responding party 14 days from the date of referral to submit any response he may wish to make. This period can be extended by agreement between the parties and the adjudicator. Setting a period of 'within 14 days' for the responding party is a restraint on the process. If the adjudicator is to comply with the procedure agreed by the parties, there is an obligation here for him to allow the 14-day period agreed. It would however have been better if this provision had not been made. In practice it is almost impossible to give the referring party a reasonable period to reply to the responding party's submission, and hold a meeting if one should prove necessary, within the remaining 14 days after the response is served. It is the practice of one of the authors to seek a 7-day extension of time on receipt of the referral so that the referring party has a proper opportunity to respond and a sensible time is made available before and after the meeting that will in many adjudications prove to be necessary. It has never been refused and the referring party almost invariably wants time to make a proper written reply.

Clause 5.5 states that the adjudicator has complete discretion as to how to conduct the adjudication and shall establish the procedure and timetable, subject to any limitation that there may be in the contract or the Act. He shall not be required to observe any rule of evidence, procedure or otherwise, of any court.

The adjudicator's discretion is obviously subject to any restrictions that there may be in the Act or the contract. It is arguable that this provision was almost unnecessary. If the adjudication procedure were silent on these matters, the adjudicator would still have complete discretion as to how to conduct the adjudication. There is express provision that the adjudicator is not required to observe any rule of evidence, procedure or otherwise, of any court. Whatever procedure the adjudicator may wish to adopt, adjudication is a judicial process. He will have to apply the facts to the law to determine the rights of the parties under the contract.

Clause 5.5 continues with the customary non-exhaustive list of powers. The adjudicator may:

(a) ask for further written information;
(b) meet and question the parties and their representatives;
(c) visit the site;
(d) request the production of documents or the attendance of people whom he considers could assist;
(e) set times for (a)–(d) and similar activities;
(f) proceed with the adjudication and reach a decision even if a party fails:
  (i)   to provide information;
  (ii) to attend a meeting;
  (iii) to take any other action requested by the adjudicator;
(g) issue such further directions as he considers to be appropriate.

Clause 5.6 allows the adjudicator to obtain legal or technical advice having notified the parties first. There is however no mention here of having to obtain an estimate of the likely cost and notifying the parties of that.

Clause 5.7 provides that any party may at any time ask that additional parties be joined in the adjudication. Joinder of additional parties is subject to the agreement of the adjudicator and the existing and additional parties. An additional party has the same rights and obligations as the other parties, unless otherwise agreed by the adjudicator and the parties.

Part of any agreement to joinder, by necessity, must take account of a timetable that will permit the adjudicator to hear and deal with all the inter-related matters going to the collective dispute. It is unlikely that any situation where there is joinder will permit completion of the adjudication in 28 days and the parties to the first adjudication or adjudications will have to take this into account when deciding whether or not to agree to a proposed joinder situation.

Clause 6.1 relates to the decision. The prime task of the adjudicator is to reach his decision and so notify the parties within the time limits in paragraph 5.1.

There is an important additional power in paragraph 6.1. This allows the adjudicator to reach a decision on different aspects of the dispute at different times. He will therefore be able to conclude any dispute in stages, summarising the whole position in his final decision. This is a sensible procedure in a minority of cases but, in the context of a procedure where the response does not come in until 14 days after the referral, it is probably only useful where there are separate and distinct issues within the same dispute, some of which may require longer consideration and an extension of time while others can be dealt with without delay.

Clause 6.2 provides that the adjudicator may direct the payment of such simple or compound interest at such rate and between such dates or events as he considers appropriate.

The right to interest is already contained in some contracts. Where no such right exists this is a useful provision. There may also be the additional burden of considering interest in accordance with the Late Payment of Commercial Debts (Interest) Act 1998 if that matter comes within the ambit of the adjudication.

Clause 6.3 deals with the situation where the adjudicator fails to reach his decision and notify the parties in the due time. Either party may then give seven days' notice of its intention to refer the dispute to a replacement adjudicator appointed in accordance with the procedures in paragraph 3.3. Clause 6.4 provides that notwithstanding any failure of the adjudicator to reach and notify his decision in due time, if he does so before the dispute has been referred to a replacement adjudicator under paragraph 6.3 his decision shall still be effective.

Clause 6.4 continues by providing that if the decision is not notified to the parties then it is of no effect and the adjudicator shall not be entitled to any fees or expenses. The parties are however responsible for the fees and expenses of any legal or technical adviser appointed under paragraph 5.6 subject to the parties having received such advice.

This is a practical provision dealing with the adjudicator's decision that arrives late. If no other adjudicator has been appointed the decision is nevertheless effective. This obviously deals with the situation where the decision may be late by an odd day through some mishap or oversight.

Clause 6.5 provides that the parties shall each bear their own costs and expenses incurred in the adjudication.

By clause 6.5 the parties are also made jointly and severally responsible for the adjudicator's fees and expenses, including those of any legal or technical adviser appointed under paragraph 5.6, but in his decision the adjudicator is permitted to direct a party to pay all or part of his fees and expenses. If he makes no such direction the parties are required to pay them in equal shares.

It would be unwise for any adjudicator not to direct in his decision who is responsible for the payment of his fees and expenses. There is however some wisdom in having a default position where the adjudicator's decision is silent on the matter.

Clause 6.6 allows the adjudicator, at any time until seven days before he is due to reach his decision, to give notice to the parties that he will deliver his decision only on full payment of his fees and expenses. Any party may then pay these costs in order to obtain the decision and recover the other party's share of the costs in accordance with paragraph 6.5 as a debt due.

There is nothing in law that requires any person in commerce to give credit. There was reluctance, when the Act and the Scheme were drafted, to allow any lien on the decision. The ability of the adjudicator to secure payment of his fees and expenses before he delivers the decision is therefore important to him. The CIC, which originally followed the ICE as regards liens, has now omitted this provision as a result of the decision in Scotland in *St Andrews Bay* v. *HBG*[39] where the adjudicator sought to impose a lien and the court found an obligation to notify the decision by the 28th or other properly extended day.

Clause 6.7 states that the parties are entitled to the relief and remedies set out in the decision and to seek summary enforcement of them, regardless of whether the dispute is to be referred to legal proceedings or arbitration. This clause also provides that no issue decided by an adjudicator may subsequently be laid before another adjudicator unless so agreed by the parties.

Clause 6.8 provides that in the event that the dispute is referred to legal proceedings or arbitration, the adjudicator's decision shall not inhibit the court or arbitrator from determining the parties' rights or obligations anew.

The provision on summary enforcement of the decision may be of assistance to the courts

---

[39] *St Andrews Bay Development Ltd* v. *HBG Management Ltd* (20 March 2003).

or an arbitrator as well as the party seeking enforcement. The parties are bound by the decision until such time as the dispute is heard anew.

Clause 6.9 allows the adjudicator on his own initiative, or at the request of either party, to correct a decision so as to remove any clerical mistake, error or ambiguity provided that the initiative is taken, or the request is made, within 14 days of the notification of the decision to the parties. The adjudicator is required to make his corrections within seven days of any request by a party.

The adjudicator thus has authority to correct any clerical mistakes, errors or ambiguities within 14 days of the notification of the decision to the parties. Without these express powers it would be arguable that there was no authority to correct such slips. This is therefore a sensible provision in this adjudication procedure.

Clause 6.9 is effectively a copy of the provision in the Arbitration Act 1996 for the correction of an arbitrator's award. There is however no specificity as to what the words 'clerical mistake, error or ambiguity' actually mean. These words can be either construed as limiting correction first to clerical mistakes, second to clerical errors and third to ambiguities or as extending the second category to errors of any description. It appears from the decisions of the court in *Bloor* v. *Bowmer & Kirkland*[40] and *Nuttall* v. *Sevenoaks*[41] that the court may take the wider view where there is no provision at all in the adjudication provisions for the correction of a decision. We examine this further in Chapter 11.

Our unrepentant view is that an adjudicator should not trouble himself with such definitions. If he has made an error he should correct it; he should generally allow the parties to make submissions to him before making the correction and should always do so if the error is of some magnitude. His decision then properly reflects his findings rather than leaving the situation of an obvious error that can only be sorted out by further dispute resolution processes. If a party objects to the correction having been made, it becomes a matter for the court in any enforcement proceedings to decide if the adjudicator has exceeded his jurisdiction by making the correction.

Civil Engineering Contractors' Association sub-contracts

The provisions in the CECA sub-contracts do not differ greatly from the ICE main contracts that they complement. They also rely on the ICE Adjudication Procedure for conduct of the adjudication. The main important difference is the interface between the sub-contract form and the main contract. There is no matter of dissatisfaction procedure under the sub-contract itself. A dispute comes into being without such a procedure. The link with the main contract is where the sub-contractor considers he is entitled to a payment greater than the amount determined by the contractor. Where in the opinion of the contractor this gives rise to a matter of dissatisfaction under the main contract, he gives notice to the sub-contractor and the matter is pursued under the main contract. Obligations are then imposed on the main contractor to keep the sub-contractor fully informed and on the sub-contractor to provide information and to attend meetings to resolve the matter of dissatisfaction. The parties have contracted on the basis that no dispute shall arise until the matter of dissatisfaction is resolved under the main contract. This attracts the same comments and criticisms that we have considered earlier when reviewing the main contract form.

---

[40] *Bloor Construction (UK) Limited* v. *Bowmer & Kirkland (London) Limited* [2000] BLR 314.
[41] *Edmund Nuttall Limited* v. *Sevenoaks District Council* (14 April 2000).

Clause 18

18 (1) If any dispute or difference shall arise between the Contractor and the Sub-Contractor in connection with or arising out of the Sub-Contract, or the carrying out of the Sub-Contract Works (excluding a dispute concerning VAT but including a dispute as to any act or omission of the Engineer) whether arising during the progress of the Sub-Contract Works or after their completion it shall be settled in accordance with the following provisions.

   (2) (a) Where the Sub-Contractor seeks to make a submission that payment is due of any amount exceeding the amount determined by the Contractor as due to the Sub-Contractor, or that any act, decision, opinion, instruction or direction of the Contractor or any other matter arising under the Sub-Contract is unsatisfactory, the Sub-Contractor shall so notify the Contractor in writing, stating the grounds for such submission in sufficient detail for the Contractor to understand and consider the Sub-Contractor's submission.

      (b) Where in the opinion of the Contractor such a submission gives rise to a matter of dissatisfaction under the Main Contract, the Contractor shall so notify the Sub-Contractor in writing as soon as possible. In that event, the Contractor shall pursue the matter of dissatisfaction under the Main Contract promptly and shall keep the Sub-Contractor fully informed in writing of progress. The Sub-Contractor shall promptly provide such information and attend such meetings in connection with the matter of dissatisfaction as the Contractor may request. The Contractor and the Sub-Contractor agree that no such submission shall constitute nor be said to give rise to a dispute under the Sub-Contract unless and until the Contractor has had the time and opportunity to refer the matter of dissatisfaction to the Engineer under the Main Contract and either the Engineer has given his decision or the time for the giving of a decision by the Engineer has expired.

This may give rise to a further dispute if the sub-contractor thinks that the main contractor is not justified in forming the opinion that the payment matter is also a matter of dissatisfaction under the main contract. This dispute, by the terms of the clause itself, has to wait for the completion of the matter of dissatisfaction procedure under the main contract. It seems inevitable that in these circumstances the sub-contractor will suffer a delay of up to one month before adjudication can even be commenced. The same challenge that a matter of dissatisfaction constitutes a dispute and there is therefore entitlement to immediate adjudication, applies under the sub-contract as it does under the main contract. These matters provide the initial jurisdiction problems in any adjudication. Given the apparent failure to allow access for the sub-contractor to adjudication at any time we repeat our view expressed earlier that the adjudicator should proceed with the adjudication if so requested by the sub-contractor and it then becomes a matter of enforcement.

   (3) (a) The Contractor or the Sub-Contractor may at any time before service of a Notice to Refer to arbitration under sub-clause 18(7) by notice in writing seek the agreement of the other for the dispute to be considered under the Institution of Civil Engineers' Conciliation Procedure (1994) or any amendment or modification thereof being in force at the date of such notice.

      (b) If the other party agrees to this procedure any recommendation of the conciliator shall be deemed to have been accepted as finally determining the dispute by agreement so that the matter is no longer in dispute unless a Notice of Adjudication under sub-clause 18(4) or a Notice to Refer to arbitration under sub-clause 18(7) is served within 28 days of receipt by the dissenting party of the conciliator's recommendation.

The sub-contract maintains the conciliation procedure in the same way as the main contract. The conciliation procedure might conflict with adjudication. The sub-contractor could issue a notice to proceed to adjudication and the contractor could in response issue a notice to proceed to conciliation. It would then require agreement of the sub-contractor to proceed to conciliation.

(4) (a) The Contractor and the Sub-Contractor each has the right to refer any dispute under the Sub-Contract for adjudication and either party may at any time give notice in writing (hereinafter called the Notice of Adjudication) to the other at any time of his intention to refer the dispute to adjudication. The Notice of Adjudication and the appointment of the adjudicator shall, save as provided under sub-clause 18(10)(b), be as provided at paragraphs 2 and 3 of the Institution of Civil Engineers' Adjudication Procedure (1997). Any dispute referred to adjudication shall be conducted in accordance with the Institution of Civil Engineers' Adjudication Procedure (1997) or any amendment or modification thereof being in force at the time of the appointment of the adjudicator.

    (b) Unless the adjudicator has already been appointed he is to be appointed by a timetable with the object of securing his appointment and referral of the dispute to him within 7 days of such notice.

    (c) The adjudicator shall reach a decision within 28 days of referral or such longer period as is agreed by the parties after the dispute has been referred.

    (d) The adjudicator may extend the period of 28 days by up to 14 days with the consent of the party by whom the dispute was referred.

    (e) The adjudicator shall act impartially.

    (f) The adjudicator may take the initiative in ascertaining the facts and the law.

(5) The decision of the adjudicator shall be binding until the dispute is finally determined by legal proceedings or by arbitration (if the Sub-Contract provides for arbitration or the parties otherwise agree to arbitration).

(6) The adjudicator shall not be liable for anything done or omitted in the discharge or purported discharge of his functions as adjudicator unless the act or omission is in bad faith and any employer or agent of the adjudicator shall similarly not be liable.

(7) (a) All disputes arising under or in connection with the Sub-Contract, other than failure to give effect to a decision of an adjudicator, shall be finally determined by reference to arbitration. The party seeking arbitration shall serve on the other party a notice in writing (called the Notice to Refer) to refer the dispute to arbitration.

    (b) Where an adjudicator has given a decision under sub-clause 18(4) in respect of the particular dispute the Notice to Refer must be served within three months of the giving of the decision, otherwise it shall be final as well as binding.

The remainder of the clause and the use of the ICE Adjudication Procedure is the same as the main contract form. The comments made under the main contract form apply here.

Government contracts

There are 1998 versions of GC/Works/1, 2, 3 and 4 contracts. They provide a sensible approach to jurisdiction. There is no attempt to qualify contracts to cover only construction contracts as defined in the Act. The adjudicator may deal with disputes, differences or questions arising under, out of, or relating to the contract. This is drafted as widely as any arbitration clause. It enables all disputes likely to be encountered in a construction project under this form of contract to be dealt with by adjudication in the first instance.

Condition 59: Adjudication

(1)   The Employer or the Contractor may at any time notify the other of intention to refer a dispute, difference or question arising under, out of, or relating to, the Contract to adjudication. Within 7 Days of such notice, the dispute may by further notice be referred to the adjudicator specified in the Abstract of Particulars.

   . . .

(4)   The PM, the QS and the other party may submit representations to the adjudicator not later than 7 Days from the receipt of the notice of referral.

(5)   The adjudicator shall notify his decision to the PM, the QS, the Employer and the Contractor not earlier than 10 and not later than 28 Days from receipt of the notice of referral, or such longer period as is agreed by the Employer and the Contractor after the dispute has been referred. The adjudicator may extend the period of 28 Days by up to 14 Days, with the consent of the party by whom the dispute was referred. The adjudicator's decision shall nevertheless be valid if issued after the time allowed. The adjudicator's decision shall state how the cost of the adjudicator's fee or salary (including overheads) shall be apportioned between the parties, and whether one party is to bear the whole or part of the reasonable legal and other costs and expenses of the other, relating to the adjudication.

It is unfortunate that there is an earliest date for the adjudicator to reach his decision. If the philosophy of adjudication is to reach early decisions to avoid delay to the works, there are instances where a decision could be made much earlier than the ten days stipulated here. For example, a dispute concerning the quality of brickwork ought to be capable of resolution in much less time than ten days. There is also a limitation here on the adjudicator's fee. In his decision he is required to state how the fee is to be apportioned between the parties. The fee may be calculated on the basis of salary plus overheads. Alternatively, the fee is at a quoted rate.

(6)   The adjudicator may take the initiative in ascertaining the facts and the law, and the Employer and the Contractor shall enable him to do so. In coming to a decision the adjudicator shall have regard to how far the parties have complied with any procedures in the Contract relevant to the matter in dispute and to what extent each of them has acted promptly, reasonably and in good faith. The adjudicator shall act independently and impartially, as an expert adjudicator and not as an arbitrator. The adjudicator shall have all the powers of an arbitrator acting in accordance with Condition 60 (Arbitration and choice of law), and the fullest possible powers to assess and award damages and legal and other costs and expenses; and, in addition to, and notwith-standing the terms of, Condition 47 (Finance charges), to award interest. In particular, without limitation, the adjudicator may award simple or compound interest from such dates, at such rates and with such rests as he considers meet the justice of the case—
   (a)   on the whole or part of any amount awarded by him, in respect of any period up to the date of the award;
   (b)   on the whole or part of any amount claimed in the adjudication proceedings and out-standing at the commencement of the adjudication proceedings but paid before the award was made, in respect of any period up to the date of payment;
         and may award such interest from the date of the award (or any later date) until payment, on the outstanding amount of any award (including any award of interest and any award of damages and legal and other costs and expenses).

Not only does the adjudicator have authority to take the initiative in ascertaining the facts and the law but also both the employer and the contractor are under a duty to enable him to do so. It must of course always be remembered that there is a duty upon the adjudicator to put matters that he ascertains in carrying out this initiative to the parties before reaching his decision (*Balfour Beatty* v. *Lambeth* and *RSL* v. *Stansell* mentioned earlier.)

The adjudicator is also given jurisdiction to take into account the way in which the parties have behaved during the course of the contract in complying with any procedures in the contract. Whether this will have any real effect is doubtful. If a notice is expressed to be a condition precedent in the contract and no notice has been issued, it will be of no effect in any event. This is merely following the contract. The adjudicator does not really have the power to penalise a party who has been slipshod in the administration of the contract.

The jurisdiction and powers of the adjudicator are considerably widened by this clause. For an adjudicator to be invested with the same powers as an arbitrator under Condition 60 of the contract is much wider than anything ever intended by the Act. The minimum criteria of the Act are satisfied and there is nothing to prevent the parties from increasing the scope of matters that can be dealt with by adjudication as this clause does.

(7)   Subject to the proviso to Condition 60(1) (Arbitration and choice of law), the decision of the adjudicator is binding until the dispute is finally determined by legal proceedings, by arbitration (if the Contract provides for arbitration, or the parties otherwise agree to arbitration), or by agreement: and the parties do not agree to accept the decision of the adjudicator as finally determining the dispute.

(8)   In addition to his other powers, the adjudicator shall have power to vary or overrule any decision previously made under the Contract by the Employer, the PM or the QS, other than decisions in respect of the following matters–

(a)   decisions by or on behalf of the Employer under Condition 26 (Site admittance);

(b)   decisions by or on behalf of the Employer under Condition 27 (Passes) (if applicable);

(c)   provided that the circumstances mentioned in Condition 56(1)(a) or (b) (Determination by Employer) have arisen, and have not been waived by the Employer, decisions of the Employer to give notice under Condition 56(1)(a), or to give notice of determination under Condition 56(1);

(d)   decisions or deemed decisions of the Employer to determine the Contract under Condition 56(8) (Determination by Employer);

(e)   provided that the circumstances mentioned in Condition 58A(1) (Determination following suspension of Works) have arisen, and have not been waived by the Employer, decisions of the Employer to give notice of determination under Condition 58A(1); and

(f)   decisions of the Employer under Condition 61 (Assignment).

In relation to decisions in respect of those matters the Contractors's [sic] only remedy against the Employer shall be financial compensation.

Rather than simply declare the rights of the parties under the contract, this provision permits the adjudicator to vary or overrule decisions made under the contract by the employer, project manager or quantity surveyor. There is a limitation on the matters listed in paragraph (8)(a) to (f). These are not to be varied or overruled by the adjudicator but he can decide that the remedy shall be financial compensation where he would otherwise have varied these decisions.

(9)   Notwithstanding Condition 60 (Arbitration and choice of law), the Employer and the Contractor shall comply forthwith with any decision of the adjudicator; and shall submit to summary judgment and enforcement in respect of all such decisions.

(10)   If requested by one of the parties to the dispute, the adjudicator shall provide reasons for his decision. Such requests may only be made within 14 Days of the decision being notified to the requesting party.

(11)   The adjudicator is not liable for anything done or omitted in the discharge or purported

discharge of his functions as adjudicator, unless the act or omission is in bad faith. Any employee or agent of the adjudicator is similarly protected from liability.

Reasons must be given for the decision if so requested. It is clear from the drafting of paragraph 10 that this may be after the decision has been given and the adjudicator who has produced an unreasoned decision should have his reasons ready and be prepared to answer such a request pretty quickly. It is not acceptable for him to have to rework his decision to come up with his reasons. What happens if the answer that he comes up with after going through his reasons in detail is different from that which he has already sent to the parties?

NEC Engineering and Construction Contracts

These contracts attempt to cover the bare minimum required by the Act. They follow the policy in the civil engineering contracts on matters of dissatisfaction. It is therefore debatable that these contracts do not include the statutory right to refer a dispute to adjudication at any time. The lack of prescription in this contract must regard the adjudicator as having all the powers necessary to enable him to reach his decision as required by the timetable.

At the time of writing the adjudication provisions are set out in supplement Y(UK)2/ APRIL 1998. The forms of contract have not been revised. This supplement includes amendments to the earlier form which are set out below.

We have considered clauses 90.1 to 90.5 in Chapter 7.

Clause 90.6 provides for the party referring the dispute to the adjudicator to include with his submission information to be considered by the adjudicator. This clause then goes on to say that any further information from a party to be considered by the adjudicator is provided within 14 days of referral.

This seems to suggest that the referring party itself can provide further information in the 14-day period. This could cause difficulty unless the adjudicator brings some order to the proceedings and directs a response within seven days and a reply seven days after that.

Clause 90.7 requires the parties and the project manager to proceed as if the action, failure to take action or other matters that are the subject of the referral to adjudication were not disputed until the adjudicator has given his decision on the dispute.

Clause 90.8 covers the Act's requirements that the adjudicator acts impartially and that he may take the initiative in ascertaining the facts and the law.

Clause 90.9 covers the Act's requirements that the adjudicator reaches a decision within 28 days of referral or such longer period as is agreed by the parties after the dispute has been referred and that the adjudicator may extend the period of 28 days by up to 14 days with the consent of the notifying party.

Clause 90.10 requires the adjudicator to provide reasons for his decision. In addition to notifying these to the parties he is required to do this to the project manager as well.

Clause 90.11 confirms that the decision of the adjudicator is binding until the dispute is finally determined by the 'tribunal' or by agreement. The definition of the 'tribunal' is left blank in the contract data section of the standard form for insertion by the parties. The choices will be either legal proceedings or arbitration in conformity with section 108(3) of the Act..

Clause 90.12 provides the contractual immunity required by section 108(4) of the Act.

Clause 91 is entitled 'Combining procedures'.

Clause 91.1 provides that if a matter causing dissatisfaction under or in connection with a sub-contract is also a matter causing dissatisfaction under or in connection with this contract, the subcontractor may attend the meeting between the parties and the project manager to discuss and seek to resolve the matter.

Clause 91.2 provides that if a matter disputed under a sub-contract is also a matter disputed under the main contract, the two disputes can be submitted to the adjudicator at the same time.

Clause 92.1 confirms that the adjudicator settles the dispute as independent adjudicator and not as arbitrator. It also confirms that the decision is enforceable as a matter of contract and not as an arbitral award. This clause also confirms that the adjudicator can review and revise any action or inaction of the project manager or supervisor in relation to the dispute. It requires communications between one party and the adjudicator to be communicated to the other party. Lastly, it relates the adjudicator's decision to the 'compensation event' procedure in this form of contract.

Clause 92.2 deals with the situation where the adjudicator, who is by this form of contract named in the contract data, resigns or is unable to act. The parties are then required to choose a new adjudicator but if they fail to do so there is a provision in this clause for a nominator. There is no nominator specifically named in the contract data, it being left to the parties to insert their own choice. This does create a difficulty if no nominator is named. If the parties cannot agree the name of an adjudicator, a party cannot be deprived of its right to refer to an adjudicator and in these circumstances we have no doubt that, as a last resort, the nomination procedure in the Scheme should be used. The new adjudicator is appointed when the dispute is submitted to him.

One interesting drafting point is that the appointment of the adjudicator named in the contract does not refer to the NEC Adjudicator's Contract but the appointment of the replacement adjudicator does.

The sub-contract follows the pattern of the main contract.

The Institution of Chemical Engineers Adjudication Rules (The Grey Book)

Subsequent to the procedure relating to the adjudicator's appointment, which we deal with in Chapter 7, the IChemE procedure is split into three sections: procedure on appointment, procedure for the adjudication, and the decision. These sections set out in considerable detail the powers and duties of the adjudicator and also the duties of the parties. We deal with the first two of these here and the provisions as to the decision in Chapter 11.

As noted elsewhere we have not been given permission to reproduce these clauses verbatim which is rather unfortunate as they give excellent guidance in respect of procedural matters. We set out a précis below.

Clause 6 is entitled 'Procedure on appointment'.

Clause 6.1 provides that the adjudicator directs each party in writing to produce the following 'within specified times which shall be reasonable':

- a statement setting out a detailed view of the dispute including a summary with supporting documentary evidence and a copy of the contract;
- details of representative(s) at any meeting;
- the address of any place which the adjudicator may need to visit; and
- details of witnesses of fact and/or opinion including the subject to be covered by each.

Clause 6.1 covers both the referral and the response by the responding party. It sets out in detail the requirements of these documents. In Chapter 7 we considered the appointment, which under this form is the signing of the form of agreement which appears in Annex B to this form of contract, by either of the parties. This fulfils the requirement that the adjudicator's appointment be confirmed. Clause 6.1 does not deal with the usual situation that the referring party will send the referral notice almost as soon as the nomination is made. It appears that the requirement that the adjudicator writes to the parties on appointment, that is set out in clause 6.1, will generally go by default and these requirements will follow receipt of the referral.

Clause 6.2 sets out further matters that the adjudicator is to direct:

- that the parties must copy all documents sent or given to the adjudicator to the other party at the same time (unless otherwise agreed); and
- that the adjudicator is to be informed if a party intends to discuss the dispute with the other party and the adjudicator is subsequently to be provided with an agreed written account of any conclusions reached. Discussions between the parties on possible settlement of all or any part of the dispute are not be reported to the adjudicator unless and until a settlement has been concluded.

The provision in clause 6.2 that requires the adjudicator to direct that a party informs him of its intention to discuss the dispute with the other party is an interesting one. We suspect that this is a little over the top and may well be considered by parties to be an intrusion into their privacy.

The requirement placed on the adjudicator to direct these matters does seem rather over the top when they are set out in detail in the contract itself.

Clause 6.3 deals with the situation where the parties reach an agreement on all or any part of the matter(s) under dispute. They are required to send the adjudicator a statement to that effect, requesting the adjudicator to terminate the relevant part of the adjudication and render accounts of the fees and expenses due for payment by the parties.

Clause 6.4 contains the required time period of 28 days and the provisions for extension of this period required by section 108 of the Act. This clause also defines the date of referral of the dispute as being the date upon which the adjudicator received all the documents and information referred to in sub-clause 6.1(a) from the referring party.

Clause 7 is entitled 'Procedure for the Adjudication' and is a detailed guide to what any adjudicator has power to do in order to reach his decision. It includes the following provisions. Clauses 7.1, 7.2 and 7.3 provide as follows:

- The adjudicator has sole discretion as to the conduct of the adjudication. The adjudicator establishes the timetable for the adjudication subject to any limitation that there may be in the contract or the Act.
- The adjudicator is not required to observe any rule of evidence, procedure or otherwise of any court.
- The adjudicator is to consider the matters set out in the notice and any other matters which the parties and the adjudicator agree should be within the scope of the adjudication.

Clause 7.4 provides that the adjudicator has the power to:

- decide any question of interpretation of the contract between the parties that is relevant to the dispute;
- ask for further written information;
- meet and question the parties either separately or together;
- require any party to make available for inspection any premises or item pertinent to the dispute but he must allow representatives of the parties to be present if they so wish;
- require the production of documents or the attendance of people;

- direct the preservation and, if necessary, storage of any property of or under the control of either party;
- call meetings with the parties' representatives. Locations and times to be chosen by the adjudicator after consultation with the parties. If a party objects unreasonably to the adjudicator's proposed location, the adjudicator may proceed to hold the meeting and advise the objecting party of the outcome;
- impose reasonable time limits on the parties for replies to requests by him for information. In the event of a party failing to respond in accordance with the time limits imposed, to proceed with the determination in accordance with evidence available, and to draw such inferences as seem to be appropriate arising from such failure;
- set times for the previous and similar activities;
- proceed with the adjudication and reach a decision even if a party fails to provide information, to attend a meeting or to take any further action requested by the adjudicator;
- give a decision on different aspects of the dispute at different times;
- issue such further directions or take such actions as he considers to be appropriate.

This is an unusually exhaustive list and is in our opinion very useful for the sake of the understanding of the parties as to the adjudicator's role. The IChemE is to be congratulated on its approach. It must of course be remembered that any such list can only be non-exhaustive and, as is set out in the final item, the adjudicator can in appropriate circum-stances give further directions or take other actions that he considers to be appropriate for the adjudication in question.

Clause 7.5 allows the adjudicator in addition to engage advisers on any matter to assist in reaching a decision. He must however give the parties notice before making such appointment and if possible provide an indication of likely costs. This clause also states that the appointment of advisers does not alter the fact that all decisions in the case are the adjudicator's alone.

Clause 7.6 allows the adjudicator to open up, review and revise any decision (other than that of an adjudicator unless agreed by both parties) made under or in connection with the contract and which is relevant to the dispute. He may order the payment of a sum of money, or other redress. In particular, he may open up, review and revise any decision taken or certificate given by any person referred to in the contract unless the contract states that the decision or certificate is final and conclusive.

Chapter 10

The Process

Section 108 of the Act approaches adjudication in a rather different way from that in which the Arbitration Act 1996 deals with arbitration. There are none of the detailed procedures or lists of powers that can be found in the Arbitration Act that govern the arbitration process. These are left for the Scheme or the contract between the parties and any adjudication rules that may apply.

All section 108 says about the adjudication process, once the dispute has been referred to the adjudicator, is:

- that the adjudicator must reach a decision within 28 days of referral or such longer period as is agreed by the parties after the dispute has been referred;
- that the adjudicator may extend the period of 28 days by up to 14 days, with the consent of the party by whom the dispute was referred;
- that the adjudicator must act impartially; and
- that the adjudicator may take the initiative in ascertaining the facts and the law.

There may be provisions in the contract or in any applicable rules which set out more than this but, unless there are, the adjudicator is totally at liberty as to the procedure that he may adopt to enable him to reach his decision within the time allowed. Most adjudication rules in fact do no more than set out a non-exhaustive list of actions that the adjudicator may take and the Scheme does not differ in this respect.

Initiating the adjudication process

The notice of adjudication and the referral to the adjudicator

The parties to the contract are in dispute and one of them has decided to exercise his right to refer that dispute to adjudication. The first thing he has to do is issue his notice of intent to refer the dispute to adjudication ('the notice of adjudication'). The notice must comply with the minimum requirements of the contract or any rules adopted by the contract. The Scheme is quite specific as to its content. It is vital that the referring party makes every effort to make his notice of adjudication as specific as he can. The primary reason for this is that the notice sets out the details of the dispute that is to be referred and thus sets the adjudicator's jurisdiction. The notice also enables the prospective adjudicator, whether he be named in the contract, agreed by the parties or nominated, to understand the nature of the dispute. Where an ANB is involved it will need to have a clear idea of the nature of the dispute so that it has the opportunity of nominating the best person for the job of adjudicator. If the notice is too widely drawn it may be considered not to identify any dispute.

All the various adjudication clauses and the Scheme set out requirements as to the nature of the notice of adjudication. They require varying degrees of detail but in principle what is required is sufficient information for the other party to identify the disputed matters and for the most appropriate adjudicator to be identified.

The Chartered Institute of Arbitrators uses an application form for the nomination of an adjudicator which, in addition to requiring that the notice of adjudication be provided, includes a list of typical disputes that the referring party has to tick as appropriate. This eases the problem for the appointing body in identifying the right adjudicator. The other ANBs have their own permutations of forms necessary for their purposes to make an appointment or nomination. It is important that the forms are completed correctly and accompanied by the correct fee. Most ANBs do not consider that time is running until they have received a properly completed form and the appropriate fee.

The next stage after the adjudicator is either agreed or nominated, is for the referring party to send its referral notice to the adjudicator and to the other party. If the parties are pragmatic and have not allowed the dispute to fester and develop in size and complexity, the referral notice may be very brief. The practical approach is to be well prepared and to exercise the right to adjudication on a dispute that contains a discrete issue or issues. Adjudication is most likely to succeed in terms of the timetable and result if this approach is adopted.

Quite often, however, the matters referred include a complete final account where many hundreds or thousands of variations are in dispute between the parties; there is a prolongation and disruption claim with the associated claim for loss and/or expense. There are divergent views on the practice of submitting a dispute of such a nature to adjudication. Some consider that this almost amounts to an abuse of the adjudication process and should be actively discouraged. Others have a rather more pragmatic view, however sizeable the dispute, and provided that the parties are prepared to allow sufficient time for the adjudicator to consider and reach a decision on the totality of the dispute properly, what is wrong with a process that will produce a decision that could well be accepted by the parties or used by them as a means to settle the dispute? Even if the result is a decision that does not lead to a resolution of the dispute, the adjudication may well result in a narrowing of the issues and as a result reduce the costs of a later arbitration or court proceedings.

If the referral notice is succinct and to the point, it will allow the adjudicator to set the most appropriate procedure for the adjudication. Human nature being what it is, the more likely scenario is one where the referring party will have done a considerable amount of preparation beforehand and will submit what would be, in arbitration terms, a statement of its case.

In the early days of statutory adjudication great concern was expressed with regard to the referring party setting an ambush for the other party in this way. This does of course sometimes happen but most adjudicators are alive to the problem and can take steps to deal with the unfairness that can result. If the referring party produces a referral that involves matters of considerable complexity, the adjudicator may well in any event be unable to deal with the matter properly unless time is extended, and there is no doubt that responding parties are only too willing to allow extensions greater than the 14 days to which the referring party is limited, to give them the time to make a proper response. We deal with the possibility of ambush that may arise out of the possible imbalance in preparation time, in more detail later in this chapter.

One matter that has become evident with regard to the referral is the view held by some that the referral notice should set out the dispute in its entirety, with the full detail of both parties' contentions being included. This procedure is all very well when the referring party is professionally represented, but the practicalities of adjudication are that many referring parties do it themselves; this after all is the great attraction of adjudication, and the last thing that they will consider doing is putting the other side's case. However beneficial such an approach is to the adjudication process and in particular in ensuring that the dispute has properly crystallised in the way set out in the various judgments of the court, the practicalities of adjudication are that this ideal situation seldom happens. While we have sympathy with the concept of the referring party presenting both sides of the case from the point of view of avoiding problematical jurisdictional situations and making the adjudicator's life easier, we suggest that any adjudicator who decided that he would not adjudicate unless the referral was presented in this way would find very few adjudications to do.

It is vital however that the referral relates to matters that have already been debated between the parties. See *Sindall* v. *Solland*[1] and *Beck Peppiatt* v. *Norwest Holst*[2]. The addition of new material may be considered to create a new dispute that is not within the jurisdiction of the adjudicator, see *Nuttall* v. *Carter*[3].

On receipt of the referral

Whatever the nature of the referral notice the adjudicator will have to make sure that he has time to read it in detail on the day he receives it. Anything of any complexity will require the allocation of a reasonable amount of time on the first day to digest it and to prepare a plan of action for the forthcoming 28-day period.

The adjudicator may be of the view that he would ideally like to get his fees agreed with the parties. This may be possible and in many cases the procedure will require that an adjudicator's agreement accompanies the referral notice, which will provide the opportunity to resolve the issue of fees from the outset. The adjudicator should not allow this to interfere with his duty to complete the adjudication within the very restricted timescale. ANBs will take a very dim view if a party that had previously applied for a nomination approached them with a complaint to the effect that the person nominated as adjudicator would not proceed because he wanted his fees agreed beforehand. The more usual approach, save in cases where the appointment is by agreement and terms are settled between the arbitrator and the parties, is for the adjudicator to state what his fees will be and rely on his entitlement to a reasonable fee.

It is vital that the adjudicator stamps his authority on the proceedings from the outset. He should therefore acknowledge receipt of the referral notice immediately by fax or e-mail as is specifically required by some forms of contract. If the referral notice is short he should consider it immediately, decide what the procedure will be and include his directions with his acknowledgement. If the referral notice is of any length or complexity he should merely acknowledge receipt and inform the parties that he will be issuing his directions as soon as he has considered and decided what the appropriate procedure should be. This does not mean that the adjudicator can sit back; it is of paramount importance that he issues his directions at the earliest possible moment to avoid vital time being lost.

He must at this time satisfy himself that the dispute is of the same nature as he believed it to be when he told the parties or the ANB that he would take it on. He must also check that the matters referred are within the ambit of the notice of adjudication. Anything that is not identified in the notice of adjudication and is included in the referral is not within the

---

[1] *Sindall Ltd* v. *Solland and Others* (15 June 2001).
[2] *Beck Peppiatt Ltd* v. *Norwest Holst Construction Ltd* [2003] BLR 316.
[3] *Edmund Nuttall Limited* v. *R.G. Carter Limited* [2002] BLR 312.

adjudicator's jurisdiction and, if he makes a decision on such things, that decision is unenforceable. It does no harm to repeat that it is vital that the adjudicator does have an understanding of the subject matter of the dispute. If he is being asked to adjudicate on something wholly outside his expertise he should tell the parties immediately, put it down to experience and withdraw.

The authors have no sympathy with the view that has been expressed by some members of the arbitral community to the effect that anyone who can arbitrate can also adjudicate on any matter that is put before them. This totally misses the point of adjudication. If an adjudication is to succeed, it is vital that the adjudicator not only has a grasp of the principles of dispute resolution and the law of contract but that he is also well up to speed with the current practical aspects of the adjudication process and that part of the construction industry out of which the dispute emanates. The adjudicator who takes on disputes in areas in which he does not have experience is likely to be found out quickly, both in that particular adjudication and in the marketplace. Where the modern arbitrator and the adjudicator are similar is that both must be prepared to be innovative in approach, a good manager, particularly skilled in time management, and command respect because of their competence.

The adjudicator must also find out what the contract is, whether it has any particular procedural requirements and whether there are any adjudication rules that apply. If there are, the adjudicator must ensure that any procedure he sets up complies with these requirements.

Before he issues his directions the adjudicator will need to ask himself a considerable number of questions, among which are the following:

- What do the contract and/or the applicable rules say?
- How will this affect his approach to the adjudication?
- How long will be needed to reach and to write the decision after all the information is to hand?
- Having considered the referral, are the issues involved of a nature that he should seek an extension of time?
- What is the available period for information gathering, allowing for the submissions of the parties and for his own investigations? Any dates that he sets must ensure that the period that he needs for this purpose is not eaten into in any way. The first rule for any adjudicator must be: never get backed into a corner that you can avoid; the time limits are going to impose immense pressure without making it worse for yourself.
- Is his diary clear for the requisite number of days immediately before the expiry of the 28-day period? If not, the period in question should be lengthened and the period for information gathering reduced as a result.
- Is the referral notice adequate? Does it set out the position of the referring party in an understandable fashion or does it need amplification?
- Is the question framed in a way that he can answer? It is vital that the question does not ask him to make a decision in a form that will usurp the authority of the architect, the contract administrator, the engineer or engineer's representative. There may be jurisdictional problems if a question is put to the adjudicator that he does not have authority to answer. At this juncture he should consider taking the initiative and telling the parties the question he is prepared to answer.
- If the question needs amplification, should the referring party be given the opportunity to set out something further in writing or should the adjudicator convene an immediate meeting and set about ascertaining the facts (and possibly the law) himself?

- If it appears appropriate to get something further in writing, how long should be allowed before it must be submitted?
- Are there any documents that appear vital to the adjudication that are not included in the referral notice?
- Does it appear that a site visit will be necessary and are there likely to be any restrictions in obtaining access to the premises?
- How is he going to organise a site visit? Should he allow the parties some flexibility in respect of dates or is he going to state a date and time (after ensuring access is available) and expect them both to fit in?
- Does it appear that any tests or experiments will be required?
- Does it appear that legal or technical advice will be needed? Is he required to notify the parties or obtain agreement from them to do this? How long will it take?
- Is he likely to be using his own knowledge or expertise? If he is going to use his own knowledge or expertise or obtain advice, the parties must be given the opportunity to comment. How will this affect the programme?
- Should the responding party be given the opportunity to respond in detail in writing? Do the applicable rules require him to allow such a response? If so, what period should be allowed? Do the contract or rules specify a period?
- Should this submission be restricted in length? Do the contract or rules require such a restriction?
- Should the responding party's response be restricted to a fact-finding meeting?
- What periods of time should be allowed for the parties to make further submissions?
- A meeting with the parties may well be needed. Should he set a provisional date at the outset or should he wait and see how things turn out before doing anything?

The adjudicator must, as noted above, in considering each of these points and any others that, in the circumstances, appear appropriate, ensure that he takes into account any rules that govern the adjudication.

Once the adjudicator has considered these matters and has decided what he is going to do, he must tell the parties what he intends. It is for the individual adjudicator to decide how he will do this but, in keeping with the spirit of adjudication, the method preferred by most adjudicators is a simple letter addressed to both parties setting out precisely what he is going to do and wants them each to do, and by when. Alternatively, it could be more formal and be set out as directions reminiscent of the arbitral process. This may not always suit the impression of the more informal nature of adjudication, but even the letter, however informal, should be framed in a way that indicates to the parties that the adjudicator means business and will expect them to comply in order that he can reach his decision in the allotted period.

In the case of a dispute that requires his attendance at site, a telephone call to each party, in conference mode if that is felt to be more appropriate, followed up by a written confirmation, might well be a way of getting things moving quickly, but communications by fax will generally be as quick as anything and avoid possible misunderstandings.

Progress of the adjudication

After the issue of the initial directions

The ideal adjudication will take the following path. After the referral the adjudicator receives a response followed, if appropriate, by a reply from the referring party. These can be written or oral as suits the circumstances of the case. He then has the time necessary to satisfy himself with regard to any matters upon which he requires clarification, either from the parties or by his own efforts. He will have time to take legal or technical advice as necessary. The parties will have sufficient time to comment upon the results of the adjudicator's own investigations and any advice received. The adjudicator will then have time to give proper consideration to all the information that he has to hand. All this should be completed in sufficient time to allow the adjudicator to set his conclusions down on paper and to ensure that he deals with everything that he has been asked to.

To achieve this it is vital for the adjudicator, once he has issued his directions, to keep a checklist which has every action set out on it, with a note of who has to do it, and when it must be done by. If he doesn't do this there is the danger that a date will be missed and the whole process will then be thrown into disarray.

Each direction must be reviewed frequently and adjustments may well have to be made as the reactions of the parties are received. The referring party will probably make every effort to comply but it is not inconceivable that the responding party will be less than helpful. In this event it is for the adjudicator to make it quite clear that he has his job to do and that he will be making his decision in the allotted time whether or not the responding party complies with his directions. This should at the very least bring such a party to its senses and to a realisation that there is no benefit whatsoever in not co-operating with the process. We look at what the adjudicator should do in the face of non co-operation by a party a little later in this chapter.

Whatever the adjudicator does, he must at all times remember his obligation to reach his decision within 28 days or any agreed extension to the time and he must subordinate everything else, other than his duty to act impartially and as fairly as the time limitations permit, to achieving this target.

Submissions by the parties

All standard form contracts, rules and the Scheme are drafted to enable an adjudicator to be appointed and for the dispute to be referred to him within seven days of the issue of the notice of adjudication. The requirements for this referral notice can vary from minimalistic to comprehensive. In most cases it is likely that the referring party will have done some homework beforehand and even where the requirements for the referral notice are slight there will be a substantial submission. Whatever the requirements are, the 28-day period will not commence until any specific requirements for the referral have been complied with properly.

Where the submission of the referring party appears long and complex, the immediate reaction of the adjudicator may be to seek more time than the normal 28-day period. The adjudicator may have to use some persuasive powers on the referring party in this circumstance, perhaps indicating that he will draw adverse inferences if additional time is not allowed and the responding party is not allowed a proper opportunity to answer all the points made. Any sensible referring party will be more than happy to extend the period but the adjudicator must be prepared for those with less amenable attitudes.

Save in the case of the CEDR Rules, there is no specific requirement in respect of the nature of the responding party's submission. This leaves the decision as to how the responding party is to acquaint the adjudicator of its position totally in the adjudicator's hands.

The responding party will, in all likelihood, wish to make a written submission and this is probably the preferable course in all normal instances. If the adjudication relates to a simple discrete issue there will be ample time for a response from the responding party, still allowing the adjudicator time to ask his own questions if he so desires and to prepare his decision.

If, however, the parties have allowed the dispute to escalate and there is a complex web of issues involved, the adjudicator may have to consider the imposition of procedures that may appear rather radical to those more used to the timescales allowed for arbitration and litigation. This is where the concept of adjudication must be seen as a separate function from the more traditional forms of third party dispute resolution. The normal procedure in litigation or arbitration is for the parties to set their stalls out formally by the exchange of pleadings or statements of case. This allows the parties to know what the other party's case is all about and to identify those matters that are in dispute and those that are agreed.

In adjudication there is no time for such detailed procedures. If the parties want such procedures it would be far better if they followed the arbitration route with the detailed checks and balances which that procedure provides. The spirit of adjudication is a speedy decision and the adjudicator should be looking for ways in which he can achieve such a result while endeavouring to get as near as he can to an answer that will place the parties in a position to finally resolve the dispute.

It is possible to take the responding party's submissions orally at a meeting or series of meetings but any party responding to a detailed written referral document of almost any kind will invariably want to respond in writing.

Whatever he does, the adjudicator should make every effort to allow for a response by the responding party. The adjudicator should avoid, if at all possible, having to decide the dispute on the basis of the referring party's submission and his own investigations. There is one particular reason for this and that is the overriding requirement upon the adjudicator to act fairly albeit this must be within the restrictions of the time limits imposed by the Act. In the law this concept is described as the 'rules of natural justice' which we looked at in the context of adjudication in the last chapter.

The time allowed for this response will be extremely limited. A cut-off date for receipt of information and evidence must always be borne closely in mind by the adjudicator. He should never accept the situation where the responding party's submission arrives too late for him to deal with it or more particularly for the referring party to respond to. What happens if it raises more questions than it answers? In any event it is not unusual for the responding party's response to raise points that are not covered in the referring party's original submission, and the adjudicator must allow time for these to be answered if they are within his jurisdiction. It must therefore be incumbent upon any adjudicator to allow a sufficient time after the responding party's response for the referring party's reply and his own investigations, when he sets his original programme and sends out his initial directions.

Parties do have a habit of not meeting deadlines and there could be a few critical days when the adjudicator is busy elsewhere and suddenly everything is delayed. If the delay has been on the part of the referring party, the adjudicator would, it is suggested, be right to tell that party that more time should be allowed and that they should agree to an extension of time. This can of course be done unilaterally. Any reluctance on the part of the referring party to agree to such a delay can in appropriate circumstances be met by the comment that adverse inferences could be drawn in respect of the refusal.

If the boot is on the other foot and the responding party has produced its response late, the referring party may well allow more time as it will probably want to consider and respond to it. The referring party is likely to make things more difficult for itself if it refuses an extension of time in these circumstances.

There will be times, however, when things conspire against the adjudicator and there just is not any time for a party to make a written response. The adjudicator may have decided that the matter is of such complexity that he wishes to obtain the responding party's submissions orally as has been suggested above. He may be aware of a large number of points made by one party that have not been answered by the other. Whether the answers are satisfactory or not, he is in difficulties without them. In these circumstances it is almost inevitable that the adjudicator will have to take proactive action to fill in the gaps, and convene a meeting between himself and the parties.

One matter that has developed as a problem for adjudicators since the decision of the court in *Nuttall* v. *Carter*, mentioned earlier, relates to questions surrounding the nature of the dispute referred and how restricted the responding party will be in making its defence. It is clear that the dispute that is referred must have been that which has crystallised between the parties. This means that new disputes cannot be referred and they must be aired between the parties in order that the adjudicator has jurisdiction. A referring party cannot raise new issues but what if the responding party wishes to develop its defence? It is possible that the dispute has reached a stage that the parties can take no further. When the referral is received the responding party realises that there are defences that it has not put to the referring party beforehand. Can these be put in the response? If *Nuttall* v. *Carter* is followed to the letter it would appear that they cannot. But what about fairness? Is it right to find that a claim succeeds because the responding party is prevented from making a proper defence? There are two principal scenarios. Firstly, the referring party can agree to the introduction of the new defence and an extension of time is agreed to allow the new matters to be addressed properly. Secondly, if the referring party refuses to do this the responding party puts its contentions to the referring party outside the adjudication and creates a new dispute which can be referred to adjudication. The responding party may then seek to resist enforcement of any award against it pending the finalisation of the second adjudication.

Meetings with the parties

The timescale within which the adjudicator has to work is such that often he will, as noted above, have to take proactive steps to obtain as much information as possible. This will ensure that his decision, when made, has the best chance of reflecting the contractual rights and obligations of the parties. One of these steps is likely to be meeting the parties.

It is self-evident that, unless the circumstances of the adjudication are most unusual, the parties themselves will have most, if not all, of the evidence that the adjudicator will need in order to reach his decision. The parties may be able to provide him with all he needs in writing by means of their submissions. He will not require a meeting in this event.

It may be that neither party wishes to elaborate on their written submission and resists attempts of the adjudicator to act in any way that might be considered inquisitorial. In this event the adjudicator will have to do his best on the basis of the written submissions, however unsatisfactory they may be. If the parties are set on an adjudication that is entirely on the basis of documents, we suggest that there is little if anything that the adjudicator can do other than go along with it. He could force the situation and insist on a meeting, but this should not be done if such a move would be counterproductive. It would be most appropriate for the adjudicator to set down any questions that he may have in writing and request that the parties provide him with their answers in writing as well. Where the adjudicator feels that he has been unable to obtain answers to everything that he would have liked, he may, if asked to give reasons, be likely to use the formula, 'on the basis of the submissions made to me' rather a lot, as he may be reaching conclusions that he feels may well have been different had he had the opportunity to make some more extensive investigations of his own.

More usually, the parties are receptive to the adjudicator's expressed desire for a meeting and where appropriate he should not be afraid to hold one. The authors would suggest that it is not described as a hearing; that sounds too formal and may put the parties in the wrong frame of mind. It should be described as a meeting and tend towards some level of informality as this will encourage rather less defensive attitudes on the part of the parties, and the meeting may be rather more productive as a result. This does not mean that the adjudicator should let the parties have a completely free hand; in that way lies disaster. It is vital for the adjudicator to maintain total control, particularly as disputing parties can express every ambit of emotion and there is nothing worse than a meeting that degenerates into a 'slanging' match.

That said, as long as the adjudicator remains entirely in control, it is often the case that a frank discussion between the people who are at the heart of the dispute, even of the 'you did, I didn't' variety, can lead the adjudicator to a much better understanding of the differences than the most expert of cross-examinations will reveal. Any risk of the meeting getting out of control is, in the authors' experience, quickly quelled by a quiet comment from the adjudicator such as, 'gentlemen, you are no longer assisting me'.

Another reason for holding a meeting is something that is well known in the field of mediation. That is the technique of 'venting'. It is often found that if the parties have the opportunity of telling a third party what is troubling them, blowing off steam and giving vent to their frustrations if you like, it can assist in reaching an agreement. This is a process that is also not unknown in arbitration, especially in smaller cases where the parties themselves are involved without representation. It is often done on documents alone but many arbitrators convene a short hearing to allow some face-to-face presentation to make up for any inadequacies in the documentary submissions and for the parties to make their feelings known. In all these cases the parties are often more likely to accept the end result if they feel that they have had an opportunity to acquaint the decision maker with their feelings on a face-to-face basis.

There is no reason why a similar technique should not be used in adjudication. There is considerably less time for the preparation of submissions, especially for the responding party, and it may well be that the parties will be more inclined to accept the decision as resolving the dispute if they have had a chance to make some form of oral presentation.

Before convening the meeting the adjudicator should have familiarised himself totally with the submissions of the parties and have identified the questions that he believes need answering. There is nothing wrong with setting out an agenda of the matters he wants to discuss and the questions that he wants to be answered. He should have made the parties aware of his agenda and if possible asked them if there are any matters that they want to be included in the discussions. This can be quite important if witnesses to the matters that are the subject of the dispute attend, either by invitation of the adjudicator or of the parties' own volition. There is little point in the adjudicator arriving primed to examine the witnesses in detail if what the parties really want is to carry out a cross-examination of the other side's witnesses themselves. In the time that the adjudicator has had available for preparation it is in any event far more likely that the parties themselves will know the right questions to ask to bring out the facts, than will the adjudicator himself.

It is for the adjudicator to decide whether he will allow the meeting to proceed in a way similar to a traditional formal arbitration hearing, and allow the parties to present their respective cases and then to ask questions himself, or whether to take a more inquisitorial stance from the start. Each individual dispute will have its own differing characteristics and personalities involved and an appropriate procedure will have to be devised accordingly.

Investigating the facts and the law

There is no obligation upon the adjudicator to take any positive action other than to set the procedure for the adjudication. He can just sit back, require the parties to make submissions to him, and, once these are received, reach his decision by a process of analysis of the documents before him.

The adjudicator will generally be able to ascertain the facts from the submissions of the parties, the relevant documents and from any questions that he puts to them. If he needs to carry out tests or investigations, or have them carried out on his behalf, he should have the technical expertise to ensure that these are done properly or, if done by others, to understand the results.

Investigating the law may be slightly different and more difficult for the technical adjudicator who is not familiar with that specific area. He may just take the submissions of the parties as to the contract and deal with them as best he can; in everyday terms he comes to a reasonable commercial view. The problem with this is that he may well not be properly fulfilling his obligations as adjudicator. As we have noted previously, the adjudicator's job is to ascertain the rights and duties of the parties under the contract. Where the parties have agreed that a commercial view can be taken then he may be able to act in this way, for example the TecSA Rules, but otherwise he may be skating on thin ice.

The adjudicator is obliged to apply the law in reaching his decision on the contractual rights and duties of the parties. A losing party to an adjudication will be understandably upset if the adjudicator reaches a decision that is patently wrong in respect of established law. The winning party could also be rather concerned if it is faced with a reference to arbitration or the courts that could have been avoided had the adjudicator applied a well established and well known point of law correctly.

It is therefore paramount that an adjudicator is able to apply the established law in a capable manner to the dispute he is called upon to resolve. Whether he does this from his own knowledge or by obtaining legal advice is for the individual adjudicator in any specific circumstances that he finds himself. One thing that has become clear over the first years of adjudication is that it is vital that adjudicators do understand the law of contract and if there is one area that has been a source of complaint it is that certain adjudicators have shown inadequacies in the knowledge and application of the law of contract.

It is not for a practical guide such as this to attempt to offer any guidance in respect of the law. In any event it is not a requirement for an adjudicator to have an exhaustive knowledge of the subject. What the adjudicator needs is an awareness that certain concepts exist and to be able to recognise when he may be getting out of his depth and as a result that it might be appropriate to obtain legal advice. A few pointers follow.

First, and most importantly, the adjudicator must have an understanding of the way contracts are put together. He must be able to understand the concepts of offer, acceptance and consideration. He must be able to identify when a contract is formed and what terms it contains. He must be conversant with how the actions of the parties before and after the contract is formed are admissible as evidence in construing a contract. Given that contracts in the construction industry are very often made using standard forms which are supplemented by large numbers of supporting documents, the adjudicator must understand the rules governing the primacy of these documents in the event of conflict.

We would also go so far as to say that an adjudicator must have some idea of the concept of rectification and the parameters that apply in a situation where the matters put to him encompass the argument that the contract as written does not reflect the agreement that was reached. The adjudicator does not have authority to deal with such matters in disputes

arising under the contract; however, some contracts may give him such authority.
be able to identify those points over which he has jurisdiction so that he can elimir
over which he does not.

Section 108(1) of the Act limits the right of a party to refer a dispute to adjudication to one
that arises under the contract. This means that a dispute must arise under the contract and
not in connection with it unless the contract so provides. An adjudicator should therefore be
able to differentiate between them. If he cannot do this, he is at risk of incurring the parties in
considerable costs that could later be shown to be abortive. A losing party might well seek to
avoid a decision being enforced by showing that the adjudicator exceeded his jurisdiction by
dealing with a dispute that does not arise under the contract. One point that the adjudicator
must be aware of, however, is that whilst negligence is a tort and a dispute in this respect
does not arise under the contract, what is commonly called 'professional negligence' can
often in fact be a breach of contract.

An adjudicator should also be aware of the basics of the law of evidence and specifically to
recognise relevance and hearsay. He must be able to understand the concept of giving 'due
weight' to evidence put before him.

There is one set of circumstances in which the adjudicator will have to take active steps to
ascertain the law and that is when he is adjudicating under a regime that places specific
obligations upon him in this respect. The Scheme is one such. Paragraph 12(a) of the Scheme
requires that the adjudicator 'shall reach his decision in accordance with the applicable law
in relation to the contract'. It can be that the submissions of the parties to an adjudication that
is conducted under the Scheme do not even remotely touch on the legal rights and duties of
the parties in any ordered way. The Scheme does, after all, only operate when the parties
who have entered into the contract are not using a standard form and, unless the parties are
represented, they are thus likely to be rather less sophisticated and knowledgeable of legal
and contractual concepts. If the dispute requires the adjudicator to interpret the terms of the
contract he will have an obligation to make his decision in accordance with the law
applicable to the contract. If he does not, he is in breach of the terms of his undertaking as
adjudicator.

Taking advice

The adjudicator may take advice of a technical or legal nature. It is to be hoped that adjudicators will not generally need to take technical advice. It is almost inevitable that on
occasion the non-legally qualified adjudicator will feel that he must take legal advice in
order to deal properly with the matter put to him for his decision.

It has already been said that the adjudicator is making a statement of the parties' rights
and duties under the contract. If he does not have sufficient legal knowledge himself to deal
with a specific dispute he may well consider it appropriate to take legal advice. There is a
requirement under the Scheme that the adjudicator shall reach his decision in accordance
with the applicable law in relation to the contract. He should therefore do everything that he
can to reflect the applicable law. He may have detailed submissions from the parties in this
respect and have to choose between them if they differ. In this case he may not need to take
legal advice. Alternatively, there may be no such submissions and if he is not legally qualified he might well feel safer taking legal advice rather than risking making his own
interpretation of the situation, which may subsequently turn out to be patently wrong.

It must be a cardinal rule for any adjudicator, unless time just does not permit, to allow the
parties to comment upon any matters that he ascertains for himself before he reaches his

decision. This must include any specialist advice that he obtains. The person giving the advice may take up a fair amount of the time that is available for the adjudication and the earlier that such advice is sought the better. There is then the greatest possible chance of the parties having an opportunity to comment upon the advice as they choose. Even the time factor may not always be too much of a problem. A suitably worded confirmation of the reasons why the advice is needed ought, if framed in a suitable way, to result in the parties agreeing to allow further time. The adjudicator must be wary not to fall into the trap of applying any findings that his researches may uncover to his decision without allowing the parties the opportunity to consider and make submissions upon them (*Balfour Beatty* v. *Lambeth*[4] and *RSL* v. *Stansell*[5]).

There are certain situations where the adjudicator may not be in a position to make arrangements to obtain advice immediately after he has read the referring party's submission. The referring party's submission may need amplification; it may for instance not include the contract between the parties. There may be a requirement placed upon the adjudicator to inform the parties of his intention to take advice before he takes it. In some instances there may be a requirement to obtain agreement from one party before he does so. Cost estimates may have to be given. It may in fact be as a result of the responding party's response that the adjudicator decides that he should take advice. The adjudicator may well find that he is unable to obtain the advice that he considers that he needs within the set timescale. In this event the adjudicator is very much in the parties' hands. He should make every effort to persuade the parties that they should extend the time necessary and it may be that they both agree. If one agrees and the other does not, it is for the adjudicator to draw the appropriate inference. If neither agree, the adjudicator will just have to get on with the adjudication and accept that his decision will not be as satisfactory as it otherwise might be.

Whenever advice is sought it is on the basis that the adjudicator will take account of that advice in reaching his decision. Legal or technical advice is not a substitute for the adjudicator reaching the decision himself. It is not part of the task he is there to perform to 'subcontract' out of parts of the process. In the final analysis the decision is that of the adjudicator and nobody else.

Obstacles in the process

The reluctant party

As distinct from arbitration, which is a consensual process in which the parties choose to have an arbitration clause in their contracts, adjudication provisions in contracts are required by legislation. The right to refer a dispute to adjudication can be exercised by one party to a contract alone. An arbitrator has to make every effort to ensure that a party who does not initially take part in the arbitration has been given every opportunity to do so before he proceeds in that party's absence. The position in adjudication is different; the time limit is paramount. There must be an obligation on the adjudicator to attempt to contact the responding party, but, given the timescale involved, he must fulfil his primary obligation, to reach his decision, even if he is unable to get any reaction from the responding party.

The failure of the responding party to take any action means that the referring party may

---

[4] *Balfour Beatty Construction Limited* v. *The Mayor & Burgesses of the London Borough of Lambeth* [2002] BLR 288.
[5] *RSL (South West) Ltd* v. *Stansell Ltd* (16 June 2003).

have problems of enforcement but this is not a matter that should concern the adjudicator, at least not directly. The adjudicator must of course ensure that he carries out his duties in accordance with the contract or any applicable rules and there may be specific requirements with which he has to comply as a result. If he fails to act as required by the contract or rules, his decision will possibly be unenforceable.

The adjudicator has two choices when faced with an unco-operative party. He can proceed simply on the basis of the referring party's submission or he can decide to investigate the facts and the law. Proceeding in accordance with the referring party's submission effectively means accepting it at face value. Even the most inexperienced adjudicator must recognise the difficulties that this will ultimately present. Claims are generally put together in the most favourable light for the referring party and a totally uncritical acceptance of the referring party's contentions will, in all likelihood, mean that the decision will be rather more in favour of the referring party than it perhaps should be.

The adjudicator may also come to the view that the claim is exaggerated. In this case it would be remiss of the adjudicator not to endeavour to acquaint the referring party of the position and give him an opportunity to respond, but as with all matters in adjudication the timescale is paramount and this may not always be possible.

This situation arose in an arbitration where the arbitrator came to the view that the claim was vastly exaggerated and a 'try on'. In *Fox* v. *Wellfair*[6] the arbitrator reduced the claim substantially and on appeal was found to have misconducted himself in failing to put his alternative evidence to the claimant's expert.

The adjudicator should always remember the possibility that what appears to be an exaggeration may in fact be an inadequacy in the evidence that can be remedied. If this is the case he will have rendered the referring party a service and that party may request the adjudicator to make a 14-day extension of time to remedy the inadequacy.

If the adjudicator is required to reach his decision in accordance with the applicable law, as he is when operating under the Scheme, he will have to take specific action in this direction as he will not have the responding party's views as to what the applicable law is. This may well mean taking advice. It is almost certain that the unco-operative party will contest the decision and consequently it is very important that the adjudicator does not give that party an obvious opportunity by producing a decision that clearly does not properly reflect the applicable law.

The drawing of inferences by the adjudicator

We have on a number of occasions suggested that the adjudicator will draw adverse inferences from the actions or non-actions of a party. As a general rule if a party refuses to do something that quite evidently should be done to assist the adjudication process, it can be reasonably assumed that that party believes that the action, if taken, will be to its own disadvantage.

This is where the adjudicator would be perfectly within his rights to seek to uncover what that party might be trying to hide. Time may, however, preclude this, and in any event it must be remembered that the adjudicator does not have the same powers to order discovery as does an arbitrator and the information may remain concealed. The adjudicator would then be quite within his rights to draw the inference that the information that is not offered or that cannot be obtained would be damaging to that party's cause.

---

[6] *Fox* v. *Wellfair* (1981) 19 BLR 52, CA.

The 'commercial' decision

The first edition of the ORSA (now TeCSA) Adjudication Rules allowed the adjudicator, 'if it appears to him to be impossible to reach a concluded view upon the legal entitlements of the parties within the practical constraints of a rapid and economical adjudication process' to make his decision on the basis that it 'shall represent his fair and commercially reasonable view'. ORSA, who took a bold step in allowing adjudicators to make commercial decisions, have reviewed this provision and it has now been amended to read:

> 'Where it appears to the Adjudicator impossible to reach a concluded view upon the legal entitle-
> ments of the Parties within the practical constraints of a rapid and economical adjudication process,
> his decision shall represent his fair and reasonable view, in light of the facts and the law insofar as
> they have been ascertained by the Adjudicator of how the disputed matter should lie unless and
> until resolved by litigation or arbitration.'

This is the only such provision of which the authors are aware. It is unlikely that any such provision will appear anywhere else.

A commercial decision has connotations of allowing a departure from the establishment of the strict contractual rights and obligations of the parties. It also suggests a lower standard of proof than 'on the balance of probabilities' and even a decision which may not relate to the evidence.

One other question relating to commercial decisions is whether the adjudication process has to be 'fair'. We have already looked at this point in some detail in Chapter 9. As a process it has to be as fair as it can be in the time available and comply as far as is possible with the rules of natural justice. It is not, however, within the remit of any adjudicator to try to make a fair decision in the face of the contractual provisions, which may on the surface appear extremely onerous on one party to the benefit of the other. If that is the contract that they have entered into, the adjudicator should always remember that the contract agreement will, or should, have reflected the risks involved. He should also remember that he does not actually have the power to do anything other than find in accordance with that contract.

What can be adjudicated?

Disputes relating to design

One area that it is suggested will cause difficulty is where the adjudicator assumes a design responsibility and thus takes it away from the design team. It is unlikely that many adjudicators will fall into this trap. The typical example is where a builder has been instructed to demolish and rebuild a wall. He applies to the adjudicator for a counter instruction to the effect that the wall should remain. This is not the adjudicator's function. If the architect/ contract administrator/engineer has instructed that the wall should be removed, that instruction must be complied with unless the architect/contract administrator/engineer has no power to order such an instruction. The adjudicator must not take away any design responsibility from the members of the design team. They are responsible for design and must remain so. The adjudicator has contractual immunity and if he gives an instruction for work to be done or retained that ultimately proved defective, the employer has nowhere to seek redress. The questions that the adjudicator must answer are whether the wall as built complies with the specification or has the architect/contract administrator/engineer issued an instruction that properly accords with the contract. If the adjudicator finds that the wall does comply with the specification, the wall still comes down but the employer pays for the demolition and reconstruction in addition to the first time it was put up. If the adjudicator finds the instruction was not issued in accordance with the contract, the architect/contract administrator/engineer will still have the opportunity to issue an instruction that does comply and the wall can still be taken down.

The end result is that the employer may think he is paying for the wall twice. All that has really happened, however, is that the onus of seeking redress has passed from the contractor to the employer. The employer will have to make the decision as to whether it is really worthwhile going through the litigation or arbitration process in the hope of recovering the money. This is what contractors have always had to do. The boot is now on the other foot. Employers, and contractors in a similar position in relation to sub-contractors, will argue that putting the money with the contractor, or sub-contractor, means that there is a risk that the party holding the money will go bust before the money can be recovered. This is true, but isn't this the risk that contractors and sub-contractors have always run? In any event, there is a substantial argument that runs to the effect that the employer or contractor had the whip hand when letting the contract and they always had the choice not to enter into the contract if they felt that there was a chance that the other party was financially unsound.

There is one other reason for the adjudicator to avoid getting involved in design and that is that his immunity is contractual only. A subsequent owner of the building would not be restricted from suing the adjudicator if something that he instructed either to be built or to remain later proved defective.

When a notice of adjudication is framed in such a way as to invite the adjudicator to give an instruction, the prospective adjudicator has the option to refuse the nomination. If it were the referral notice that offends in the absence of a prior indication in the notice of adjudication, the adjudicator would, in our view, be right to decline to proceed at least in respect of those parts of the referral not covered by the notice of adjudication. These points actually go to whether the adjudicator would have jurisdiction to do what is being asked of him, as well as what liability he might have if he carried out a task that was beyond his remit. If the adjudicator were simply to resign in these circumstances there would be nothing to prevent the parties putting the same question to any number of other adjudicators until they found one who would act. This would not of itself make the question or the decision valid. Most matters concerning design and quality can be answered by a declaratory-type decision which says yes the wall is in accordance with the contract or no it is not. They can also be answered on the basis of reducing the matter to a sum of money. The wall was in accordance with the contract and the contractor is entitled to the cost of rebuilding it. It is simply a matter of procuring the right questions from the parties on which to make a decision. There is nothing to prevent the adjudicator writing to the parties and stating 'having read your notices and submissions these are the questions I am prepared to answer and this is where I get my authority to do so'. Most pragmatic parties will be persuaded by such an approach.

One further point relating to design is where the dispute relates to work that must be 'to the architect/contract administrator/engineer's reasonable satisfaction'. This is not a major problem for the adjudicator who is familiar with the normal standards that would apply to work instructed by an architect/contract administrator/engineer. It might be more difficult for an adjudicator who is, for example, rather more familiar with the costing of building work. This may be an instance where the adjudicator should say at the outset on receiving the referring party's submissions that the adjudication should go to someone who is more familiar with the specific issues involved. If such an issue comes up as a limited part of a wider dispute it may be that technical advice from someone experienced in the specific matter that is outside the adjudicator's own experience can be obtained very quickly and the problem is resolved in that fashion.

Disputes relating to extensions of time

We have already identified that the adjudicator would be very unwise to usurp the design responsibility of the contract administrator. This is equally so in the case of disputes relating to extensions of time. The adjudicator must be exceedingly careful not to usurp the certifier's responsibility. All he should do is to make a decision that the contractor is entitled to an extension of time of x weeks for the reasons stated. The adjudicator cannot grant the extension of time; all he can do is declare the entitlement. The contractor will then be absolved from the responsibility of paying liquidated and ascertained damages. Extensions of time of themselves only give relief from liquidated and ascertained damages; they are not, unless the parties have so agreed, a prerequisite or condition precedent to entitlement to loss and expense. In answering questions concerning extensions of time, the adjudicator should not stray into the area of relief from liquidated and ascertained damages or loss and expense unless he is asked to do so. Where an adjudicator's decision declares that there is an entitlement to an extension of time of x weeks, it will be for the architect/contract administrator/engineer to grant that extension of time and thus provide relief from liquidated and ascertained damages. The parties are bound by the decision until such time as it is heard anew in arbitration or litigation or until they agree it is final. Such a decision is a defence where the employer then seeks to deduct damages where his architect/contract administrator/engineer has declined to grant the extension of time.

Claims under professional indemnity insurance

Professional service contracts fall within the remit of the Act and therefore disputes which arise under them are subject to the right to adjudication. The establishment of claims against professional men or others carrying out similar services and the rebuttal of such claims is a process that may involve a great deal of investigative work. Such claims are, in the authors' view, matters that will create considerable difficulties in accumulating the necessary evidence if the timescales set out in the Act are not relaxed. It may be that both parties will relax the timescales but, if such extensions are not allowed, the adjudicator will just have to do his best in the time available. He will certainly have to take the utmost care in the production of his decision. Professional indemnity (PI) insurers back most professionals. The adjudicator has no jurisdiction to decide that a PI insurer meets the claim as a result of the decision. The contract of insurance itself is exempt from the operations of the Act by virtue of the Exclusion Order. The insurers will nevertheless be faced with 'picking up the pieces' as a result of any decision that goes against the insured party.

It is reasonable to say that insurers were somewhat late in taking grasp of the effects that adjudicators' decisions might have on the insured and therefore on the contracts of insurance. There are anecdotal reports of insurers making unrealistic and unreasonable demands on the insured if they are to hope that the appropriate cover remains in place following the decision. There are reputedly insurers demanding notice of any adjudications that occur on contracts whether or not the insured is a party to the dispute. These matters do not affect the adjudicator. The adjudicator must perform the task he is there to perform and not be influenced by whether or not insurers will meet the claim.

The note of caution here is for all those who are covered by professional indemnity insurance. They should be aware of any requirements of their insurers as to whether or not they are actually insured for the effects of an adjudicator's decision and what they must do to ensure they remain insured when a notice to refer is received.

The 'unadjudicable' dispute

There have been many debates about the 'unadjudicable' dispute. The first draft of the Scheme, which reached the House of Lords' committee stage, had a provision to deal with the unadjudicable dispute.

Paragraph 32 of this draft Scheme stated:

> The adjudicator may terminate proceedings in respect of any matter if, or to the extent that, he decides that the matter is not suitable for resolution by the procedures under the scheme. This will constitute all or part of his award, and a party may not then lay the same matter before another adjudicator unless agreed with all other parties.

This matter was then revisited in the consultation document, which was the forerunner to the current Scheme. Under the heading of 'Issues to resolve', Question B39 stated:

> What should happen if the adjudicator indicates that he is unable to make a decision?

Two answers were given to this question, neither of which, to the authors, seemed satisfactory. Response B39.1 stated:

> parties may treat the procedure as in failure and start again with another adjudicator. To make the appointment any party may choose whether to rely on any existing agreement or use an appointing body specified in the Scheme.

Response B39.2 stated:

> in such cases only reasonable expenses incurred by the adjudicator need to be met, and parties are not obliged to pay fees.

The situation we are now left with is less satisfactory than that proposed in the first draft of the Scheme. This would have offered total control over the ambush situation and the unadjudicable dispute. To the authors' knowledge there is no situation that would allow the adjudicator the powers expressed in the first draft of the Scheme.

The adjudicator is required to make a decision within the 28-day period or any extended period. The construction industry has got adjudication, warts and all. The industry and the adjudicators must make the system work. For this to happen adjudicators must not be looking for reasons not to adjudicate; they must provide decisions.

The basic principle that the authors would espouse is that there are few disputes that are unadjudicable. A decision can always be made on the evidence presented whether or not this evidence is extensive. An adjudicator may be faced with a submission that includes a great number of documents submitted as evidence that may be contained in a large number of lever arch files. The response from the other party may be similarly lengthy. It is not appropriate for the adjudicator to throw his hands up and say that it is impossible to reach a decision in the given timescale on receipt of such submissions. The submissions themselves will probably be relatively short. The voluminous documents that accompany the submissions are generally by way of back-up evidence. They cannot be totally ignored but a pretty good view of their content can be gleaned from a consideration that is not as exhaustive as it might be were there more time.

People should not put themselves up as adjudicator if they are incapable of identifying and understanding the basic contentions contained within a lengthy written submission. It is perfectly within the capability of a person experienced in the ways of the industry to consider a lengthy submission and to pick out the matters of importance within it. If the adjudicator cannot do this he should not be offering his services in this field.

It is obvious in these circumstances that the adjudicator will have to cut corners and probably make intuitive steps. He will not be able to analyse the submissions in detail but will have to take a view based on the balance of probabilities.

One example of this that might be considered to cause difficulties is the disruption claim by a contractor. For an entitlement to costs resulting from disruption it is necessary to show that the disruption suffered has resulted directly from the actions of the other party. It is relatively easy to make a 'judgement' that the referring party has been prevented from proceeding efficiently with the contract works, for example because there have been a considerable number of variations, or because there have been deficiencies in the provision of information, or that a sub-contractor has been disrupted because of a delay in the completion of earlier works. It is rather more difficult to decide the extent of the disruption and the financial entitlement that results. This is where the adjudicator is going to have to make use of his powers to ascertain the facts. He will have to do everything he can to find out what actually happened. He will have to work very hard. He will have to interview site staff, head office staff and, if allowed, members of the professional team and the employer's own staff. He will have to put all this information together and come to his decision. He must not concern himself with the possibility that a fuller forensic investigation in a later arbitration or litigation may show up something different. He cannot be criticised if he makes every effort to ascertain what went on and reaches his decision on that basis.

One thing that the adjudicator must realise in dealing with matters such as disruption is that he makes his decision on the balance of probabilities. He must get away from the mindset that is sometimes apparent in the approach of the professional team to disruption claims by contractors or by contractor's staff when in receipt of a claim from a sub-contractor. It is not unknown for a surveyor, in carrying out his contractual obligation to 'ascertain' the amount of direct loss and/or expense, to require that the organisation making the claim establishes liability to a standard that is to all intents and purposes 'beyond reasonable doubt'. This is in fact a higher standard of proof than is required in a civil action or arbitration. If the adjudicator is satisfied from the information presented to him that the disruption that is evidenced by the referring party was, on the balance of probabilities, caused by the responding party, he should frame his decision accordingly.

Unless the adjudicator is empowered to make his decision on the basis of a 'commercial view', the decision must always be based on the evidence and weighed using the balance of probabilities; it cannot be an intuitive leap. The acid test is always: 'has the claiming party proved its case on the balance of probabilities?'

One point when considering disputes that may at first appear to be unadjudicable, is to remember the opportunity that the adjudicator has to draw inferences. A party may refuse to allow additional time for a proper examination of their claim or perhaps for tests to be carried out. This refusal may appear unreasonable to the adjudicator. If it does the adjudicator is at liberty to draw the inference that the fuller examination or the tests might well show that the case put up by the party who is refusing to co-operate is not as convincing as it may at first appear.

The large and detailed claim for loss and/or expense

A specific type of claim that might in some eyes come within the category of disputes that are possibly unadjudicable is a claim for loss and/or expense.

There are two court decisions that provide some guidance relating to the way an arbitrator should deal with such claims, which are of assistance to adjudicators. In *McAlpine* v. *Property and Land Contractors*[7] Judge Humphrey Lloyd QC said:

> 'Furthermore "ascertain" means "to find out for certain" and it does not therefore connote as much use of judgment of an opinion as had "assess" or "evaluate" been used. It thus appears to preclude making general assessments as have at times to be done in quantifying damages recoverable for breach of contract.'

These words were examined by Dyson J in one of the many *How* v. *Lindner*[8] judgments:

> 'Judge Lloyd applied this approach when answering a question of law that had been raised in relation to a claim in respect of plant. He said that in ascertaining direct loss or expense, the actual loss or expense incurred must be ascertained and not any hypothetical loss or expense that might have been incurred by way of assumed or typical hire charges or otherwise.
>
> I do not understand Judge Lloyd to be saying that there is no room for the exercise of judgment in the process of ascertainment. I respectfully suggest that the phrase 'find out for certain' might be misunderstood as what is required is absolute certainty. The arbitrator is required to apply the civil standard of proof.'

The only standard form contracts that actually require an 'ascertainment' of loss and expense are the JCT 98, IFC 98 and JCT with Contractors Design 98 Contracts. The adjudicator has to reach his decision in accordance with the contract and when asked to determine loss and/or expense under these forms he should take into account the contract requirement that this be ascertained but in doing so he should bear in mind the comments of Dyson J.

The requirements of the other standard forms do not include the word ascertain in connection with the determination of entitlement relating to delay or disruption:

- In the ICE contracts the words 'loss and expense' do not appear. The term that is used is 'cost' which appears in various places. It can be cost 'that the Engineer considers fair'. The cost 'as may be reasonable' or the cost 'as determined'.
- The NEC Engineering and Construction Contract uses the word assess.
- The IChemE contract uses the word evaluate.
- GC/Works says that loss and expense shall be 'determined'. This must, in the eyes of the drafters of that contract, mean something less than 'ascertain' as that was the word used in the 1977 edition of this contract.
- The JCT Minor Works 98 form says that the architect shall value loss and expense 'on a fair and reasonable basis'.
- DOM/1 states that the recovery of direct loss and/or expense shall be 'the agreed amount'.
- DSC/C states that any direct loss and/or expense shall be 'an amount agreed by the parties'.

The adjudicator may have to consider making a request for more time if faced with a detailed and lengthy claim, but we would suggest that none of these formulae for the

[7] *Alfred McAlpine Homes North Limited* v. *Property and Land Contractors Limited* (1995) 76 BLR 59.
[8] *How Engineering Services Limited* v. *Lindner Ceilings Floors Partitions plc*, CILL July/August 1999 p. 1521.

determination of the amount a referring party is entitled to recover in respect of loss and expense, including the main JCT 98 contract and its fellows, should give an adjudicator cause to think that the dispute is unadjudicable.

The adjudicator who decides 'I cannot decide'

Can an adjudicator say, 'my decision is that I cannot decide'?

On the basis of the discussion set out in the paragraphs above, the authors have some difficulty in formulating a situation where an adjudicator comes to a decision that he cannot decide, but it is a question that is often asked. This is not in the least bit helpful.

There is in the authors' view no reason why the adjudicator should make such a decision. His duty is to state the rights and duties of the parties, albeit on a temporary basis. The adjudicator will not have performed the task he has been employed to do and any such decision might well be void. What this situation really amounts to is that the claiming party will have failed to prove his case. Where the weight of argument and evidence is equal between the parties, the claiming party will still not have done enough to prove his case and therefore loses. That is the decision the adjudicator should make rather than saying 'I cannot decide'.

The parties will realise the strengths and weaknesses of their respective positions and as a result settle or abandon their differences or reframe their strategy for the next 'round' in arbitration or litigation.

There is clearly a conceptual difficulty in deciding that a claiming party loses in such circumstances, especially if the chances are that additional evidence will be available. The real nub of the difficulty in such situations is the lack of time for the adjudicator. If more time is made available it will always be possible to adjudicate. The difficulty will only occur when one of the parties will not grant further time to reach a decision. The adjudicator is then faced with the situation that if he adjudicates on the basis of the documents that he has, it may be suggested that he has shown bias to one of the parties. If he is biased he has failed to act impartially or with fairness.

The solution to this lies in the relationship between the parties and the adjudicator. The three parties (or more) are in a contractual relationship. The terms of that contract will require that the adjudicator reaches a decision within 28 days. The terms will also include the ability to extend time by 14 days with the consent of the referring party and by a longer period with the consent of both parties. It therefore forms part of the contract that there will be occasions where time needs to be extended before a decision can be reached. It will form part of the implied term as in the contract between the adjudicator and the parties that the parties will do all things necessary to allow the adjudicator to adjudicate and reach a decision. There should be no difficulty in implying this term into the contract between the adjudicator and the parties. This is a common implied term in contracts; where one party is to perform a task, the others must facilitate the performance of that task. If one of the parties seeks to prevent the performance of this task by not allowing sufficient time, this is a breach of the implied term and will amount to repudiation of the contract. It would amount to repudiation as the adjudicator is prevented from carrying out the task that he was employed to do. The adjudicator can treat the contract as at an end and would be entitled to his fees and expenses up to the date of the repudiation. When faced with this the reluctant party will probably allow the time needed rather than abandon the adjudication and be faced with a similar scenario in a further adjudication.

Some concerns about adjudication

In the first edition we said that a number of matters had been raised by those wh ... that adjudication was likely to cause more problems than it solved. In view of the support for adjudication by the courts, many of these fears have been addressed but it is still worth reviewing these points.

The first point that was identified was the fact that the losing party may be unable to recover monies that have been paid because of an adjudicator's decision. This would happen, for example, where the winning party goes bankrupt before the subsequent arbitration is complete. As we have said already, a building contract is an allocation of risk. Contractors have worked for years under the threat that they may be entitled to recover money for work done but may be unable to do so. The payment of monies to a contractor who may go bust is doing no more than shifting the risk. The court has however stated that where there was a real doubt as to the receiving party's ability to repay, a stay of execution might be granted pending final determination of the proceedings[9]. In at least one case[10] the court has decided that the adjudicator's decision should not be enforced due to the financial position of the winning party.

Another objection, as already discussed, is where the employer may have to pay twice for work condemned by the architect/contract administrator/engineer but found to be in accordance with the specification by the adjudicator. Instead of the contractor only being paid once, although he has carried out the work twice, and having to seek recompense later, the employer is the one who pays twice. Ultimately, the end result should be the same but instead of the employer benefiting from any reluctance of the contractor to open up such matters after completion of the works, the contractor may benefit in the same way.

Another concern expressed was that the adjudicator's decision would produce a less accurate result than litigation or arbitration. Our comment remains the same. It is unlikely, in the timescales allowed, that the result will be as accurate as the full forensic investigation that is carried out in arbitration or litigation. This is true, but who, save for very few, can afford the luxury of a full forensic investigation? The vast majority of arbitrations and matters dealt with by the courts settle before the arbitrator or the judge has pronounced on the matters in issue. Where is the accuracy then? At least with adjudication the party that may be less well financed and thus forced to settle at a disadvantage will have the benefit of the conclusions of an impartial third party. The party is likely to receive a recompense that is closer to its actual contractual entitlement than it would have been if it was forced to settle due to having inadequate funds to continue with an arbitration or litigation.

We also considered the possibility that the adjudicator may undermine the authority of the architect or other person supervising the works. This is a matter for each individual practice. Design matters have been considered at some length earlier in this chapter but what must be remembered is that one of the reasons that adjudication was introduced was a perception that architect and engineers do not necessarily always carry out their functions in the totally impartial way that the contract assumes.

While these matters are all still of understandable concern, the introduction of adjudication has meant a change of culture in the construction industry. Adjudicators may sometimes get their decisions wrong but the courts have expressed a clear support for the process. Adjudication has had a definite effect on the cash flow of the industry and this has been with no obvious evidence that it has done anything other than what it was intended to do.

---

[9] *Herschel Engineering Limited* v. *Breen Property Limited* [2000] BLR 272.
[10] *Ashley House Plc* v. *Galliers Southern Ltd* (15 February 2002).

The adjudicator and subsequent legal action

In concluding this chapter on adjudication practicalities one particular question that is asked is worth considering: 'Is an adjudicator likely to get involved in later legal action?' The answer, after 170+ cases that have been through the courts, is that it is unlikely.

There is one case, *Woods Hardwick* v. *Chiltern*[11], where the adjudicator was involved in the proceedings to enforce his decision in which he provided evidence in support of the application to enforce. This did not assist the party seeking enforcement in that the adjudicator's involvement was seen as evidence of bias.

---

[11] *Woods Hardwick Limited* v. *Chiltern Air Conditioning* [2001] BLR 23.

Chapter 11
The Decision

Section 108 of the Act includes the word 'decision' four times. In none of these instances is there any indication of the form that the decision is to take. The Arbitration Act 1996, by comparison, does set out certain requirements relating to an arbitrator's award, but even these are very limited. By the Arbitration Act 1996 an arbitrator's award must be in writing, be signed and contain reasons unless agreed otherwise by the parties. It must also state the seat of the arbitration and the date that the award is made. You will look in vain for even such limited provisions in the Act regarding the adjudicator's decision.

The references to the decision of the adjudicator in the Act are as follows:

108(2)(c)   'require the adjudicator to reach a decision within 28 days ...'
108(3)   'The contract shall provide that the decision of the adjudicator is binding until ...
            The parties may agree to accept the decision of the adjudicator as finally determining the dispute.'
108(6)   'For Scotland, the Scheme may include provision conferring powers ... relating to the enforcement of the adjudicator's decision.'

There is absolutely nothing there that requires or guides the adjudicator as to the way he should formalise the decision. All that is required of the adjudicator is that he reaches a decision. The Act does not even require that the adjudicator puts his decision in writing or that he delivers his decision to the parties at all, although this must be implied for the process to have any effect. This situation reinforces the point that any procedures relating to the adjudication arise out of the contract between the parties and not directly from the Act itself.

The requirements of the various adjudication rules applying to construction contracts are dealt with in detail in Chapter 6. It is, however, worth noting here that there is little, if any, guidance as to the form that the adjudicator's decision should take in any of the published rules or in the Scheme.

As with all other aspects of the adjudication process, the adjudicator must make himself fully aware of the requirements of the contract in this regard. These requirements will obviously include any applicable rules. He must follow the required procedures to the letter. If he does not, he is likely to be in breach of his own contract with the parties. Not that he will necessarily suffer as a result, due to the immunity provisions that the contract is required to contain, but, if an adjudicator fails to fulfil any express or implied duty that he has under his own contract with the parties, the result may not be the quick efficient resolution of the dispute that ought to be the desired end to the adjudication process.

If, however, the contract, any applicable adjudication rules or any adjudicator's agreement entered into upon appointment are silent as to the form that the decision must take, it is for the adjudicator alone to decide how he wishes to convey his decision to the parties.

These points apart it is apparent that the adjudicator is, in most circumstances, very much on his own when he reaches the point of having, in the words of the Act, to reach his decision, and as a consequence, to draw the adjudication process to a close in respect of that particular dispute.

It is essential, first, to consider what it is that the adjudicator is trying to achieve with his decision.

In the broadest of terms the purpose of the decision is to acquaint the parties of the adjudicator's decisions on the various issues that have been put to him, both clearly and in a form that is enforceable by the court if necessary. A supplementary purpose that has, in our opinion, clearly developed from the success of the adjudication process in producing decisions that generally mean that the dispute goes no further, is an endeavour to put the parties in the position of being able to resolve the dispute. It must of course always be remembered that, as distinct from arbitration, the adjudication process is not a dispute resolution process in itself, either party being able subsequently to take the dispute on to arbitration or litigation if desired. The most the adjudicator can do is to produce a decision that enables the parties to accept it and thus resolve the dispute.

It is self-evident that there is no point in the adjudicator just reaching a decision and then doing nothing with it. He must convey that decision to the parties. He may even consider the possibility of doing this orally. If the dispute relates to a clearly defined argument concerning a specific matter, there is nothing to stop the adjudicator meeting the parties on site, hearing what they each have to say and telling them there and then what his conclusions are. The parties may not even want a written record of that decision, but this is very unlikely.

This approach is of limited application. However desirable the 'referee' approach is in keeping the project moving, in the majority of disputes that require the services of an adjudicator the parties will want to see his decision in written form. Even in the case of the instant decision on site it is likely that some written confirmation of that decision will be required by at least one of the parties.

The majority of adjudicator's decisions, if not all, will therefore be conveyed to the parties in written form and, as discussed above in this respect, the adjudicator will be very much on his own unless a specific form of decision is required by either the contract or by his agreement.

Before considering the form that the decision should take, it is necessary to consider in a little more detail what it is that the adjudicator is trying to communicate to the parties through his written decision.

His first purpose, as has already been noted, must be to convey the nature of his conclusions to the parties. He must do this in a manner which ensures that each party knows exactly what that decision is, how it affects them individually, what they have to do and what they can expect the other party to do as a result. It must be clear as to its effect. In order to achieve this it should be set out in such a way that it is readily understandable. It should deal with all the issues that the parties require to be decided. Each issue that is readily separable should be dealt with individually; this is particularly important where there may be some dispute as to the adjudicator's jurisdiction to deal with some aspects of the claim. It should fulfil any specific requirements that relate to the decision, for example, if the contract or applicable rules require that reasons be given, this should be done.

In addition, the adjudicator should always be mindful of the fact that the unsuccessful party in the adjudication, who is likely as a result of the decision to be required to pay money or to carry out some specific act, may well be unwilling to do so. The successful party will then have to take steps to enforce the decision and it is therefore vital that the written decision does not put any obstacles in his way.

Enforcement has little to do with the adjudicator and much to do with the court or, if there is an arbitration agreement in the contract that does not exclude enforcement of an adjudicator's decision as an arbitral matter, the arbitrator. The adjudicator's decision should clearly state the details of the decision reached, and clearly identify such things as what it is, how it came to be made and the contract to which it relates, and what has to be done by whom and when. If it does not it could, at worst, be nothing more than a worthless piece of paper. At best, it may be something that will require the adjudicator himself to be called as a witness to verify the nature of the document and possibly what it means.

Adjudication is, after initial negotiation between the parties themselves, save for the possibility of mediation or conciliation, likely to be the first formal stage of the process to resolve the dispute. If it can also be the last stage, the adjudicator will have rendered the disputing parties a great service. The adjudicator should therefore always bear in mind that he might, by the way in which he carries out his duties, prevent the dispute from going any further. Much of this will of course be done in the stages of the adjudication prior to him sitting down to write out his conclusions, but the way in which he imparts his decision to the parties may well have a significant effect.

The process of making the decision

Decision-making is a process that we all go through on an infinite number of occasions every day of our lives. Nine times out of ten this process is a very simple one. For instance, we decide that we want a cup of coffee. This is purely instinctive and we don't have to think about it very much. We know that we want a cup of coffee and we have one. On the tenth occasion we have alternatives to consider and these can range from simple choices between two alternatives, to complex matters involving a great number of permutations.

Decisions do, however, all boil down to a simple process. That process is as follows:

- we sort out the facts that have a bearing upon the decision we have to make;
- we consider any influences that will also have a bearing;
- we look at the alternatives; and
- we choose the appropriate one in the light of all the circumstances applying at that particular time.

Decision-making as an adjudicator in respect of a dispute, however complex it may be, is little different from that.

The adjudicator has to ascertain the facts. Where facts are disputed he has to decide which version is correct. Once those facts are ascertained, the adjudicator has to find out whether any constraints apply to the parties that may have a bearing on those facts. For example, those facts may show either that the terms of the contract have been complied with or they may show the reverse. The common law, that is statute and case law, may also have a bearing upon the issues. The adjudicator makes his decision by applying his findings in respect of compliance with the contract and/or the common law to the facts he has ascertained.

There is only one occasion where this process can be varied and that is where there is an agreement between the parties that the adjudicator can diverge from ascertaining the legal (contractual) entitlements of the parties. Then his decision can be formulated upon another basis, for example so that his decision represents his fair and reasonable view, as is the case in Rule 17 of the TeCSA Adjudication Rules.

This provision at first glance appears eminently practical. What could be better than sorting out a dispute on a fair and reasonable basis? The problem is that unless the parties have agreed that precisely the same provision is a term of their contract, or they agree that the adjudicator's decision settles the dispute, a subsequent arbitration or litigation brought by a dissatisfied party would most probably come up with a different answer after a proper and detailed examination of the evidence.

Save in this last instance, where there is some leeway, the adjudicator has to decide the disputed facts and also the disputed law. He has then to apply the decided law to the decided facts and the decision follows. However complex the dispute, this must be the process followed. There is no other way of ascertaining the rights and obligations of the parties under the contract that they have entered into.

While an adjudicator should always endeavour to reach his decision fairly, he must always be wary of making a decision simply in the interests of apparent 'fairness'. We looked at this in Chapter 10 but it is worth repeating here. Contracting parties have, or should have, entered into their contract with a pretty good idea regarding their rights and obligations. A construction contract is, after all, an allocation of risk. Take, for example, a term of a contract that appears to be to the great disadvantage of one of the parties and one of the major planks of that party's submission is the unreasonable effect of the term. Unless the situation is caught by the unfair contract terms legislation, it will be presumed that the apparently disadvantaged party will have taken the item into account when reaching the original agreement. Any attempt to be 'fair' in this situation by shading or modifying the effect of the agreement between the parties will not cause problems in enforcement but will almost certainly result in dissatisfaction on the part of the party whose contractual rights have been affected, and will mean that the dispute may well be taken further.

Deciding the facts

This process can be analysed as follows:

- What facts are alleged by the applicant?
- Which of these facts is disputed by the non-applicant?
- Of each disputed fact which party has the burden of proof?
- In respect of each fact that is disputed what evidence is adduced by the party who has the burden of proof?
- Is the evidence in respect of any of the disputed facts incontrovertible?
- For those alleged facts not supported by incontrovertible evidence, what evidence does the party without the burden of proof have to rebut the allegation?
- Which party's evidence does the adjudicator prefer?
- The adjudicator has then made his decision as to the facts.

The words 'burden of proof' are mentioned above. This may, in the eyes of some, appear to be unnecessarily legalistic in the context of construction adjudication. There is, however, no doubt that in the process of ascertaining the rights and obligations of the parties, it is a vital aspect. 'He who asserts must prove' is almost a truism but, if a party's assertion is upheld without being properly proved, the result must be unfair, wrong and should, if possible, be reversed. The party that makes the assertion must prove his assertions, and thus bears the burden of proof.

The burden of proof encompasses some established legal principles and it is vital that the adjudicator in establishing the rights and obligations of the parties has some understanding of these principles. We examine these below.

The burden of proof

The legal or ultimate burden of proof is the obligation imposed by law to prove a fact that is in issue between the parties. It is necessary to be careful in the situation where several facts are in issue as the legal burden is often distributed between the parties. For example, the applicant may allege that the other party has delayed him but the other party says that the applicant has contributed partly or entirely to the delay by his own actions. Who then has the burden of proof?

A useful starting point (but only a starting point) is, as noted above, 'he who asserts must prove'.

The legal burden of proving a contract, its breach and consequential loss, lies with the applicant. The legal burden of proving a defence that goes beyond a simple denial of the applicant's assertions lies with the other party.

An allegation that is in negative terms has to be considered carefully. An allegation by the party in receipt of a claim for payment to the effect that 'the other party did not carry out the work properly, so I am not going to pay him' might be thought by the lay recipient of such an allegation to place upon him the burden of proving that he did carry out the work properly. It does not. The other party has to prove that he did not carry out the work properly. The adjudicator must look at the words, rather than the substance and effect.

There are certain facts which may automatically be in a party's favour in legal proceedings. These are called 'presumptions'. An example in the construction industry is where the applicant is an architect seeking fees for services rendered. The architect enjoys a presumption that he will be paid. The non-applicant says that the architect operated on the basis that if planning permission was not obtained, he would not be paid. In this instance, if no planning permission is granted, the non-applicant has the burden of proving that no fees should be paid for the services in preparing the planning application.

The procedure that the adjudicator should adopt in such a case is to ask, 'against whom does the presumption operate?' It operates against the person who bears the tactical burden of proving that the presumption is wrong.

Some basic presumptions are as follows:

- that the work will meet the 'usual' standard for that type of work
- a material will do the job expected
- when no programme is agreed, the performance will be within a reasonable time.

Matters that are in the general knowledge of all need no proof and are open to no evidence of rebuttal. In legal terminology this is known as 'judicial notice'. An adjudicator will be quite right to use matters that are of general knowledge in reaching his decision without telling the parties, but he must be absolutely certain that it is general knowledge and not specialised or knowledge that is particular to him.

The burden of proof is removed from a party where there is a formal admission by either an express admission or a submission. It is also removed where there is a failure to rebut an allegation.

This last is a problem area for the adjudicator faced with submissions from non-legally qualified parties. In arbitration, the statements of case are usually prepared with the normal conventions of legal pleadings in mind. Where a statement of defence is prepared, care is generally taken to ensure that every allegation is identified and responded to by an admission as to the fact it is true or by a denial. This makes the position quite clear for the arbitrator. He can work on the basis that unless there is a denial the alleged fact is accepted.

In adjudication things are unlikely to be so clear-cut. Allegations may be made that are not answered. Should the adjudicator follow the convention that they are accepted or not? It is suggested that it is inappropriate to operate on strict legal principles. Where, upon perusal of the papers, it appears that there is no reaction whatsoever to an allegation, the adjudicator should, at the very least, seek clarification. He should be careful not to accept an allegation of this nature without question just because it is not denied. The application of the strict legal principles of pleading in the High Court is not acceptable in adjudication and not only should the adjudicator avoid falling into this trap by his own doing, he should be wary of any attempt to bulldoze him into accepting them. They are not the rules of this game!

Equally, the adjudicator must appreciate that he treads a narrow path. He must, if he seeks clarification, do no more than that. He must avoid at all costs the possibility that what he does is perceived by one party as being biased in favour of the other, as this could result in his decision not being enforced.

Probability

Probability is a test that is applied to individual statements made. For instance, in respect of a claim for labour costs the adjudicator must be satisfied that it is probable that the operative in respect of whom costs are claimed was working on the activity claimed for in the light of the progress of the works.

One final matter is what happens where it is impossible to decide between conflicting evidence. This happens particularly in connection with the evidence of two witnesses who state that there were diametrically opposed results from a meeting. One says there was an agreement, the other says that there was not. Normally there will be evidence that corroborates one or the other. But what if there is no such evidence and the subsequent actions of the parties give no clue? The adjudicator does not know which witness to believe but he must make a decision between them. What is the adjudicator to do? The answer lies once again with the burden of proof. The party who asserts must prove. If the adjudicator cannot separate them the party making the assertion must fail.

Consideration of the subsequent actions of the parties as corroboration must be used circumspectly, especially if the dispute relates to a contract or an alleged agreement. This is an example of an area where the need for the adjudicator to have a basic idea of the concepts attaching to the formation of a contract is so important. There is a long line of cases including the decision of Lord Hoffman in *Investors Compensation Scheme* v. *West Bromwich*,[1] that confirm that in construing a contract it is not appropriate to take the subsequent actions of the parties into consideration.

It is not for a practical guide such as this to go into the detail of contract law. It is important, however, that anyone taking on the task of adjudicator should obtain some familiarity with such matters, enough at the very least to ring warning bells and encourage him to seek advice. Decisions made that are totally contrary to the settled law are unlikely to result in a resolution of the dispute to which they relate.

Assessing the evidence

At the time the adjudicator starts to pursue his enquiries, it is likely that all he has before him will be the notice of referral. He will, in most adjudications, seek the responding party's views in some form of written statement. He will no doubt have accompanying documents.

---

[1] *Investors Compensation Scheme* v. *West Bromwich Building Society* [1998] 1 WLR 896.

The assertions in the statements presented by the parties and even the written evidence may well conflict. If they do, these conflicts will have to be resolved by the adjudicator. In setting his timetable at the outset of the adjudication the adjudicator will always have to bear in mind this possibility and do his best to allow time to undertake any oral examination of those making the assertions.

In this event the adjudicator will be able to call those making the conflicting statements in front of him and try to make up his mind as to which one he believes. It is not always possible, however, to decide from the demeanour of a witness whether he is telling the truth or not. In fact, sometimes the most nervous of witnesses can be shown to have been telling the truth where the barefaced liar appears to be very convincing.

What the adjudicator has to do in these circumstances is the same function as anyone who has to consider evidence. He has to assess and test it for consistency and probability. The basic approach is to have a framework of incontrovertible facts. This should always be done at the outset of any consideration of evidence. This framework could be said to form all the edge pieces of a jigsaw and all the inside pieces have to be fitted into place. Some pieces of evidence will fit directly into the pieces at the edge that are already in place. These are consistent with what is already known. The added interest and problem for the individual assessing evidence can be seen as a jigsaw that has got mixed up with another and there are pieces that at first glance appear to fit but on further examination they do not. These pieces may well be placed in the hole because they appear probably to fit but they may have to be rejected later as the whole picture develops because they cannot be reconciled with the other evidence.

Some examples of the tests that must be applied to individual pieces of evidence are now considered.

Consistency

Consistency in respect of evidence given by a single witness is tested by considering the relationship of the evidence to the following:

- The incontrovertible facts
- The other parts of the witness' own evidence
- Evidence as to what the witness has said other than in his own statement
- Evidence of what that witness has done
- The evidence of other witnesses
- The documents.

All the above apply equally to a general statement submitted by a party in the adjudication.

The balance of probabilities

This is where another legal concept comes into play. What standard of proof is the adjudicator to apply to test whether the assertions made are true?

If the adjudicator is not to apply a hurdle greater than that applied in civil proceedings in the court, he must go no further than seeking to ascertain on the balance of probabilities whether the assertions made are true. One of the better examples of this concept relates to the two separate trials that American footballer OJ Simpson underwent in the USA. He was accused of murdering his wife. In the criminal trial he was found not guilty on the basis of his guilt not being 'beyond reasonable doubt'. In the civil trial that followed he was found to be guilty because the standard of proof required was 'on the balance of probabilities'.

Often this is a simple matter. It can, however, be an area where the adjudicator has to be very careful. The adjudicator may, having examined the papers in the time available to him, come to the conclusion that the evidence shows that there is a possibility that the claim is well founded. The responding party must have caused some of the loss claimed; the adjudicator is, however, not sure how much. His experience tells him that in similar circumstances it is common for this to be the case. He feels from his own experience and on the balance of probabilities from the evidence that he has seen, that there must be some blame on the responding party. There is, however, no real evidence to back up this con-clusion. If he has the time he will be able to take active steps to ascertain what actually went on when considering a disruption claim. If there is no time for such an investigation the adjudicator might suggest to the referring party that the period be extended by 14 days, as he is permitted to do without needing the responding party's agreement. If the referring party refuses to allow the extension, the adjudicator may well draw the inference that the inves-tigations would not produce an answer that was favourable to that party. The adjudicator would be well within his rights to draw such inferences, and 'fairness' probably suggests that he should inform the parties that he is doing so. That at least gives the party against whom he is drawing the inferences a chance to do something about it.

The ascertainment of the facts by the adjudicator himself

Section 108(2)(f) of the Act requires that the contract shall enable the adjudicator to take the initiative in ascertaining the facts and the law. Note that the provision does no more than enable the adjudicator to do this; it does not require him to do so.

The adjudicator may be satisfied that the materials presented to him by the parties provide him with sufficient information to reach his decision without anything more. He may not have any time to do more than this. There may, however, be a situation where the adjudicator considers that he must find out more before he can fulfil his obligations as adjudicator properly.

It is worth mentioning here the wording of the Scheme, paragraph 12(a) of which requires that the adjudicator *shall* reach his decision in accordance with the applicable law relating to the contract. This is a very onerous requirement and begs the question regarding who is fit to undertake such a responsibility in a 28-day period. This point is looked at in rather more detail elsewhere.

The adjudicator is fully at liberty to ask the parties to provide further information, but he is also allowed by section 108(2)(f) to make his own investigations. The time factors may mean that this is a far more sensible means of getting to the root of things than merely waiting for submissions, which may not arrive in time to be considered properly. Making investigations may well include using information that is already within his own knowledge.

In arbitral proceedings the basic concept of 'natural justice' or fairness applies and it is improper for the arbitrator to use specialised knowledge or particular knowledge without inviting argument from the parties. This is an area in which the adjudicator, operating under a different regime, has to be very careful. He may in fact have been engaged as adjudicator for the very reason of his own knowledge and experience. Does the concept of fairness go out of the window given the tight timescales? Is the adjudicator obliged to inform the parties of matters within his own knowledge that will affect his decision? Common sense would suggest that the answer must, in all normal circumstances, be yes. It is vital to do this when the information results in a conclusion that differs from that which might reasonably be reached from the information provided by the parties. This is particularly important when the adjudicator has to give reasons. It would certainly not assist in the resolution of the dispute if it becomes quite clear from the reasons, that the adjudicator has used a private theory that flies in the face of the evidence or even the law of contract. If the time period for the adjudication precludes informing the parties of the results, it is suggested that it is appropriate to seek the approval of the parties to the extension of the time for reaching the decision. Reasons for requesting the delay will probably have to be put forward. If a party refuses to allow extra time they have only themselves to blame if the decision goes against them.

Anyone who has operated in a particular trade or profession will have formed views concerning the way in which work is done or contractual clauses should be applied from their own experience and practice. The adjudicator should beware of applying privately developed conclusions without carefully considering whether evidence before him that may conflict with those conclusions should be ignored. For example, if the adjudicator has views which appear to him to be perfectly reasonable, being based upon years of experience, but he finds that they conflict with the evidence before him, then he should do his very best to ensure that he makes the time to appraise the parties of those views and gives them the opportunity to argue against them if they wish.

The above six paragraphs are repeated verbatim from the first edition. They are as true now as when they were written in 1999. As already noted in Chapter 10, adjudicators have been found to have acted unfairly when they have applied their own knowledge and experience or used information that has not been seen by the parties (*Balfour Beatty v. Lambeth*[2] and *RSL v. Stansell*[3]).

Deciding the law

The adjudicator must endeavour to ascertain the contractual rights and obligations of the parties. Adjudication arises out of the contract. The parties have, by entering into that contract, agreed to be bound by it. That contract operates within the overall framework of the law of the land. The adjudicator will not be doing his job properly if he fails to take these matters into account.

It may be that a provision similar to that in the original ORSA Adjudication Rules applies and the parties have agreed that the adjudicator can apply commercial reasonableness to his decision. The law may become less important but the adjudicator should be careful not to fly in the face of any agreement that is evident from the contract between the parties.

It is worth in this context repeating the points made with regard to fairness, in the introductory paragraphs to the earlier section of this chapter headed 'The Process of Making the Decision'. A decision that seeks to be 'fair' to the claimant party may fly in the face of the interpretation of contract law or the common law as developed by the courts. The example given earlier related to an attempt to be fair by altering the effects of a contract clause that, a party argues, creates unfairness. Seeking to redress the balance may in fact be acting unfairly towards the other party who is likely already to be making recompense for the apparent unfairness in the contract price.

---

[2] *Balfour Beatty Construction Limited* v. *The Mayor and Burgesses of the London Borough of Lambeth* [2002] BLR 288.
[3] *RSL (South West) Ltd* v. *Stansell Ltd* (16 June 2003).

The process of deciding the law is similar to that discussed above for deciding the facts, but it needs its own analysis. The questions that the adjudicator must ask himself are:

- What law applies?
- From where does that law arise?
- How does that law affect the parties?

Answering these three questions will certainly require an inquiry into the contract that the parties have entered into, in order to ascertain what effect it has on the parties.

The adjudicator may receive an analysis of the effects of the relevant contract from one or both the parties, that includes all the relevant authorities (textbooks, previously decided cases and the like) that support the contentions of the party putting them forward. In this case the adjudicator will have to review these submissions and will decide the law that applies by a process of analysis. He must of course be particularly careful to question the conclusions if such submissions come from one side alone. It is not unknown for an extract from a court judgment, taken in isolation, to appear to say something entirely different from what it does when taken in context.

In a situation where there is no formal submission of the law from either party the adjudicator will have to decide the law without their assistance. There will, however, in all likelihood be general statements regarding each party's interpretation of what the other party has failed to do which it is alleged they were required to do by the contract. In that event the adjudicator has to answer all the same questions; what law applies, from where does that law arise and how does that law affect the parties, and if the parties' submissions are of little or no help, he will have to do some investigation on his own account. There is, of course, no absolute requirement on the adjudicator to do this under the Act but, if he does not, he may well be less likely to produce a decision that reflects the true contractual position and puts the parties into the position of being able to resolve the dispute.

This may well entail an inquiry into the implied terms of the contract and the effect of any applicable case and statute law. The adjudicator himself may well need to consult the authorities in order to ascertain with any certainty what the law is in relation to the assertions of the parties.

If in this process the adjudicator ascertains law that the parties have not made him aware of, as confirmed by *RSL* v. *Stansell*, he must ensure that he makes the time to put this to the parties and allow them to comment.

Once he has done all this, the adjudicator is in a position to decide the law that applies to the dispute or, more to the point, the rights and obligations of the parties.

Pulling the strands together

The adjudicator has now completed his jigsaw of the facts and has as complete a picture of what did or did not happen as he is going to get. He has decided what the law is and he knows what rights and obligations each party has in respect of the matters that are in dispute.

There are two ways of reaching each individual decision that makes up the decision as a whole, the linear method and the intuitive method.

In the linear method the adjudicator takes each issue and works logically through each applicable decided fact and applies the decided law to those facts, ultimately reaching his decision on each issue.

The alternative, the intuitive method, is probably a far more likely course in adjudication given the timescale. During the assembly of all the details that relate to the dispute, the adjudicator will have intuitively decided the answer to the various questions that he ultimately will have to decide. He just knows which party is going to succeed in respect of each issue. The danger here is that the conclusions he has reached do not logically flow from the facts and the law. If an adjudicator reaches his decision in this way it is vital that he returns to basics and reviews everything using the linear method. If he does not do this he may get away with it in a decision that does not include reasons, but if he has to give reasons they may well not stand up to scrutiny in these circumstances. It is a well known phenomenon among those who make decisions in complicated matters that they start out convinced as to the answer. By the time they have analysed all the evidence and the submissions and reasoned it all through in detail, the actual answer is the reverse of what they originally believed to be the case. The adjudicator must test and test again until he is satisfied that his conclusions follow from the evidence, through his reasoning, even if not set out formally, and into his decision.

If there is one consistent cause of complaint about adjudicators' decisions it is that it is not possible to find the link between the conclusions reached and the facts and law that have been identified earlier in the decision. It can be most disconcerting for parties when, having asked for reasons, they receive a decision that recites the fact that, for example, work was carried out later than programmed and then reaches a conclusion that the contractor should be recompensed for that occurrence on the basis that the entitlement arises from the fact of the lateness without setting out the adjudicator's reasons for deciding that the employer was liable for the delay.

Once he has completed this process, the adjudicator has finalised his decisions on each individual issue making up the overall dispute and is in a position to complete his written decision.

Draft decisions

Some adjudicators like to present the parties with a draft of their decision before finalising it. Others do not. If a draft decision is sent to the parties it should be on the strict understanding that it is only for the purpose of allowing the parties to identify obvious mathematical errors and other obvious slips or errors that might cause confusion once the decision is finally released. The danger is that the losing party will see this as an opportunity to try to reopen all the arguments. Although the adjudicator has not formally reached his decision at this point the logistics of allowing the other party to respond to the further submissions and dealing with all the new material are impossible, and any attempt to make further submissions other than on specifically identified matters must be strongly resisted if this course is taken.

Writing the decision

The title of the written decision

There is a tendency in some quarters to refer to the adjudicator's decision as an award. This has connotations of the arbitration process and this, in the authors' view, is not appropriate. It is important to remember that adjudication is a separate and distinct process and it should as a result have its own separate and distinct terminology and avoid anything that is reminiscent of arbitration. Neither the Scheme nor the Act mention the word 'award'.

It is recommended that the written result of the adjudicator's considerations should therefore be entitled the 'Adjudicator's Decision'. That decision may well, of course, include the award of a sum found to be due.

The general format of the written decision

The decision must be set out in a logical and understandable format. The sections that a decision conveniently falls into are as follows:

- The introductory section which sets out what the document is, who the parties are, the nature of the contract, how the adjudicator was appointed and the like.
- The section that identifies the detail of the issues that the parties require the adjudicator to decide and sets out the adjudicator's conclusions in respect of each one.
- The decision itself, which summarises the conclusions reached and sets out what each party has to do and when. This will also, if appropriate, set out the adjudicator's conclusions relating to the costs of the adjudication.
- Finally, the concluding words, which will include the adjudicator's signature and the date.

We shall now look at these aspects individually in some detail.

The decision – the introductory section

The decision must, as noted above, state what it is. It does not have to take the form of an arbitrator's award. An arbitrator's award will, for example, commence with the formulaic words 'In the matter of the Arbitration Act 1996 and in the matter of an arbitration between...'. We are aware that many adjudicators use this concept in the heading to their decision for example using 'In the matter of the Housing Grants, Construction and Regeneration Act 1996 and in the matter of an adjudication between...'. In our view this reference to the Act in the headings to an adjudicator's written decision is not appropriate in an adjudication which is, after all, a process governed by the contract, and even if the Scheme applies it is not the Act that applies but the Scheme as implied terms of the contract.

Based on advice received from the Technology and Construction Solicitors Association there should be included in the title of any action concerning a decision 'In the Matter of an Adjudication' or 'In the Matter of a Proposed Adjudication' as appropriate. This is however what is required in an action in the courts and decisions have been and continue to be enforced that do not include these formulaic words.

An early point relates to the terminology that is used to describe the parties. In arbitration the party commencing the arbitration is called the claimant and the other party is called the respondent. These terms could quite easily be used in adjudication but it is the authors' view that this again creates a document rather reminiscent of an arbitrator's award. As noted in Chapter 1, the authors are of the opinion that the appropriate terminology is 'referring party' or 'applicant' and 'responding party' or 'respondent'.

As far as the written decision itself is concerned, the authors' view is that as this decision is a matter that arises out of the contract, it is quite appropriate to use the contract terminology to describe the parties. Thus, if the referring party is the contractor in the contract and the responding party is the employer, those are the terms that should be used to describe them in the decision.

This convention becomes a little more unwieldy if the dispute arises between a sub-contractor and a sub-sub-contractor, but with the availability of macros in modern word processing programs this is easily dealt with if the words have to be used on numerous occasions in the decision. In any event, if this all becomes too much, the names of the parties could be used. One other way sometimes seen is to use initials as shorthand; Bloggs Contractors Limited would for example be identified as 'BCL' or 'Bloggs' throughout the written decision.

Another area concerning the form of the adjudicator's decision relates to what are commonly known as 'recitals' in an arbitrator's award. The purpose of such recitals is to set out in readily accessible form the following information:

- the contract and the agreement to arbitrate;
- the nature of the dispute referred to the arbitrator; and
- the method of appointment.

The inclusion of these points establishes that the arbitrator has jurisdiction to make the award. The attitude of arbitrators to recitals is quite varied, ranging from minimalist to long, flowery descriptive passages including what might be described as quasi-legal terminology. Some arbitrators even go to the length of setting out in great detail the submissions of the parties in the view that this is the best and safest way of identifying the nature of the dispute. This is, in the authors' view, unacceptable in arbitration and should certainly never be considered in adjudication. It is worth remembering the wise words of the authors of *Commercial Arbitration* (2nd edition), Mustill & Boyd, who suggest on page 383 that the inclusion of such information, if not absolutely accurate, may be a source of confusion and dispute.

There is thus little benefit in saying any more than is absolutely necessary in a decision. A decision is a working contract document that is there to convey the adjudicator's conclusions to the parties. The decision needs to be identifiable for what it is if the successful party needs to take action to enforce it. Save in the situation where the responding party has taken no part at all in the proceedings, both parties should be fully aware of everything that has gone on during the progress of the adjudication.

The document must obviously be identifiable as 'The Decision' that includes decisions on all the issues referred to the adjudicator. It must state the names of the parties and the contract out of which it arises. It should state what the dispute is that the adjudicator has been asked to decide and it should give the adjudicator's name and address. It should also identify that there is an adjudication provision in the contract or that there is no such provision and that the adjudication is governed by the Scheme. In order to reduce the possibility of challenge due to any question arising regarding the adjudicator's appointment, the decision should include brief details of how this came about.

A matter for consideration is whether these introductory parts need to set out, in any detail, the particular nature of the dispute. It is suggested not; this will be done later in the decision when each aspect of the dispute is detailed. A brief reference to the general nature of the dispute will be necessary as it may well be that there are a number of separate disputes arising out of one contract, each with its separate written decision. It must be remembered that these will not all necessarily be dealt with by the same adjudicator.

All the necessary information in the above respects can be readily accommodated on the title page of the decision covering the following points:

- Heading – 'Adjudicator's Decision'
- Name and address of adjudicator
- Name and address of referring party
- Name and address of responding party
- Details of contract
- Nature of dispute
- Date.

The decision – the substantive aspect

The substantive aspect of the decision is that part that deals with each individual issue and sets out the conclusions that the adjudicator has reached. There may of course be only one issue, in which case this part of the decision will be extremely short, especially if reasons are not required.

Implementation and enforcement are the most important requirements. If the adjudicator's decision cannot be implemented or if one party refuses to implement it or it cannot be enforced, for reasons within his powers, the adjudicator is wasting his and both parties' time.

The decision must relate to the issues put to the adjudicator for him to decide upon. An adjudicator is not fulfilling his obligations if he fails to deal with something that the referring party has referred to adjudication, unless it is not within his jurisdiction. The other side of the coin is that any decision that he makes will not be enforced if it is not something that has been the subject of the notice of adjudication and indeed part of a dispute that has crystallised between the parties before the adjudication commenced.

The decision must set out the obligations and liabilities of the losing party in a clear and understandable fashion. If it does not do this, the winning party may well be disadvantaged in that the decision will not be implemented by the losing party in the way that the adjudicator intends, and the court may find it impossible to enforce.

A clear simple decision is far less likely to be challenged than one that goes into intricate detail and offers the opportunity of more than one interpretation of the decisions reached.

The adjudicator must ensure, as discussed earlier, that he does not usurp the authority of the contract administrator in any way, save of course where the dispute itself relates specifically to the jurisdiction of the contract administrator.

While the decision is binding on the parties, the dispute may be reopened in arbitration or in the court. The obligation of the adjudicator relates to the matters referred to him. He should avoid the temptation to attempt to second guess what may transpire later. The adjudication process stands alone. The decision has been reached in a very short time, probably on the basis of far less information than will be available should the matter come before an arbitrator or the court. The adjudicator's decision will in all probability be considered to be an irrelevance in later proceedings.

The basic concept to be borne in mind in our view is the 'KISS' principle (keep it simple – stupid!). Reasons may be required or the adjudicator himself considers that reasons should be given (of which more later), but all the parties really need to know is what it is that they each have to do and when. The simpler it is the better.

If this concept is followed, the arguments of the parties, the interpretation of the contract clause that relates to the issue and the facts that have been decided should be kept brief. These will all have had to be considered by the adjudicator and it is vital that he goes through this process before putting pen to paper in writing his decision, but there is no need to set them out in detail.

If the issues are in any way complex, there may be merit in setting out a summ
each party has to do and if a monetary award is made, what the amount to be p<
terms.

Claims will very often include mention of VAT. What the adjudicator has to remember is
that he has no authority in this area. That is a matter between the parties and the Customs
and Excise. If an adjudicator is asked to deal with VAT he should do so. He cannot decide
the amount. The most that an adjudicator should do in all normal circumstances is to declare
that any VAT that is properly due on any sums awarded should be paid. The party who is
receiving money then issues a VAT invoice and the paying party has the right to object to the
amount of VAT charged and seek the ruling of the Commissioners for Customs and Excise if
he wants to.

The only exception to this is the case where a householder or other party who is not
registered for VAT has had to effect repairs upon the failure of a contractor to carry them out.
In that event it is often the case that the householder has paid VAT and seeks to recover it as
a part of the claim. The adjudicator when faced with this situation is quite entitled to include
recovery of the VAT element of the claim in his decision.

The decision – reasons

The next question to consider is whether the decision should include reasons. As already
noted, the adjudicator may be required to give reasons by the contract or by his agreement
with the parties. He may be operating under the Scheme and one party may ask for reasons.
In that situation the decision whether to give reasons in his decision is taken out of his hands.
He has no option but to do so.

Where he does have an option regarding the giving of reasons, there are a number of
points to be considered for giving reasons and against including them.

The arguments for the giving of reasons include:

- the losing party may accept the decision more readily and the winning party accept that it
  has been awarded a lesser amount than it had hoped for if they understand the thought
  process behind it; particularly if the adjudicator has taken legal or technical advice, the
  detail of that advice may well assist in persuading the parties to accept the decision;
- the giving of reasons will impose on the adjudicator a discipline, which will assist in
  ensuring that his decision stands up to scrutiny if necessary.

While they are not reasons for the substantive issues, an adjudicator may set out the reasoning behind his decision to proceed in the face of a challenge to his jurisdiction and also
the reasons for any action that he has taken which might appear to the losing party to be a
failure in procedural fairness. At least the court will then have some idea of the adjudicator's
position, and enforcement may be less likely to be refused as a result.

The arguments against reasons include:

- an unreasoned decision is less likely to be picked over for mistakes;
- giving reasons may emphasise the limited information the decision is based on and
  encourage parties to seek to open up the dispute afresh at arbitration or litigation;
- the time to prepare the written decision may be less and thus the parties get the decision
  sooner and possibly at a lesser cost.

Reasons should not be given merely because the adjudicator feels that they will assist in later
proceedings. It must be remembered that in the event of the dispute being reconsidered

afresh in arbitration or litigation, the adjudicator's decision is almost certainly an irrelevance.

If the adjudicator decides of his own volition to give reasons, he should be certain that the end result of doing so will benefit the parties to the dispute. He should not do it for his own benefit. He must at all costs resist the temptation to show off how clever he has been in interpreting all those difficult problems that the parties have given him to solve.

The only real reason for the giving of reasons is that it will assist the parties, particularly the losing one, to understand why the adjudicator has reached his particular conclusions. As a result it may make it less likely that that party will seek to open up the dispute afresh in arbitration or through the courts. Human nature being what it is, there is a considerable danger that the reverse situation will apply and the reasons given will be used by the losing party to encourage him into an arbitration or litigation to get back the money awarded in the adjudication.

When adjudication under the Act started, the received wisdom was for an adjudicator not to give reasons unless specifically asked to do so. Things have moved somewhat since then and many adjudicators now give brief reasoning as a matter of course unless specifically asked not to. The arguments for and against remain the same but it is generally the case that losing parties do want to, and deserve to, know why they have lost.

There is one other aspect to the giving of reasons. A trend that has become apparent is that the adjudicator's decision can create a climate for discussions to settle the dispute. A winning party may feel that it has not received enough and seeks to achieve a higher settlement. Equally a losing party may seek to reduce the amount awarded in the decision. Rather than take the dispute afresh to arbitration or the courts, they commence further negotiations. Reasons in the decision, if clearly articulated, will give both parties a lead in these discussions.

In *Joinery Plus* v. *Laing*[4] Judge Anthony Thornton QC gave a detailed résumé of his views concerning reasons:

> 39. The statement by the arbitrator that he was only giving reasons for a limited purpose or was only giving limited reasons has little if any practical effect. If an adjudicator gives any reasons, they are to be read with the decision and may be used as a means of construing and understanding the decision and the reasons for that decision. There is no halfway house between giving reasons and publishing a silent or non-speaking decision without any reasons. There is no way in which reasons may be given for a limited purpose and which are only capable of being used for that purpose.
>
> 40. Indeed, although adjudicators are, under some adjudication rules, permitted to issue a decision without reasons, it is usually preferable for the parties for reasons to be given so that they can understand what has been decided and why the decision has been taken and so as to assist in any judicial enforcement of the decision. A silent decision is more susceptible to attack in enforcement proceedings than a reasoned one. Moreover, judicial enforcement, being a mandatory and compulsory exercise imposed by the state, should only be ordered by a court once it has been satisfied that the underlying adjudication decision is valid, is in accordance with law and complies with all applicable contractual and statutory procedures. For that purpose, a court will always be greatly assisted by cogent, albeit succinct, reasons.

Even if no reasons are given on the face of the decision, it is vital that the adjudicator is very clear as to the reasoning behind his conclusions. He must have reached his conclusions logically and in accordance with the facts of the matter in dispute. It is very easy to jump to a conclusion on the basis of experience. That conclusion may be right but it may just as equally

---

[4] *Joinery Plus Ltd (in administration)* v. *Laing Limited* (15 January 2003).

be wrong, and will be so if it is not relevant to the facts in the dispute. It is vital to test conclusions fully. If the adjudicator's conclusions have been reached by a logical progression through the documents, the reasoning is probably already there. If the conclusions have been reached intuitively, care must be taken to ensure that the conclusion fits the facts. This is especially important to bear in mind in relation to the short period that is available to the adjudicator. This aspect has been examined earlier in this chapter where the mechanics of decision-making are looked at more fully.

If reasons are to be given, how should this be done? What is still probably the best advice regarding reasons was given by Sir John Donaldson, as he then was, in his judgment in *Bremer Handelsgesellschaft* v. *Westzucker*[5]. He was referring to arbitrators but this advice applies to adjudicators just as much:

> 'All that is necessary is that the arbitrators set out what, in their view of the evidence, did or did not happen. They should explain succinctly why, in the light of what happened, they reached their decision and what the decision was.'

That is really all there is to it. 'Tell the story in a simple and concise fashion' is another way of putting it. Whatever else is done it is vital to ensure that the reasons relate to the matters in dispute set out in the submissions of the parties. The reasons should be logical and the conclusions reached must actually follow from the reasoning given.

If any legal or technical advice has been taken, in giving his reasons the adjudicator should set out the results of that advice together with the conclusions reached. He must let the parties see the advice and allow them to comment (*RSL* v. *Stansell* once again) albeit that the Scheme has no requirement to allow the parties to do this.

Reasons utilise the evidence given to form a bridge between the issues and the conclusions. They have to relate to both the evidence given and the conclusions reached. It is vital to ensure that a logical connection of this nature is made and can be seen in the reasons given. If reasons are included in the decision, sufficient time must be taken to ensure that they fulfil these objectives. It is all too easy in the rush to complete the task within the given time limit, to know exactly how the conclusions have been reached but fail to explain it properly or clearly in the reasoned written decision.

One other matter that needs consideration is what the adjudicator should cover when he gives reasons. An arbitrator has historically only been required to give reasons for his decisions on the law, albeit this may well be different under the Arbitration Act 1996. This is not appropriate for the adjudication process. When the parties ask for reasons what they want is the adjudicator's reasoning relating to all the decisions he has made, whether they be in respect of fact or law. In any event it is extremely unlikely that an adjudicator will, in the 28 days allowed, be able to analyse the contract and the law to give the sort of reasons that an arbitrator gives. At all times the adjudicator must bear in mind the point made earlier in this chapter that his aim must be to prepare a decision that the parties will accept and as a result settle the dispute.

The adjudicator should always be prepared for the situation in the Scheme where the parties have the option of asking for reasons but no time limit is stated. It may be that on receiving an unreasoned decision, one or both parties may ask for reasons. This situation really does mean that the adjudicator should have his reasons set out at least in draft so that there is not an embarrassing wait while he tries to formulate reasons that tie in with his unreasoned decision. Late requests can be made under the Scheme but a sensible adjudicator

---

[5] *Bremer Handelsgesellschaft mbH* v. *Westzucker GmbH (No 2)* [1981] 2 Lloyd's Rep 130.

will be prepared for this and will have pre-empted the situation by getting the parties to agree to him setting a final date for such requests if he can.

The decision – interest

The basic principle of law in England and Wales has always been that there is no entitlement to interest as damages where the only breach alleged is lateness in making payment. This is now changed by the passing of the Late Payment of Commercial Debts (Interest) Act 1998, which has been subject to phased commencement but is now fully in force for all contracts currently entered into.

This Act makes a considerable change to the recovery of interest upon amounts that remain unpaid. Interest will be statutorily recoverable on outstanding amounts at a rate that may be as much as 8% above base rate. It will, however, only apply to established debts. Adjudicators will no doubt be asked to apply its provisions to sums that they find to be due where there is no provision in the contract for the recovery of interest.

One thing to remember in this respect is that parties to contracts may agree to oust or vary the statutory right to interest, but only if they provide 'a substantial contractual remedy' for late payment. The JCT Standard Forms of Contract include the figure of 5% over base. As far as the authors are aware this has not been a matter raised by referring parties, but were it to be raised adjudicators would have to decide whether this is a substantial remedy. We would not hazard a guess as to the answer in individual circumstances, but it is certainly arguable that 5% over the bank base rate is a substantial remedy in the case of a commercial organisation that can borrow at a lesser figure. Equally, it must be remembered that the figure was originally set by the government at 4% but it was increased to 8% on the basis that it wished to prevent the smallest of businesses suffering because they have to borrow overdraft finance.

A further aspect that may have to be considered is the statutory right set out in the Act to suspend performance. We have no doubt that the right to suspend performance will be argued as a substantial remedy. It is not for us to predict the circumstances in which such arguments may be used. It does seem to us that a contractor who was found to be unable to recover interest to which he was statutorily entitled because he had a right to stop work, might feel that the system was working against him, especially if he had decided not to invoke his right to suspend work in the interests of the contract as a whole.

It may be, however, that there is a contractual right to interest as there is in the JCT and, save for arguments if the amount is substantially below the statutory figure, that will be the provision that the adjudicator will have to apply to any monies that he decides should be paid. Should the question of simple or compound interest arise, the adjudicator must again be guided by the contract between the parties.

A question arises with regard to interest, which we commented on in our discussion of paragraph 20 of the Scheme in Chapter 5: should the adjudicator award interest if it is not claimed? We noted there that there are differing views on this. Where interest is not claimed, the adjudicator must be very careful not to run the risk of making a party's case. At the very least he should notify the parties that he is considering his award of interest and allow them to make submissions on the point to avoid any accusation of unfairness.

There is another aspect of interest in its broadest sense that the adjudicator may be called upon to consider, and that is interest paid on the borrowings needed to fund the operations of the party who claims it. This is commonly known as financing charges. Financing charges are allowable as part of a claim where it can be shown that the failure to pay by the party against whom the claim is being made had the direct effect of requiring the claimant to borrow money to fund its activities. Financing charges in this event are considered to be what is known as special damages. They have to be a specific head of claim and cannot be wrapped up in a claim for general damages. The validity of financing charges as a head of claim relies upon there being evidence to show that these charges have been incurred and that the reason for them being incurred is the direct responsibility of the party against whom the claim is made.

The adjudicator should not consider financing charges unless they form a part of the claim or the issues put to him for his decision. If put, such a claim should be treated on its merits. If a party has incurred them they will surely form a part of the claim. Financing charges should not be allowed unless it can be shown that there is a proper entitlement. This entitlement may well not be spelt out in the contract between the parties. The adjudicator is then reliant upon the decisions made in the courts as to how the wording of contracts in this respect is to be interpreted. It is not appropriate for us to go into a detailed discussion of the principles of the award of financing charges here, and if the adjudicator has any concerns about dealing with such matters he should take the appropriate advice.

The decision – costs

Insofar as costs are concerned the adjudicator will, as in all other matters, be bound by the contract. If the contract sets out a provision such as 'the parties shall each pay their own costs', the decision is made for him. This is certainly in the general spirit of adjudication as a contractual procedure.

The parties may give the adjudicator power to deal with their costs during the adjudication. This occurs where the referring party claims its own costs in the referral and the responding party does likewise in the response. If the referring party claims its costs and the responding party disputes this, there is no agreement and the adjudicator has no power to award the parties' costs. If the referring party seeks its costs and the responding party is silent, the position is not as it would be in the case of court proceedings where a matter is deemed to be admitted unless specifically denied. There is no agreement and thus no power for the adjudicator to award party costs in this case.

Certain adjudication provisions include the requirement that the referring party pays the costs of both parties come what may. We discussed the case of *Bridgeway* v. *Tolent*[6] in Chapter 1, in which such a provision was not found to be unlawful.

The contract may, however, be silent on the question of costs or alternatively the Scheme may apply and this is also silent in respect of the parties' costs.

There are two conflicting judgments on this question, both at first instance. In *Cothliff* v. *Allen Build*[7] it was held that as the Scheme is silent on the matter, the adjudicator has the option to award costs in certain circumstances. If the parties make no mention of costs, then the adjudicator himself cannot raise it.

This was however not followed in *Northern Developments* v. *J&J Nichol*[8], which appears to be a more generally accepted interpretation of the position regarding the Scheme and costs where it was held that there is no implied term in the Scheme to the effect that an adjudicator

---

[6] *Bridgeway Construction Ltd* v. *Tolent Construction Ltd* (11 April 2000) unreported.
[7] *John Cothliff Limited* v. *Allen Build (North West) Limited* [1999] CILL 1530.
[8] *Northern Developments (Cumbria) Limited* v. *J&J Nichol* [2000] BLR 158.

may award that one party pay the costs of the other, unless there is an express or implied agreement between the parties that he may do so.

Where the Scheme does not apply and the contract is silent on the matter, the adjudicator would do well to require that the parties each pay their own costs if for no other reason than the interests of finality. It is almost inevitable that a party who is required to pay the costs of the other will contest the level of costs claimed. All this does is to replace one dispute with another. This cannot be a sensible situation and it is one that should be avoided if at all possible. It is our view, in the interests of finality, if the adjudicator decides to make one party pay the costs of the other, that it is vital to include a lump sum in the decision, preferably having called for an estimate of the total beforehand. Anything else means that there will be further argument in respect of the costs, and the adjudication is unlikely to result in a settlement of the dispute.

We are aware of strongly held views to the contrary, that matters of the quantum of costs should be left out of the decision, but we do find some difficulty with this view. The adjudicator is required to deal with all matters in dispute in his decision. It is not clear to us where the adjudicator gets his authority from without a specific dispensation from the parties to deal with this point after the decision is made.

The adjudicator may reach the view that the actions of one party have been such that the costs of the other party have been increased unnecessarily and that there should as a result be some recompense. Nothing can be done by the adjudicator in such a situation unless there is power for him to deal with the parties' costs. If there is, the submission of a claim for costs shortly prior to issuing the decision is the appropriate way to do this.

The adjudicator may not want to reveal his hand at this point and submissions from both parties are generally called for.

The other aspect of costs is the adjudicator's own fees and expenses. Again, the contract may have specific requirements. If it does not, the adjudicator will have to decide whether one party pays the whole amount or whether it will be shared between them in some proportion. The choice between these is a matter for the adjudicator but it seems customary for the referral party and the response party each to seek that the other party pays the adjudicator's fee in full, and winning parties will be rather upset if the adjudicator does not follow the customary course that costs follow the event. This matter should be dealt with finally in the decision, as the adjudicator will obviously know the total fee he is going to charge before he releases the decision.

Stakeholders

The original Latham proposals suggested that the use of stakeholders should only be permitted if both parties agree or the adjudicator so directs. There is no provision in the Act that requires such a procedure. The principle of using stakeholders also conflicts with one of the intentions of adjudication, which is to improve cash flow in the industry.

The contract may, however, include such a provision. This procedure does provide a useful safeguard where there is concern that the adjudicator's decision may be reversed subsequently by an arbitrator or the courts and the party receiving the money is seen as being in danger of going into liquidation. As discussed elsewhere in this book, the authors do not necessarily subscribe to this concern. If a building employer or sub-contractor is worried about the financial stability of an organisation next down the contractual chain, it is seriously arguable that they had the opportunity to investigate before entering into the contract. If they entered into a contract with an organisation that is unstable financially, it was probably because they obtained a lower price and they should take the consequences of their decision.

Even if there are no such provisions in the contract, the party that is likely to have to pay as a result of the decision may apply during the course of the adjudication for any monies that it may be required to pay to be lodged with a stakeholder. The adjudicator does not have the power to do this if there is no provision to that effect in the contract.

If the adjudicator has the power under the contract to order monies to be held by a stakeholder, he may order this in his decision. There are some hybrid forms of contract that require that any monies due as a result of an adjudicator's decision must be deposited with a stakeholder. It is questionable as to whether clauses of this type comply with the payment provisions of the Act.

The decision – final matters

One point that has to be considered is the inclusion within the decision of any specific enforcement provisions that will be required by the contract or adjudication rules that apply.

The formal enforcement provisions of the Scheme for Construction Contracts are reliant on an amended version of section 42 of the Arbitration Act 1996 as set out in paragraph 24 of the Scheme. In order for this to work the adjudicator has to state in his decision that the whole decision or part of it must be complied with peremptorily. The adjudicator also has to give permission for the application to the court to be made under the revised section 42(2)(b) of the 1996 Act, and the prudent adjudicator will avoid the possible problem of objections that he is no longer empowered to give permission after he has completed his decision by covering this point in his decision as well.

A suggested wording to cover this situation is as follows:

'I order that this decision be complied with peremptorily. Either party is, subject to giving notice to the other and to me, permitted to apply to the court for an order requiring compliance with this peremptory order.'

The Scheme appears to have been written in anticipation of there being difficulties in enforcing adjudicators' decisions if there was no such specific provision. Since *Macob* v. *Morrison* this does not appear to trouble the court and decisions are being enforced by another route. This provision of the Scheme is not being utilised greatly any more. Some adjudicators use a wording requiring peremptory compliance but this may be more habit than anything else. If a referring party seeks a peremptory order there is no reason for the prudent adjudicator not to comply with this request. In general the inclusion of a date or period for compliance is sufficient.

As far as the time for compliance with the decision is concerned, there needs to be some specificity. 'Immediately' may create difficulties. 'Forthwith' is a good legal phrase that means as soon as possible. Often adjudicators state the period of seven days, which is generally not unreasonable and lines up with section 114(4)(a) of the Act.

It is worth repeating the point made in Chapter 5 when considering paragraph 20(b) of the Scheme, regarding the question of whether or not an adjudicator has power to order anything. It may be on a pure interpretation of the duties of an adjudicator that he is limited to declaring what the contract says. We repeat our view that there is nothing wrong with actually telling the parties what they have to do.

Once all other matters have been dealt with, the adjudicator will have to draw the decision to a neat and tidy close. The decision should be signed and dated. It may be witnessed but there is no specific requirement to that effect.

The last thing for the adjudicator to do is to send the completed decision to the parties. If the tradition of arbitration is followed, the original signed decision should be sent to the applicant party, a certified copy being sent to the non-applicant and one retained. It may, however, be just as acceptable for the adjudicator to retain the originally signed document on his file and send copies to the parties.

Arbitrators have historically had a lien on their awards and can, under the 1996 Arbitration Act, require payment of their fees before releasing their award. Adjudicators do not have the same statutory protection. Some sets of rules allow for a similar lien on the adjudicator's decision, others do not. The Scheme requires that the adjudicator shall deliver a copy of that decision to the parties as soon as possible after he has reached it and we interpret this wording to mean that the adjudicator has no power to hold on to it pending payment. The existence of an overriding duty for the adjudicator to deliver his decision by the 28th day unless the parties have specifically agreed otherwise was confirmed in *St Andrews Bay* v. *HBG*[9]. The adjudicator has no choice; he has to release his decision before he is paid, and as a result if the parties do not pay him his only recourse is to sue. Since the decision in *St Andrews Bay*, the Construction Industry Council has altered its Model Adjudication Procedure to delete the lien that it previously included.

The guidelines for adjudicators issued by the RICS deprecate the imposition of a lien without the prior agreement of the parties, and in common with other ANBs the RICS takes the view that the industry expects decisions in 28 days and that their nominees will ensure that their decisions are with the parties in that timescale. Where there is no lien either in the contract or applicable rules, it is arguable that an adjudicator has an implied duty to release his decision to comply with the objects of adjudication. It is equally arguable that no one has any legal duty to extend credit and an adjudicator would be perfectly entitled to hold on to his decision until paid, but this has now been found by the court in Scotland to be unacceptable and we think it unlikely that the English court will differ.

In practice, it is the authors' view that where the work involved in reaching the decision has not been extensive, it is probably in the interests of the adjudication process as a whole to release the decision and then seek payment. However, if there has been a lot of work involved it is unreasonable for the parties to expect to do other than pay up first. It would be inappropriate for the adjudicator, where there is no contractual lien, to do other than to tell the parties at the outset that he intends to operate a lien. If they object, the adjudicator may well have to release his decision without payment first. If neither party pays he will then have to commence proceedings for his money.

Where there is a lien set out in the contract or the rules, as there still is at the time of writing in the ICE Adjudication Procedure, the adjudicator would be within his rights to withhold his decision without notice. The right to a lien is not mandatory however and it is the authors' view that it would still be incumbent upon the adjudicator to acquaint the parties as early in the process as possible that the lien was going to be operated, in order that the parties have the opportunity to avoid any delay once the decision is ready.

One last point is the question of how long the documents used for carrying out the adjudication should be kept. We have considered this in an earlier chapter but it is worth repeating here. The decision itself and the adjudicator's working papers and correspondence

---

[9] *St Andrews Bay Development Ltd* v. *HBG Management Ltd* (20 March 2003).

should be kept for as long as the period of limitation requires. That is six years if the adjudicator is engaged under a simple contract and 12 years if a deed is involved.

The other papers used can become a bit of an embarrassment from the point of view of storage. If the adjudicator is appointed under the contract as a whole rather than in respect of the individual dispute, he will need to retain them. There is also the possibility in the case of a one-off appointment that the parties will want him to adjudicate on further disputes. Again, he should retain the papers. If, however, he has no further use for the papers there is no reason why he should not request that the parties collect them, giving a time limit. This notice can be accompanied by a statement that the documents will be destroyed if not collected within the set period. Destroyed does not mean put in the dustbin. No party wants its papers blowing about a rubbish tip. Various organisations provide secure destruction services and these should be used.

Checklist for decision

Introduction

- General heading – 'Adjudicator's Decision'
- Adjudicator's name and address
- Method of appointment
- Names and addresses of parties
- Title of contract
- Description of dispute
- Date (here or at end).

Substantive decision

- Adjudication under provisions of contract or under Scheme for Construction Contracts
- Issues considered
- Decision reached on each issue
- Reasons – if required by contract or applicable rules or if asked for by a party under the Scheme
- Summary of decision
- Time for compliance
- Interest or financing charges
- Costs – (a) of parties, (b) of adjudicator
- If requested under Scheme, include formula required for enforcement.

Signature

Chapter 12
Enforcement and Appeals

Save for those who were not in favour of the legislation, there was more concern about enforcement of adjudicators' decisions than any other single issue relating to the adjudication. This has been assuaged by decisions made by the courts that adjudicators' decisions should, as a matter of principle, be enforced. Adjudication is a creature of the contract brought about as an imposed regime on contracts that are defined as construction contracts within the Act.

All the customary remedies for dealing with the enforcement of a contract are available to deal with enforcement of adjudicators' decisions.

There are at the time of writing over 170 decisions of the court that relate to adjudication. Most of these are simply about whether or not an adjudicator's decision should be enforced. Some of the decisions are the result of early applications to the court in respect of the jurisdiction of the adjudicator, but these are relatively few and not directly concerned with the subject matter of this chapter.

The effects of the decision

Section 108 (3) of the Act states that:

> The contract shall provide that the decision of the adjudicator is binding until the dispute is finally determined by legal proceedings, by arbitration (if the contract provides for arbitration or the parties otherwise agree to arbitration) or by agreement.
> The parties may agree to accept the decision of the adjudicator as finally determining the dispute.

Adjudication is thus, unless agreed otherwise, only of a temporary effect. The decision is intended to be binding on the parties for the time being and has been described as being 'temporarily binding'.

If the contract does not provide for the wording in section 108(3), the Scheme applies.

Paragraph 23(2) of the Scheme provides exactly the same wording as section 108(3) of the Act except that there is no provision for the parties to accept the decision as finally determining the dispute.

As soon as the decision is made the parties are bound by it. That is what their contract provides. How temporary is the decision? The wording of the Act and the Scheme, and therefore any contract, makes the decision binding unless and until one of the parties chooses to challenge it. If neither party wishes to challenge the decision it remains binding. The contract can be completed and the final account settled and the decision still remains binding. In this way the decision finally determines the dispute. In practice the adjudication has, in the majority of cases, formed the basis of the final resolution of the dispute between the parties. No other action is taken.

The Act also makes provision that the parties can otherwise agree. It is unlikely in practice that a party that has the decision in its favour will opt to agree to alter the effects of that decision. The only grounds on which such an agreement will be reached will be a compromise to avoid the costs of a possible litigation or arbitration.

Enforcement – initial considerations

The time to consider enforcement is before the decision is even made. The relief sought in the decision will determine the route for enforcement or indeed whether the decision is enforceable or not.

For example, where there is a dispute concerning the quality of the brickwork, the decision the adjudicator makes should declare whether or not the brickwork complies with the contract. A decision should not be sought on whether the brickwork should or should not be pulled down. The adjudicator does not have the authority to make such a decision. He is determining the rights of the parties under the contract. It is for the architect to decide whether or not he wants to keep or dispose of the brickwork when the adjudicator has determined whether or not it complies with the contract. The adjudicator does not have the power to issue such instructions; this remains the job of the architect or engineer as the case may be. The adjudicator's job is to decide the respective rights and duties of the parties.

If there is a decision which declares that the brickwork does comply with the requirements of the contract, what is there to enforce? The parties are bound by that decision. There is nothing further for anybody to do. If the architect still insists that the brickwork is taken down and rebuilt, he would need to issue an instruction to that effect and the contractor would be paid for that work which is then additional to the contract. If the payment for the work is not certified there is a new dispute which can be dealt with in adjudication. The evidence that the brickwork complied with the contract is in the first adjudication decision. Unless this is set aside in arbitration or litigation, it will remain good evidence of the fact that the brickwork complied with the contract requirements.

An adjudicator makes a decision which entitles a contractor to an extension of time. The architect then refuses to honour that decision and grant the extension of time. Whether or not the architect grants the extension of time, both the employer and the architect are bound by the adjudicator's decision. Why take any steps to enforce this decision? If at a later date the architect certifies that the contractor has failed to complete the works by a date, and ignores the adjudicator's decision, there is no entitlement for the employer to deduct liquidated and ascertained damages. If the employer then deducts a sum in respect of damages for a delay covered by the extension of time granted by the adjudicator, the decision made by the adjudicator, which was previously of no monetary value, then becomes the basis of an enforcement action by the contractor. He simply commences an action to collect the money which has been wrongly deducted. In support of his claim he has the adjudicator's decision which says he is entitled to an extension of time. The employer has no defence to this as his architect has failed to follow the decision to which both are bound.

It can be seen that not all adjudicators' decisions will have to be enforced; their very nature simply makes them stand alone as a declaration of the position under the contract.

What of those decisions which do require enforcement in some form? The decision is the product of the adjudication, which is a contractual dispute resolution process. The decision therefore has a separate existence from the contract itself and the subject matter of the dispute. In *Agromet Motobimport*[1] v. *Maulden Engineering* it was decided that:

---

[1] *Agromet Motobimport* v. *Maulden Engineering Co (Beds) Ltd* [1985] 2 All ER 436.

'an action to enforce an arbitrator's award was an independent cause of action, arising from the breach of an implied term in the arbitration agreement that the award would be honoured and not from the breach of contract which had been the subject of the arbitration'.

There is no reason that adjudicators' decisions should not enjoy a similar status. The decision ought to be the source of an independent cause of action. The decision, if no express term is provided in the contract, should carry the implied term that it ought to be honoured.

Agromet was followed in *International Bulk Shipping* v. *Minerals & Metals*[2]. The text from this case gives the policy in respect of honouring arbitrators' awards:

'The six-year limitation period for the enforcement of an arbitral award began whenever a claimant became entitled to enforce the award; in legal terms, when his cause of action arose. Conceptually, such a claim arose under a contractual undertaking to honour the award, which might render the party against whom the award was made under an immediate obligation to pay the amount of the award. Alternatively, if the claim was for damages for breach of the implied promise to pay, a reasonable period for payment should be allowed, three months at most, but that period could not be extended by reference to the attitudes of the parties and their representatives during the process of seeking to enforce payment, nor could the claimant's cause of action and right to enforce the award be deferred until the respondent had unequivocally refused to pay.

... approved the following passage from the current edition of Mustill and Boyd *Commercial Arbitration* (2nd edn, 1989) p. 418: "Time begins from the date on which the implied promise to perform the award is broken, not from the date of the arbitration agreement nor from the date of the award.''...

It cannot seriously be argued that the causes of action arose before the awards were published. It is unnecessary for present purposes to decide whether an award has the same effect as a judgment, so that a cause of action arises when the award is published. I respectfully agree, however, that conceptually the claim arises under a contractual undertaking to honour the award. This may mean that the party against whom the award is made becomes under an immediate obligation to pay the amount of award, which strictly should be construed as a liquidated debt obligation. Alternatively, if the claim is for damages for breach of the implied promise to pay, then a reasonable period should be allowed for the necessary payment to be made. This period would be co-extensive with whatever is allowed by "forthwith". If regard is had to the time needed for the payment process, then this period might be, say, 28 days. It could hardly be longer than, say, three months.'

It cannot be argued that an adjudicator's decision has the same status as an arbitrator's award in terms of being directly enforceable by application to the courts. It does not have the same effect as if it were a judgment. It should however have a similar effect if the matter arises under a contract and creates either an implied or an express term that it should be honoured. This should be conceptually helpful in the enforcement of adjudicators' decisions. The simplest method of enforcement is for the parties to simply comply with what the decision says. What we are concerned with here is the situation when one party will not comply with the decision.

The Act itself provides no assistance in seeking means of enforcement. There are many who have argued this is a serious oversight in the primary legislation. Where the Scheme applies it provides its own steps for enforcement albeit that part of this paragraph is tortuous. Paragraph 23(1) states that the adjudicator may, if he thinks fit, order any of the parties to comply peremptorily with his decision or any part of it. Paragraph 23(2) repeats the first paragraph of section 108(3) of the Act regarding the binding nature of the decision.

---

[2] *International Bulk Shipping & Services Ltd v. Minerals & Metals Trading Corp of India and Others* [1996] IRLN 45; [1996] 2 Lloyd's Rep 474.

As we discussed in Chapter 11, the formal enforcement provisions of the Scheme are reliant on an amended version of section 42 of the Arbitration Act 1996 as set out in paragraph 24 of the Scheme. In order for this to work the adjudicator has to state in his decision that the whole decision or part of it must be complied with peremptorily. The adjudicator also has to give permission for the application to the court to be made under the revised section 42(2)(b) of the 1996 Arbitration Act.

Since *Macob* v. *Morrison*[3] this does not appear to trouble the court and decisions are being enforced by another route:

'37. Thus, section 42 apart, the usual remedy for failure to pay in accordance with an adjudicator's decision will be to issue proceedings claiming the sum due, followed by an application for summary judgment.

38. It is not at all clear why section 42 of the Arbitration Act 1996 was incorporated into the Scheme. It may be that Parliament intended that the court should be more willing to grant a mandatory injunction in cases where the adjudicator has made a peremptory order than where he has not. Where an adjudicator has made a peremptory order, this is a factor that should be taken into account by the court in deciding whether to grant an injunction. But it seems to me that it is for the court to decide whether to grant a mandatory injunction, and, for the reasons already given, the court should be slow to grant a mandatory injunction to enforce a decision requiring the payment of money by one contracting party to another.'

It was this case that set the whole pattern for enforcement of adjudicators' decisions in the English courts.

Arbitration

Where there is an arbitration clause in the contract it might be thought that section 9 of the Arbitration Act 'Stay of legal proceedings' (to arbitration) would preclude any reference to the court for the enforcement of a decision. This was one of the submissions resisting enforcement made to the court in *Macob* but it was unsuccessful:

'29. But what the defendant could not do was to assert that the decision was a decision for the purposes of being the subject of a reference to arbitration, but was not a decision for the purposes of being binding and enforceable pending any revision by the arbitrator. In so holding, I am doing no more than applying the doctrine of approbation and reprobation, or election. A person cannot blow hot and cold: see *Lissenden* v. *CAV Bosch Ltd* [1940] AC 412, and *Halsbury's Laws* 4th Edition Volume 16, paragraphs 957 and 958. Once the defendant elected to treat the decision as one capable of being referred to arbitration, he was bound also to treat it as a decision which was binding and enforceable unless revised by the arbitrator.'

There is an important aside here for disputes concerning jurisdiction. The Royal Institution of Chartered Surveyors has a scheme for appointing arbitrators to deal with jurisdictional challenges. The concept of this scheme is that an arbitrator can be appointed quickly during the currency of an adjudication and a declaratory award can be made on the jurisdictional point while the adjudication is in process. This has been rarely used, as has the equivalent procedure of seeking a decision of the court as to the nominated adjudicator's jurisdiction in the early stages of an adjudication, parties seemingly preferring that the adjudicator reaches

---

[3] *Macob Civil Engineering Ltd* v. *Morrison Construction Ltd* [1999] BLR 93.

his own non-binding conclusion as to his jurisdiction, and only raising the question of jurisdiction if they are not prepared to accept the decision on the substantive issues or are unable to use the decision as a basis for compromising the dispute.

The standard forms of contract

The Act only requires that the decision will be binding until the dispute is finally determined by legal proceedings, by arbitration (if the contract provides for arbitration or the parties otherwise agree to arbitration) or by agreement. This provision must be incorporated in the contract otherwise the Scheme will apply. The Scheme only repeats the Act in respect of the binding nature of the decision. There is nothing to prevent the parties to the contract making further stipulations that may assist in the enforcement of the decision.

The various JCT contracts and their associated sub-contracts have amended their arbitration clauses to take account of enforcement of adjudicators' decisions. Significantly they exclude enforcement of adjudicators' decisions from the arbitration clause. They now also permit arbitration to commence at any time on any issue. There is no longer the limitation of waiting until the practical completion before the arbitration can be commenced.

The JCT clauses assist matters by making two provisions. The first covers the requirement of the Act in that it deals with the binding nature of the decision. The second provision requires that the parties actually comply with the decision. These can be simply viewed as provisions which relate to the nature or status of the decision and to the performance of the decision. It is a separate breach of contract to fail to comply with the decision under this contract. That separate breach would be actionable in its own right. This assists the enforcement process.

Effect of adjudicator's decision

41A. 7 .1 The decision of the Adjudicator shall be binding on the Parties until the dispute or difference is finally determined by arbitration or by legal proceedings or by an agreement in writing between the Parties made after the decision of the Adjudicator has been given.

41A .7 .2 The Parties shall, without prejudice to their other rights under the Contract, comply with the decisions of the Adjudicator; and the Employer and the Contractor shall ensure that the decisions of the Adjudicator are given effect.

41A .7 .3 If either Party does not comply with the decision of the Adjudicator the other Party shall be entitled to take legal proceedings to secure such compliance pending any final determination of the referred dispute or difference pursuant to clause 41A.7.1.

The ICE civil engineering contracts and their associated sub-contracts also exclude the enforcement of adjudicators' decisions from the arbitration clause.

This contract complies with the Act in clause 66(7) in stating the binding nature of the decision in the terms of section 108(3). In clause 66(9) it excludes from arbitration disputes arising from failure to give effect to a decision of an adjudicator and then goes further and where the parties fail to serve a notice to refer to arbitration within three months of the adjudicator's decision, the decision becomes binding.

The ICE Adjudication Procedure provides in clause 6.7 that the parties are entitled to the relief and remedies set out in the adjudicator's decision and to seek summary enforcement of them, regardless of whether the dispute is to be referred to legal proceedings or arbitration.

This contract by reference incorporates its own adjudication procedure. The terms of the adjudication procedure also form part of the contract. This gives a source of further assistance in enforcement of the decision. The parties are entitled to the relief and remedies in the decision. This paragraph then entitles them to seek summary enforcement regardless of whether the dispute is to be referred elsewhere for final determination.

The NEC Engineering and Construction Contract and the associated sub-contract set a period for notification of a challenge of an adjudicators' decision of four weeks after the decision has been made. These contracts include review of adjudicators' decisions by a tribunal. This may be a tribunal in the conventional sense where arbitration is chosen in the contract data as the final means of resolving the dispute. In this case the means of enforcement will be through arbitration. The NEC does however allow the parties to choose the tribunal and thus adjudicators' decisions can be enforced either by arbitration or the courts.

The GC Works Contracts exclude enforcement of adjudicators' decisions from the arbitration clause.

> 59. (9) Notwithstanding Condition 60 (Arbitration and choice of law), the Employer and the Contractor shall comply forthwith with any decision of the adjudicator; and shall submit to summary judgment and enforcement in respect of all such decisions.

This contract also provides an additional obligation beyond the binding nature of the decision. It is similar to the ICE contract. The parties are under a separate contractual obligation to comply with the decision forthwith. It then imports summary judgment as being the method to enforce the decision.

Litigation

The courts have proved to be an effective and rapid system of enforcing adjudicators' decisions. Adjudication is a rapid process. Delay in achieving enforcement through the court system would have detracted from the speed desired by the process.

At an early meeting of ORSA (The Official Referees Solicitors Association) now TeCSA (Technology and Construction Solicitors Association) in discussion with the Official Referees, they undertook to deal expeditiously with matters concerning adjudicators' decisions. The documents should be clearly marked for the court 'In the Matter of an Adjudication or In the Matter of a Proposed Adjudication' so that the Registry is made aware at the time of issue, of the nature of the proceedings. The enforcement of adjudicators' decisions is not subject to the Pre-action Protocol for the Construction and Engineering Disputes.

An arbitrator's award can be enforced under the provisions of section 66 of the Arbitration Act 1996 but there is no direct enforcement of an adjudicator's decision in the same way as an arbitrator's award.

Adjudicators' decisions have no direct statutory status either under the Act or the Scheme, save for the provisions in paragraphs 23 and 24 for enforcement.

Before Macob *v.* Morrison

In *Cameron* v. *Mowlem*[4] it was held that:

---

[4] A. *Cameron* v. *John Mowlem & Co plc* (1990) 52 BLR 24.

'A decision of an adjudicator given under clause 24 of DOM/1 was binding only until the determination by an arbitrator on the disputed claim to a set-off, and so was not "an award on an arbitration agreement" within the meaning of section 26 of the Arbitration Act 1950, and could not be enforced summarily under that section...

The first procedure adopted was an application under section 27 of the Arbitration Act 1950 on 20 April 1989 for leave to enforce the decision of Mr Knowles in the same manner as a judgment of the court. The application came before Judge Esyr Lewis QC who dismissed it. Cameron's appeal against that decision is the first appeal before the court. Section 26 of the Act of 1950 allows enforcement only of an "award on an arbitration agreement". Is an adjudicator's decision under clause 24.3.1 an award on such an agreement? The learned judge held that it is not, and his basis was that "the adjudicator ... does not perform an arbitral function and does not make any final award definitive of the parties' rights"...

The decision has an ephemeral and subordinate character which in our view makes it impossible for the decision to be described as an award on an arbitration agreement. The structure of the sub-contract is against that conclusion.'

Although the provisions on adjudication that are now required by statute differ from those limited proceedings under the set-off provisions of DOM/1, it is thought that the temporary nature of decisions in the legislation will not change the view expressed in *Cameron*. In that particular case one of the factors was that the sum decided as being due by the adjudicator was not in fact due.

In *Drake & Scull* v. *McLaughlin & Harvey*[5] Judge Bowsher considered the circumstances in which an injunction could be used to provide a means of enforcement of an adjudicator's decision. He held:

'The plaintiff was not disregarding the arbitration procedure, but was coming to court to enforce an order made by the adjudicator in the course of the arbitration procedure. *The Channel Tunnel Group Ltd* v. *Balfour Beatty Construction Ltd* [1992] 2 WLR 741; 56 BLR 23 considered.

There was no reason why a mandatory injunction should not be granted. A mandatory injunction which in effect anticipated the outcome of the action, would be granted only in the most extreme circumstances. However, the injunction sought was a conditional injunction which would be effective pending arbitration and which would, in any event. be subject to discharge or variation of the adjudicator's award by the arbitrator.'

This case needs to be read carefully, particularly the passage which deals with the principles of granting injunctions:

'The general principles governing the grant of interlocutory mandatory injunctions are summarised in *Halsbury's Laws of England* 4th Edn, Vol 24, page 445, para 848, as approved by Mustill LJ in *The Sea Hawk* [1986] 1 WLR 658:

"A mandatory injunction can be granted on an interlocutory application as well as at the hearing, but in the absence of special circumstances it will not normally be granted. However, if the case is clear and one which the court thinks ought to be decided at once, or if the act done is a simple and summary one which can be easily remedied, or if the defendant attempts to steal a march on the plaintiff – a mandatory injunction will be granted on an interlocutory application." '

The grant of an injunction does not provide a remedy for enforcement in every case. It is probable as in the *Drake & Scull* case that any injunction granted would consider whether both parties will be in existence at the time the matter in dispute is finally decided in litigation or arbitration.

---

[5] *Drake & Scull Engineering Ltd* v. *McLaughlin & Harvey plc* [1992] 60 BLR 102.

An application to the courts for an order for specific performance will not be a means of enforcing adjudicators' decisions. Specific performance is an equitable remedy and is therefore granted at the discretion of the courts and not as of right. The usual test is that specific performance will not be granted where damages will provide an adequate remedy. We cannot think of many instances where specific performance would be required to enforce an adjudicator's decision. There are other ways in which such matters may be dealt with and these are discussed below.

Since **Macob *v.* Morrison**[6]

In the short time that adjudication has been available, the approach of the courts to enforcement has developed rapidly and almost gone full circle. It was inevitable that the first case on enforcement would set the scene for the whole of the enforcement regime.

The first aspect of this case explored whether or not arbitration was appropriate for enforcement:

> '24. Miss Dumaresq submits that clause 27 of the contract does not apply to disputes as to the validity of an adjudicator's decision. Thus, she argues, a dispute as to whether a decision should be set aside on grounds that it was made ultra vires or in breach of natural justice is not amenable to arbitration. The jurisdiction of the arbitrator is limited to disputes as to the substantive content of the decision itself. The basis for this submission is not that an arbitrator can never have jurisdiction to decide questions as to the lawfulness of an adjudicator's decision. In my judgment, there can be no objection in principle to the parties to a construction contract giving an arbitrator the power to decide such questions. Rather, Miss Dumaresq submits that, upon the true construction of clause 27, the parties to this contract did not give the arbitrator that power.'

The court's view was simply that there was nothing that would prevent the parties having an arbitration clause drafted widely enough to permit enforcement by arbitration.

The second aspect dealt with in this judgment is whether there should be a stay to arbitration, which we have considered earlier in this chapter.

The next aspect the case dealt with was injunctive relief and summary judgment:

> '33. There was some limited discussion as to whether, section 42 apart, the appropriate procedure was by way of writ and an application for summary judgment, or by way of a claim for a mandatory injunction. Miss Dumaresq submits that an injunction is more appropriate. She suggests that summary judgment is not suitable in the context of a provisional decision, which may be revised by an arbitrator at a later stage. Mr Furst submits that the summary judgment route is the correct route to follow.
>
> 34. I do not consider that the mere fact that the decision may later be revised is a good reason for saying that summary judgment is inappropriate. The grant of summary judgment does not pre-empt any later decision that an arbitrator may make. It merely reflects the fact that there is no defence to a claim to enforce the decision of the adjudicator *at the time of judgment*.
>
> 35. I am in no doubt that the court has jurisdiction to grant a mandatory injunction to enforce an adjudicator's decision, but it would rarely be appropriate to grant injunctive relief to enforce an obligation on one contracting party to pay the other. Clearly, different considerations apply where the adjudicator decides that a party should perform some other obligation, e.g. return to site, provide access or inspection facilities, open up work, carry out specified work, etc. Nor do I intend to cast any

---

[6] *Macob Civil Engineering Ltd* v. *Morrison Construction Ltd* [1991] BLR 93.

doubt on decisions where mandatory judgments have been ordered requiring payment of money to a third party, e.g. to a trustee stakeholder as in *Drake & Scull Engineering Ltd* v. *McLaughlin & Harvey Plc* (1992) 60 BLR 102.

36. The words of section 37 of the Supreme Court Act 1981 are widely expressed viz: "the High Court may by order (whether interlocutory or final) grant an injunction... in all cases in which it appears just and convenient to do so". But a mandatory injunction to enforce a payment obligation carries with it the potential for contempt proceedings. It is difficult to see why the sanction for failure to pay in accordance with an adjudicator's decision should be more draconian than for failure to honour a money judgment entered by the court.

37. Thus, section 42 apart, the usual remedy for failure to pay in accordance with an adjudicator's decision will be to issue proceedings claiming the sum due, followed by an application for summary judgment.'

The pattern was set by this case that the correct means of enforcement of adjudicators' decisions was by an application for summary judgment. This is now covered by Part 24 of the Civil Procedure Rules which enables the court to decide a claim or a particular issue without a trial.

Time

The courts have recognised the need to deal with enforcement speedily. This was first explored in *Outwing* v. *Randell*[7]:

'9. Thirdly, I consider that the issue of a summons to abridge time was justified. There are three aspects. First, it may be rare for time to be abridged but in this court (as in commercial, mercantile and other courts) it is done from time to time. Order 3, rule 5(1) is wide and without material limitation. The editors of the *Supreme Court Practice* suggest that the guiding criterion is the avoidance of injustice, i.e. the lack of prejudice to the other party. This criterion naturally re-appears in the Civil Procedure Rules under the head of the overriding objective (see Rule 1.1). It is significant that in Rule 3.1 the list of court's typical powers of management starts with "(a) extend or *shorten* time for compliance with any rule...; (b) adjourn or *bring forward* a hearing;". I see no reason why even the standard time limits for acknowledging service and for opposing an Order 14 application cannot be abridged.

10. Next, ought they to be abridged in a case such as this? In principle in my judgment the answer is: yes. Action to enforce an adjudicator's decision is not comparable to the ordinary process of recovering an apparently undisputed debt. The Rules of the Supreme Court provide a reasonable time for the defendant in an ordinary case to take stock of its position in case there is a defence to the claim. The HGCRA (and the statutory instruments made under it) constitute a remarkable (and possibly unique) intervention in very carefully selected parts of the construction industry whereby the ordinary freedom of contract between commercial parties (without regard to bargaining power) to regulate their relationships has been overridden in a number of areas, one of which is dispute resolution. The overall intention of Parliament is clear: disputes are to go to adjudication and the decision of the adjudicator has to be complied with, pending final determination. There is no provision for a "stay of execution" (unless it is part of the decision itself), presumably since that would undermine the purpose which is finality, at least temporarily. In addition the provisions in the Scheme for the enforcement of peremptory orders via what is thought to be a quick and effective procedure reinforce the conclusion that Parliament intended that adjudicators' decisions and orders, if not complied with, were to be enforced without delay. It is clear that the purpose of the Act is that disputes are resolved quickly and effectively and then put to one side and revived, if at all, in

---

[7] *Outwing Construction Limited* v. *H. Randell & Son Limited* [1999] BLR 156.

litigation or arbitration, the hope being that the decision of the adjudicator might be accepted or form the basis of a compromise or might usefully inform the parties as to the possible reaction of the ultimate tribunal. Thus before a writ is issued to enforce the order of an adjudicator (whether or not declared to be a peremptory order) there will normally have been careful consideration of the underlying dispute, its ramifications and of the adjudicator's decision by all parties. The defendant's room for manoeuvre and its need for further time will be limited.

11. I do not consider that Mr Chambers is right in his submission that if Parliament really intended that adjudicator's orders should be given preferential treatment it ought also to have changed the Rules of the Supreme Court. Parliament does not work through every aspect of its intention expressed in legislation, much of which has to be interpreted purposively, as it is now commonly said. It may only set the direction. It is then for the courts to follow it through and, in so far as it affects the procedure of the courts, to do so in a manner which is fair and respects the interests of the parties. Parliament's intention is in my judgment sufficiently expressed in the Act.'

The general rule now is that when starting an application for summary judgment the matter should be marked 'concerning the enforcement of an Adjudicator's Decision' or similar. At the same time as issuing the Part 7 claim form, the claimant may issue an application without notice to abridge time for the defendant to acknowledge service of the claim form. That application can be heard on the same day as issuing the claim form. The Technology and Construction Court will treat these matters as requiring an early hearing.

CPR Part 8 Alternative procedure for claims

Part 8 is normally used where there is no substantial dispute of fact, such as where the case raises only the question of the construction of a document or statute, suitable to be disposed of on written evidence. (This was what used to be known as a 'construction summons'.)

It is possible to run CPR Part 24 proceedings seeking summary judgment, and CPR Part 8 proceedings seeking a declaration in tandem. The court may deal with both at once. This was the approach adopted in the first instance in *Bouygues* v. *Dahl-Jensen*[8]. Part 24 was being used to enforce the decision and Part 8 was seeking declarations to void the decision.

In *Shepherd* v. *Mecright*[9] the court decided by way of a Part 8 application that a decision was void due to a full and final settlement agreement.

In *ABB* v. *Norwest Holst*[10] the court made a declaration on whether or not there was a construction contract. Similarly in *Staveley* v. *Odebrecht*[11] the court had to make a declaration concerning the expression attached to the land.

Interestingly in *Comsite* v. *Andritz*[12] the court was initially asked to declare that the work under a sub-contract contained construction operations and was a construction contract. However the court had to examine its own jurisdiction to see if it could deal with the Part 8 proceedings where the dispute resolution clause was subject to Austrian law.

About 20 of the enforcement actions in the courts have been Part 8 proceedings. These proceedings were all in the first instance seeking a declaration. The seeking of a declaration can be used positively to make an adjudication valid, or negatively to seek that an adjudication was void.

---

[8] *Bouygues UK Limited* v. *Dahl-Jensen UK Limited* (17 November 1999) and *Dahl-Jensen UK Limited* v. *Bouygues UK Limited*
[9] *Shepherd Construction Limited* v. *Mecright Limited* [2000] BLR 489.
[10] *ABB Power Construction Ltd* v. *Norwest Holst Engineering Ltd* (1 August 2000).
[11] *Staveley Industries Plc (t/a EI. WHS)* v. *Odebrecht Oil & Gas Services Ltd* (28 February 2001).
[12] *Comsite Projects Limited* v. *Andritz AG* [2003] EWHC 958.

The use of insolvency proceedings for enforcement purposes

Where the decision involves payment of money or makes a declaration that monies are due, enforcement ought to become a much simpler proposition. There is no reason why the money should not be pursued through insolvency proceedings. A statutory demand for payment can be made under sections 123 or 268 of the Insolvency Act 1986. The party on whom the demand is made has 21 days within which to pay the amount demanded. If payment is not made within this time a petition for bankruptcy or winding up of the defendant may be made to the court. If there is a defence or the claim is contested in some way, the court will not make a bankruptcy or winding-up order. The defence however must be a real defence and not merely a ruse to avoid the proceedings and payment. This procedure will certainly apply pressure to the party who is reluctant to pay. Insolvency procedures are only useful for enforcement of liquidated monetary decisions.

This is not at all straightforward. Although the decision was not enforced because there were cross claims, the interesting point in *Parke* v. *Fenton Gretton*[13] was that adjudication created a debt which could be pursued by statutory demand.

In *Mohammed* v. *Bowles*[14] some detailed questions were considered:

'1. Is the adjudicator's decision a debt sufficient to form the basis of a statutory demand?
The simple answer is "yes".
2. If yes, what is the nature of the debt?
The nature of the debt is that it is a binding contractual obligation on the applicant to pay the sum quantified by the adjudicator's decision unless and until that decision is varied by further process either by way of arbitration or legal proceedings.
3. Does the applicant dispute the debt on substantial grounds?
In my judgment "no". He has already put his arguments on jurisdiction to the adjudicator who has rejected them. If he is unhappy with that decision his remedy is to go to court but in the absence of any such application it is not for this court to consider those arguments, although in my view they do not show an arguable case. On the substantive issues raised by the applicant whereby he seeks to argue procedural unfairness and a technical contractual point again on the evidence before me I do not consider that they are sufficient to argue that the debt is disputed on substantial grounds.
4. Are there any other grounds on which the statutory demand should be set aside?
The applicant's counsel submitted that without taking a further step and obtaining summary judgment the respondent cannot seek to enforce the decision of the adjudicator. For the reasons set out above I reject that argument.
35. 1 therefore dismiss this application to set aside the statutory demand, the respondent may petition forthwith and the applicant is to pay the respondent's costs.'

It follows that enforcement using a statutory demand remains a possibility.

Insolvency proceedings have had both positive and negative effects on enforcement of adjudicators' decisions.

Insolvent companies

Leave of the court is required to pursue a company in administration in adjudication proceedings. In *Straume* v. *Bradlor*[15] the court decided that adjudication was 'other proceedings' for the purposes of the Insolvency Act 1996:

---

[13] *Parke* v. *The Fenton Gretton Partnership* (24 July 2000).
[14] *Jamil Mohammed* v. *Dr Michael Bowles* (11 March 2003).
[15] *A. Straume (UK) Limited* v. *Bradlor Developments Limited* (7 April 1999).

'The first question, as I have indicated, is whether leave is necessary under Section 11(3) of the Act (Insolvency Act 1996). As is well known, Section 11(3)(d) provides:

"During the period for which an Administration Order is in force no other proceedings and no execution or other legal process may be commenced or continued and no distress may be levied against the company or its property except with the consent of the administrator or the leave of the court and subject, where the court gives leave, to such terms as aforesaid."

… The question for me is whether the adjudication procedure which I have outlined is "quasi legal proceedings such as arbitration or not". On behalf of the applicant it has been argued strongly that it is not. Mr Royce argued that it is the equivalent of some decision by an expert or a valuer giving a certificate. Decisions such as this, it seems to me, are largely matters of impression but I have come to the clear conclusion that the adjudication procedure under section 108 of the Act and/or clause 41 is quasi legal proceedings such as an arbitration within the classification of Vice-Chancellor Browne-Wilkinson in *Re. Paramount Airways*. It seems to me that it is, in effect, a form of arbitration, albeit the arbitrator has a discretion as to the procedure that he uses, albeit that the full rules of natural justice do not apply. The fact that it needs to be enforced by means of a further application does not stop it from being an arbitration. It is the precursor to an enforceable award by the court. It seems to me that it is "other proceedings" within section 11(3) and in my judgment accordingly leave is required.'

There is nothing that prevents companies in receivership pursuing debts or participating in an adjudication.

In *Faithful & Gould* v. *Arcal*[16] the adjudicator issued fee accounts in the name of his firm. Arcal by its administrative receivers, the two others named as defendants in the action for recovery of the fees, had been the unsuccessful referring party in the adjudication and the adjudicator's firm sought to recover fees from Arcal who were in administrative receivership. The adjudicator's firm was allowed to pursue the debt and the receivers were held liable for the fees and expenses.

In *Costain* v. *Wescol*[17] the receivers for a sub-contractor successfully used adjudication in pursuit of the company's debts and resisted a Part 8 application to void the adjudicator's decision.

Risk of bankruptcy

If an adjudicator's decision is enforced and then the recipient of the money goes into liquidation, receivership or bankruptcy, where does this leave the payer in the adjudication who has a legitimate claim in further proceedings to get the money back or to pursue a cross claim? This was examined in *Bouygues* v. *Dahl-Jensen* (see above) both at first instance and in the Court of Appeal.

At first instance there was only the mention of a possibility of insolvency being the cause of injustice, and the adjudicator's decision was enforced. The Court of Appeal explored insolvency set-off which in any event does have an effect on mutual debts. They left the first instance enforcement in place but granted a stay of execution.

In *Rainford House* v. *Cadogan*[18] a stay of execution was granted on the basis of strong prima facie evidence that Rainford House was currently insolvent. There was a similar result in *Isovel* v. *ABB*.[19]

---

[16] *Faithful & Gould Limited* v. *Arcal Limited (in administrative receivership) and Others* (25 May 2001).
[17] *Costain Ltd* v. *Wescol Steel Ltd* (24 January 2003).
[18] *Rainford House Limited* v. *Cadogan Limited* (13 February 2001).
[19] *Isovel Contracts Limited (in administration)* v. *ABB Building Technologies Limited (formerly ABB Steward Limited)* (30 November 2001).

In *Total* v. *ABB*.[20], despite the arguments that Total had been tardy in filing accounts, the adjudicator's decision was enforced save for the set-off on which there had been no adjudication as there was no valid withholding notice.

In *Lovell* v. *Legg and Carver*[21] the court rejected argument on potential insolvency of the contractor as the reason for not enforcing an adjudicator's decision on the grounds that evidence of insolvency was insubstantial.

In *Trentham* v. *Lawfield*[22] the Scottish courts refused an application to remove an inhibition on the grounds that there was significant risk of insolvency.

Set-off

Where the adjudicator's decision involves payment of a sum of money, some assistance may be obtained in the Act to ensure that payment is made. The first point of assistance is section 111(4). This follows any effective notice of intention to withhold payment where an adjudicator's decision is then made on whether or not the payment can be withheld. If the adjudicator decides the payment cannot be withheld, the payment then falls due within seven days from the date of the decision or on the date when final payment would have been made apart from the notice, whichever is the later. These provisions are as follows:

> **111** (4) Where an effective notice of intention to withhold payment is given, but on the matter being referred to adjudication it is decided that the whole or part of the amount should be paid, the decision shall be construed as requiring payment not later than–
> (a) seven days from the date of the decision, or
> (b) the date which apart from the notice would have been the final date for payment, whichever is the later.

It would be unfortunate if the final date for payment exceeded the period to be taken up by the adjudication. However, even if this is the case the period to the final date for payment will expire at some point. The money is then due. Proceedings can then be commenced to collect the money. The second point of assistance as an alternative to commencing proceedings is to invoke the suspension provisions in section 112 of the Act. There is no doubt in this instance that failure to pay the money to which there is an entitlement under the adjudicator's decision is a ground on which the works can be suspended. In fact any decision which entitles a party to money and the final date of payment is passed, would entitle a party to invoke the suspension provisions in section 112 of the Act. The threat of suspending the works may well be a more effective option than commencing arbitration or litigation to enforce the decision.

The real question is, is set-off available to resist enforcement of an adjudicator's decision? This really is a vexed question because a wide availability of set-off would create a situation where it would be simple to defeat the object of the whole legislation and nothing would ever be enforced.

There was a trend that had started to develop where set-off at enforcement had become a real possibility. We ignore insolvency set-off here as that is a special category of set-off in any event. We also ignore matters that have been dealt with during the course of the adjudication under a withholding notice. The core question is, has the Act modified in any way the common law right of set-off?

---

[20] *Total M & E Services Limited* v. *ABB Building Technologies Limited (formally ABB Steward Limited)* (26 February 2002).
[21] *Lovell Projects Ltd* v. *Legg and Carver* (July 2003).
[22] *Barry D. Trentham Limited* v. *Lawfield Investments Limited* (3 May 2002).

In *VHE* v. *RBSTB Trust*[23] the question of the assertion of a common law set-off was raised to resist summary judgment.

'36. The first subject of dispute as to the effect of section 111 is whether section 111(1) excludes the right to deduct money in exercise of a claim to set-off in the absence of an effective notice of intention to withhold payment. Mr Thomas, for RBSTB, submits that it does not. I am quite clear, not only that it does, but that that is one of its principal purposes. I was not taken to the reports or other preparatory material leading to the introduction of this part of the Act, nor to anything said in Parliament, but the see-saw of judicial decision, drafting fashion and editorial commentary in this area is familiar to anyone acquainted with construction law, and in my judgment section 111 is directed to providing a definitive resolution of the debate. The words ''may not withhold payment'' are in my view ample in width to have the effect of excluding set-offs and there is no reason why they should not mean what they say.

37. The other subject of possible dispute is the ambit of section 111(4). Clearly it requires there to have been an effective notice to withhold payment. Mr Furst, for VHE, submits that a further requirement is that the notice must precede the referral and that the ''matter'' referred to adjudication must include the effect of that notice and the validity of the grounds for withholding payment which it asserts. It may be that that was not challenged by Mr Thomas, but in case of any doubt on that score I record that in my judgment it is correct. The effect of the subsection is that, after there has been an effective notice of intention to withhold and an adjudication, payment cannot be enforced earlier than seven days from the date of the decision. There is no reason why that should be so unless the adjudication relates to the notice. Moreover that is the natural point of reference of the expression ''the matter'', with its definite article, as a matter of construction.'

It is clear from this that at least at that time, the courts were comfortable with the idea that the Act did modify the common law right of set-off. The other important point in this case is that any withholding notice should precede the referral.

In *KNS* v. *Sindall*[24] the provisions of the contract were held to have effect at enforcement. The contract had been lawfully terminated by the time the adjudication required payment. The contract provided that no further payment need be made after termination.

'There is nothing in the Act to suggest that, for example, section 111(4) means that a decision that a payment should be made, notwithstanding an effective notice of intention to withhold payment, overrides the parties' other contractual rights. An adjudicator is appointed to decide whether in the circumstances of the dispute a particular right exists and should be enforced. Unless the parties specifically agree, an adjudicator is not appointed to adapt the terms of the contract or to vary, add or take away from the terms of the contract. An adjudicator's powers are limited to those conferred by the contract and thus no more than those of a contract administrator, such as an architect, engineer or surveyor, when entrusted with the resolution of disputes. Their role is to apply the terms of the contract. An adjudicator does the same, but decisions of an adjudicator are now more immediately enforceable pending the result of litigation or arbitration. Any other interpretation would also mean that a party who, albeit apparently wrongly, questioned the existence or ambit of its obligations under the contract so that an adjudicator's decision was against it, would be worse off as a result for if had it not done so it would undoubtedly have retained its other rights under the contract. Accordingly if it had been necessary to do so, KNS's application would have been dismissed on this ground and directions given for the further conduct of the proceedings.'

In *Farebrother* v. *Frogmore*[25] the courts refused to consider a set-off at enforcement:

---

[23] *VHE Construction PLC* v. *RBSTB Trust Co Limited (as trustee of the Mercury Property Fund)* (13 January 2000).

[24] *KNS Industrial Services (Birmingham) Limited* v. *Sindall Limited* (14 July 2000).

[25] *Farebrother Building Services Limited* v. *Frogmore Investments Limited* (20 April 2001).

'I take the view that it is not right for the court to try and dismantle or reconstruct a decision. It seems to me that a party cannot pick and choose amongst the decisions given by an adjudicator, assert or characterise part as unjustified and then allege that the part objected to has been made without jurisdiction. That is not permissible under the TeCSA Rules. Either the adjudicator has jurisdiction or he does not. If he had jurisdiction, it seems to me that his decision is binding even if he was wrong to reach the conclusion he did. I take the view this award ought to be enforced in the sum found by the adjudicator and that it is not right to seek to set off the £300,000-odd which the defendant seeks to deduct from the award. I propose to make a summary order.'

In *Millers* v. *Nobles*[26], which was an application for summary judgment where there had been no adjudication, the court would not entertain any set-off without there having been a valid withholding notice in place.

In *McAlpine* v. *Pring & St Hill*[27] the court refused to allow set-off at enforcement. The matter to be set-off at enforcement was the subject of a further adjudication. The court did however refuse a stay of execution until the following adjudication had been completed.

Parsons v. *Purac*[28] was heard both at first instance and in the Court of Appeal. In both cases the set-off at enforcement was upheld.

In *McLean* v. *Swansea*[29] the court refused summary judgment on the basis that a counterclaim for liquidated damages had a reasonable prospect of success in proceedings that had already commenced.

In *Bovis* v. *Triangle*[30] the court held the following:

'1. The decision of an adjudicator that money must be paid gives rise to a separate contractual obligation on the paying party to comply with that decision within the stipulated period. This obligation will usually preclude the paying party from making withholdings, deductions, set-offs or cross-claims against that sum.
2. For a withholding to be made against an adjudicator's decision, an effective notice to withhold payment must usually have been given prior to the adjudication notice being given, or possibly the decision being given, and which was ruled upon and made part of the subject matter of that decision.
3. However, where other contractual terms clearly have the effect of superseding, or provide for an entitlement to avoid or deduct from, a payment directed to be paid by an adjudicator's decision, those terms will prevail.
4. Equally, where a paying party is given an entitlement to deduct from or cross-claim against the sum directed to be paid as a result of the same, or another, adjudication decision, the first decision will not be enforced or, alternatively, judgment will be stayed.'

The most important case on this subject is *Levolux* v. *Ferson*[31]. It was held at first instance:

'41. A party has no right to set-off claims not dealt with by the adjudicator as a defence to the enforcement of the adjudicator's decision. See *VHE Construction plc* v. *RBSTB Trust Company Limited* [2000] BLR p. 187, judgment of HHJ Hicks QC at pages 199 and 196; *Northern Developments (Cumbria) Limited* v. *J&J Nicol* [2000] BLR, judgment of HHJ Bowsher QC pages 158 and 164; *Solland International Limited* v. *Darayden Holdings Limited* TCC judgment of HHJ Seymour QC, 15 February 02.'

The Court of Appeal went further in voiding any clause that would defeat the intentions of Parliament:

---

[26] *Millers Specialist Company Limited* v. *Nobles Construction Limited* (3 August 2001).
[27] *Sir Robert McAlpine* v. *Pring & St Hill Limited* (2 October 2001).
[28] *Parsons Plastics (Research & Development) Limited* v. *Purac Limited* (13 August 2001); [2002] BLR 334, CA.
[29] *David McLean Housing Contractors Limited* v. *Swansea Housing Association Limited* (27 July 2001).
[30] *Bovis Lend Lease Limited* v. *Triangle Development Limited* (2 November 2002).
[31] *Levolux A.T. Limited* v. *Ferson Contractors Limited* [2002] BLR 341; [2003] BLR 118, CA.

'30. But to my mind the answer to this appeal is the straightforward one provided by Judge Wilcox. The intended purpose of section 108 is plain. It is explained in those cases to which I have referred in an earlier part of this judgment. . . The contract must be construed so as to give effect to the intention of Parliament rather than to defeat it. If that cannot be achieved by way of construction, then the offending clause must be struck down. I would suggest that it can be done without the need to strike out any particular clause and that is by the means adopted by Judge Wilcox. Clause 29.8 and 29.9 must be read as not applying to monies due by reason of an adjudicator's decision.'

The GC/Works sub-contract applied to this contract, the sub-contract had been determined and clauses 29.8 and 29.9 are those clauses that allow the main contractor to avoid making any payment in these circumstances.

This principle was applied in *Dumarc* v. *Mr Salvador Rico*[32].

The premise now is that set-off against an adjudicator's decision will not apply except in limited circumstances. These limited circumstances are when the contract is not a construction contract. This applies in the *Parsons* and *Bovis* cases above.

Appeals

There is in fact no entitlement or means by which an adjudicator's decision can be appealed. The decision itself is pro tem, of a temporarily binding nature. The Act requires that the decision is only binding until the dispute is finally determined by legal proceedings, by arbitration (if the contract provides for arbitration or the parties otherwise agree to arbitration) or by agreement. These provisions do not provide a mechanism for appeal from an adjudicator's decision which is in some way unsatisfactory or flawed, but a replacement process to hear the dispute de novo, from new. The dispute is heard again on its merits and decided from new on the facts and evidence presented in the new forum.

This may be by litigation as provided in the Act or by arbitration if the contract so provides, or the parties may agree on arbitration on an ad hoc basis to resubmit the dispute that was the subject of an adjudication. It may be that the parties will choose to consolidate several disputes into one set of proceedings. The words 'is binding until the dispute is finally determined' make it clear that the nature of adjudication is a temporary decision which will be replaced with a new determination from another process. The parties may agree finally that they are satisfied to be bound by the adjudicator's decision or that they will substitute a new arrangement for that decision. Where they take these steps such agreement should be recorded in writing in order to show that the dispute has indeed been compromised and to preclude any future attempt to reopen it.

Unless they do something by way of arbitration, litigation or agreement, the decision, because of lack of any action by the parties, will remain binding. The proceedings to finally determine the dispute can be commenced at any time but there can be no appeal from the adjudicator's decision. Enforcement can be resisted on grounds such as want of jurisdiction or procedural unfairness but the only 'appeal' from the decision is to commence new proceedings in litigation or possibly arbitration which involve a completely new examination of the parties' contentions, and in all normal circumstances the adjudicator's decision is irrelevant to these.

---

[32] *Dumarc Building Services Limited* v. *Mr Salvador-Rico* (31 January 2003).

Grounds for resisting enforcement

The analogous area of law in respect of challenge of decisions is in the field of expert determination. The experience in expert determination is that decisions of experts are extremely difficult to challenge. Expert determination provides decisions which are invariably permanent rather than temporarily binding. Expert determination is also a creature of contract. The courts have been reluctant to interfere with decisions which arise from this process.

The guidelines for enforcement and resisting enforcement in adjudication were established early by the courts and are summarised in *Sherwood & Casson* v. *Mackenzie*[33]:

> '24. These provisions, and the well-known background to the enactment of the HGCRA, have led Dyson J to formulate clear guidance to the approach to the enforcement of adjudicators' decisions which, although persuasive and not binding on me is guidance that I gratefully adopt. The three decisions are: *Macob Civil Engineering Ltd* v. *Morrison Construction Ltd* (1999) Building Law Reports 93; *The Project Consultancy Group* v. *The Trustees of The Gray Trust* (unreported) 16 July 1999; and *Bouygues UK Ltd* v. *Dahl-Jensen UK Ltd* (unreported) 17 November 1999. I can summarise this guidance as follows:
>
> 1. A decision of an adjudicator whose validity is challenged as to its factual or legal conclusions or as to procedural error remains a decision that is both enforceable and should be enforced (*Macob*).
> 2. A decision that is erroneous, even if the error is disclosed by the reasons, will still not ordinarily be capable of being challenged and should, ordinarily, still be enforced (*Bouygues*).
> 3. A decision may be challenged on the ground that the adjudicator was not empowered by the HGCRA to make the decision, because there was no underlying construction contract between the parties (*Project Consultancy*) or because he had gone outside his terms of reference (*Bouygues*).
> 4. The adjudication is intended to be a speedy process in which mistakes will inevitably occur. Thus, the court should guard against characterising a mistaken answer to an issue, which is within an adjudicator's jurisdiction, as being an excess of jurisdiction. Furthermore, the court should give a fair, natural and sensible interpretation to the decision in the light of the disputes that are the subject of the reference (*Bouygues*).
> 5. An issue as to whether a construction contract ever came into existence, which is one challenging the jurisdiction of the adjudicator, so long as it is reasonably and clearly raised, must be determined by the court on the balance of probabilities with, if necessary oral and documentary evidence (*Project Consultancy*).'

This list is not exhaustive as other cases have developed the policy on enforcement and possible grounds for resistance of enforcement, but it does give the basic principles.

Jurisdiction

Jurisdiction is a wide subject in itself and is dealt with in Chapter 9.

A challenge on jurisdiction grounds can survive the reference and prevent an adjudicator's decision being enforced. If the adjudicator proceeds despite objections to jurisdiction and reaches what purports to be a decision, if he had no jurisdiction at all the decision can be successfully challenged.

In *AMOCO* v. *Amerada Hess*[34] the question of jurisdiction of an expert was explored. It was held:

---

[33] *Sherwood & Casson Limited* v. *Mackenzie* (30 November 1999).
[34] *AMOCO (UK) Exploration Co and Others* v. *Ameralda Hess Ltd and Others* CHD 25 [1994] 1 Lloyd's Rep 330.

'1. In my judgement, the dispute does fall within the terms of clause 1.8.8. By clau
referred to the expert for his decision, all elements comprising a key step which are u
he should look at and the data he should take into account, was not referred to him as
element, the difference only emerged subsequently. I see no reason why it should not i   il
the difference had emerged at an earlier stage. The determination of what is and what is nc   ..ata and
what is and is not appropriate to be included in the agreed database, as provided for by clause 1.9,
seems to me to be an unagreed element suitable for his determination at the proper key step or key
steps. So the question of what is and what is not "data" and whether or not it is "within a TSC's
(Technical Sub-Committee) own agreed database" is not a suitable matter for his determination,
notwithstanding that it may involve legal elements as well as scientific.
2. Accordingly, in my judgement, it is within the remit of the expert to decide what he is entitled to
pay regard to, notwithstanding that legal questions may be involved, as an independent unagreed
element. But if that is wrong, in my judgement it is clear that that question, namely what material he
may take into consideration in deciding the unagreed element of faults in wells, is, "in respect of that
matter", in dispute. The question arises directly from the dispute concerning the unagreed matter.
As has been said in other contexts, the phrase "in respect of" has the "widest possible meaning of
any expression intended to convey some connection or relation between two subject matters to
which the order refers".'

This deals with the remit to an expert and his terms of reference. This is analogous to
adjudication. It is important that adjudicators work within the boundaries of both their
jurisdiction and powers. There needs to be particular care taken when deciding which
documents are to be considered to assist in making the decision.

Probably the most important case that goes to jurisdiction is *Nikko* v. *MEPC*.[35] This case is
dealt with under 'Errors and mistakes' below but the passage quoted also goes to jurisdiction.

Errors and mistakes

What if the adjudicator makes a mistake in a finding of fact or some error of fact in making
his decision? What if the adjudicator gets the law wrong? Save for those matters which are
obvious on the face of a decision without reasons, such error or mistake would be difficult to
find. The following cases illustrate the court's reluctance to overturn decisions made in the
field of expert determination.

In *Jones* v. *Sherwood*[36] the following extract gives guidance on mistakes:

'They say that the mistakes, if they were mistakes, would have been mistakes of law, or of mixed law
and fact, as to the true construction and effect of the provisions of appendix 1 and particularly of
paragraph 2.
It is therefore necessary to see how the law stands on the question of challenging an expert's certificate on the grounds of mistake. We are not of course here concerned with any question of fraud or
collusion on the part of the expert.
The cases have been fully analysed by Sir David Cairns in *Baber* v. *Kenwood Manufacturing Co Ltd*
[1978] 1 Lloyd's Rep. 175, 181–183, and by Nourse J in *Burgess* v. *Purchase & Sons (Farms) Ltd* [1983]
Ch. 216. The starting point for the modern statement of the law is, in my judgment, the decision in
Campbell v. *Edwards* [1976] 1 WLR 403 and in particular the passage in the judgment of Lord Denning
MR, at p. 407:

---

[35] *Nikko Hotels (UK)* v. *MEPC PLC* [1991] 28 EG 86.
[36] *Jones* v. *Sherwood Services Plc* [1992] 1 WIR 284, CA.

"It is simply the law of contract. If two persons agree that the price of property should be fixed by a valuer on whom they agree, and he gives that valuation honestly and in good faith, they are bound by it. Even if he has made a mistake they are still bound by it. The reason is because they have agreed to be bound by it. If there were fraud or collusion, of course, it would be very different. Fraud or collusion unravels everything."

That statement was, as a matter of principle and disregarding the earlier authorities, endorsed by Megaw LJ in *Baber's* case [1978] 1 Lloyd's Rep. 175, 179, and concurred in by the other members of this court in *Baber's* case. It is in line with the passage, cited by Sir David Cairns in *Baber's* case, at p. 181, from the judgment of Sir John Strange MR in *Belchier* v. *Reynolds* (1754) 3 Keny. 87, 91:

"Whatever be the real value is not now to be considered, for the parties made Harris their judge on that point; they thought proper to confide in his judgment and skill and must abide by it, unless they could have made it plainly appear that he had been guilty of some gross fraud or partiality."

On principle, the first step must be to see what the parties have agreed to remit to the expert, this being, as Lord Denning MR said in *Campbell* v. *Edwards* [1976] 1 WLR 403, 407G, a matter of contract. The next step must be to see what the nature of the mistake was, if there is evidence to show that. If the mistake made was that the expert departed from his instructions in a material respect – e.g. if he valued the wrong number of shares, or valued shares in the wrong company, or if, as in *Jones (M.)* v. *Jones (R.R.)* [1971] 1 WLR 840, the expert had valued machinery himself whereas his instructions were to employ an expert valuer of his choice to do that – either party would be able to say that the certificate was not binding because the expert had not done what he was appointed to do.'

The principal enunciated in *Jones* v. *Sherwood* is that the courts will give effect to the contractual arrangements made by the parties. Where an adjudicator makes a decision it should not be set aside lightly. A mistake of itself is not enough to set the decision aside. It is only the nature of the mistake which may provide grounds for setting the decision aside. The only grounds here are if the adjudicator departs from his instructions in a material respect. If the notice of adjudication instructs him to answer one question and he answers a completely different question, this would be grounds for challenge of the decision. Departures from the instructions which are not material, minor procedural technicalities, will not provide grounds in mistake to set the decision aside.

Nikko v. *MEPC* also involved the question of a mistake in law:

'The position is that if the parties agree to refer to the final and conclusive judgment of an expert an issue which either consists of a question of construction or necessarily involves the solution of a question of construction, the expert's decision will be final and conclusive, and therefore not open to review or treatment by the courts as a nullity on the ground that the expert's decision on construction was erroneous in law, unless it can be found that the expert has not performed the task assigned to him. If he has answered the right question in the wrong way, his decision will nevertheless be binding. If he has answered the wrong question, his decision will be a nullity.'

Even where the adjudicator is wrong in law the parties will be bound by the decision unless the adjudicator has not performed the task assigned to him. This could have serious consequences when at the time that the dispute is finally determined one of the parties is no longer in business. It is argued that one of the disadvantages of monetary decisions by adjudicators is that they cannot be reversed at a later stage if the party who has to repay the sum is insolvent. If indeed there is no right of challenge where the adjudicator is wrong in law, this unfortunate situation will prevail.

The principle identified in this case was considered and/or applied in a number of other cases[37].

---

[37] *Bouygues UK Ltd* v. *Dahl-Jensen (UK) Ltd* [2000] BLR 49; [2000] BLR 522 CA; *Karl Construction (Scotland) Ltd* v. *Sweeney Civil Engineering (Scotland) Ltd* [2001] SCLR 95; *Barr Limited* v. *Law Mining* (15 June 2001); *C & B Scene Concept Design Ltd* v. *Isobars Ltd* [2002] BLR 93, CA; *Ferson Contractors Limited* v. *Levolux AT Ltd* [2003] BLR 118, CA.

In *Mercury* v. *Telecommunications Directors*[38] doubt was cast on the binding nature of experts' decisions and the lack of right to challenge as follows:

'Reference was made to *Jones* v. *Sherwood Computer Services Plc* [1992] 1 WLR 277 where the Court of Appeal held that in a case where parties had agreed to be bound by the report of an expert the report could not be challenged in the courts unless it could be shown that the expert had departed from the instructions given to him in a material respect. In that case the experts had done exactly what they were asked to do.

What has to be done in the present case under condition 13, as incorporated in clause 29 of the agreement, depends upon the proper interpretation of the words "fully allocated costs" which the defendants agree raises a question of construction and therefore of law, and "relevant overheads" which may raise analogous questions. If the Director misinterprets these phrases and makes a determination on the basis of an incorrect interpretation, he does not do what he was asked to do. If he interprets the words correctly then the application of those words to the facts may in the absence of fraud be beyond challenge. In my view when the parties agreed in clause 29.5 that the Director's determination should be limited to such matters as the Director would have power to determine under condition 13 of the BT licence and that the principles to be applied by him should be "those set out in those conditions" they intended him to deal with such matters and such principles as correctly interpreted. They did not intend him simply to apply such meaning as he himself thought they should bear. His interpretation could therefore be reviewed by the court. There is no provision expressly or impliedly that these matters were remitted exclusively to the Director, even though in order to carry out his task he must be obliged to interpret them in the first place for himself. Nor is there any provision excluding altogether the intervention of the court. On the contrary clause 29.5 contemplates that the determination shall be implemented "not being the subject of any appeal or proceedings." In my opinion, subject to the other points raised, the issues of construction are ones which are not removed from the court's jurisdiction by the agreement of the parties.'

The point here is if the adjudicator is wrong in law he has not actually carried out the task he was asked to do. This principle was applied in *C & B Scene* v. *Isobars* by the judge at first instance[39]. He characterised an error in law as answering the wrong question and therefore there was no jurisdiction. This was overturned by the Court of Appeal[40] which favoured the analysis in *Nikko* v. *MEPC*.

Decisions can always be challenged on the basis that there was no jurisdiction. It is unlikely, particularly with the dependence on *Nikko* v. *MEPC*, that the courts will be persuaded that a simple error or mistake of law or fact can be characterised as want of jurisdiction.

It is possible that the court may be persuaded that it should not enforce decisions where the adjudicator makes a determination on the basis of an incorrect interpretation of matters which are entirely outside his jurisdiction.

It is thought that *Mercury* v. *Telecommunications Directors* is not authority for the proposition that the court can always intervene. The judges subsequently made a practical point that the court should usually decline to intervene before a decision has been reached because the application to the court has a tendency to increase the cost of the dispute. What was also not considered in this case is whether the court should decline to intervene if private systems of dispute resolution have not broken down.

In *Conoco* v. *Philips Petroleum*[41] the question of construction of the contract was revisited. It was held that:

---

[38] *Mercury Ltd* v. *Telecommunications Directors* [1996] 1 WLR 59, HL.
[39] *C & B Scene Concept Design Ltd* v. *Isobars Ltd* TCC (21 June 2001).
[40] *C & B Scene Concept Design Ltd* v. *Isobars Ltd* [2002] BLR 93, CA.
[41] *Conoco (UK) Ltd and Others* v. *Phillips Petroleum and Others*, QBD (1996).

'1. The expert had been required to place his own interpretation on the language of the procedural rules in the agreement. If his interpretation was obviously wrong, then it could be challenged on that count. If it could be demonstrated that the expert had not effectively performed his task, not only would that be a manifest error, but also an independent ground for challenging the decision.

2. However, a minor contravention of the procedure could not justify a challenge. Hence, the court should ask whether the expert had performed the task allocated to him, and, if so, whether the determination was manifestly wrong either because of the way he had interpreted the rules or on substantive grounds. An error would be ''manifest'' if it were an oversight or a blunder so obvious as to admit no difference of opinion. Whether such an oversight or blunder would have made any difference to the result was irrelevant. On the evidence, the court concluded that there had been no obvious procedural or substantive error in the expert's work.'

The interpretation placed on the procedure by the adjudicator needs to be obviously wrong. This may prove difficult where adjudicators' decisions are given with very little detail. If a matter of construction is obviously wrong that would be grounds for a challenge. This differs from the position in *Mercury* v. *Telecommunications Directors*. This case goes to two points, procedural unfairness which is dealt with below and want of jurisdiction. If the adjudicator has not performed the task assigned to him this forms a further ground of challenge. There is some sensible help here on what constitutes a manifest error. The error has to be so obvious that there is no doubt that it is wrong. Whether the error would have made any difference to the result is irrelevant. It is a matter which provides grounds for challenge.

British Shipbuilders v. *VSEL*[42] deals with both error and jurisdiction:

'Looking at the agreement, the accountants' role did not extend beyond the certification process because:

1. The agreement did not contain details of a role for the expert in resolving disputes between the parties as to their rights inter se;

2. VSEL's obligation under clause 5.5 was prima facie enforceable by specific performance or injunction and the expert had no jurisdiction in either of these respects;

3. If the expert were to have jurisdiction, clause 5.5 would have preceded clause 5.2 and 5.3;

4. VSEL had conceded that it had been intended that the court should retain jurisdiction over clause 5.5 unless and until the accountants disagreed. Consequently, if the expert went outside his remit, the court could intervene and set his decision aside. Similarly, the expert's decision could be set aside by the courts in cases of manifest error.'

The importance of remaining inside the remit is emphasised here. Where a manifest error occurs, the decision can arguably be set aside.

Dixons v. *Murray-Oboyinski*[43] follows *Jones* v. *Sherwood* and the expert's determination was upheld:

'HELD, granting the declarations sought and striking out the paragraphs in the defence which sought to challenge Mr Jackson's determination

(1) The interpretation of the agreement was left to the expert therefore whether there was any requirement in the agreement as to the basis of determination was a matter for the expert to decide.

(2) There was no evidence as to what basis of determination had been applied by the expert and therefore it could not be shown that he had failed to follow his instructions. *Jones* v. *Sherwood Computer Services plc* [1982] 1 WLR 277 applied.

---

[42] *British Shipbuilders* v. *VSEL Consortium PLC* [1997] 1 Lloyd's Rep 106, CHD 2 February 1996, TLR 14 February 1996.
[43] *Dixons Group PLC* v. *Jan Andrews Murray-Oboynski* (1997) CILL 1330; 86 BLR 16.

(3) A manifest error was an error that may easily be seen by the eye or perceived by the mind. There was no such plain and obvious error on the face of the expert's award and the terms of the agreement did not contemplate an error which after a lengthy enquiry might be made manifest. *Healds Foods Ltd* v. *Hyde Dairies Ltd* (unreported, 1 December 1994) QBD, Potter J considered.'

This illustrates again the area of difficulty where a decision is given with no reasons. If evidence cannot be provided to demonstrate that the adjudicator has failed to follow his instructions, any challenge must also fail. Similarly, errors must be plain and obvious and on the face of the adjudicator's decision, before a challenge can succeed. Unless the contract provides otherwise there should be no lengthy inquiry to seek out errors in proceedings to enforce adjudicators' decisions.

Procedural unfairness

This can be a difficult area under which to challenge an adjudicator's decision. There is a restraint in terms of the overall time available. The adjudicator has only 28 days in which to reach his decision. There is no requirement that this time be allocated equally between the parties; the allocation of time will very much depend on the nature of the dispute itself. There is no need to permit an adversarial process or procedure. The adjudicator has the authority to conduct an inquiry. Unless the adjudicator takes the lead and makes that necessary inquiry there will be problems with the timetable. It is against these restraints that any procedural unfairness will be judged. Adjudication cannot therefore be judged on the same basis of perfection that other systems of dispute resolution be judged where the time restraint is not as limiting.

In respect of arbitration awards it has been held that 'an award obtained in violation of the rules of natural justice even where there is no breach of the agreed procedure would be set aside on grounds of public policy' (*London Export* v. *Jubilee*[44]). What will be interpreted as the application of the rules of natural justice in the context of adjudication remains to be seen. It must be a factor that the process can be inquisitorial and there is a severe time restraint.

In *R* v. *Disciplinary Committee of the Jockey Club*[45], it was said:

'The remedies in private law available to the Aga Khan seem entirely adequate. He has a contract with the Jockey Club. The club has an implied obligation under the contract to conduct its disciplinary proceedings fairly. If it has not done so, the Aga Khan can obtain a declaration that the decision was ineffective and, if necessary an injunction to restrain the Jockey Club from doing anything to implement it.'

Procedural error may cover the breach of the rules of natural justice but it cannot solely be classified on that basis. There are no hard and fast rules of natural justice. In *Russell* v. *Duke of Norfolk*[46] the Court of Appeal said:

'There are, in my view, no words which are of universal application to every kind of enquiry and every kind of domestic tribunal. The requirements of natural justice must depend on the circumstances of the case, the nature of the enquiry, the rules under which the tribunal is acting, the subject matter that is being dealt with, and so forth; accordingly, I do not derive much assistance from the definitions of natural justice which have been from time to time used, but, whatever standard is

---

[44] *London Export Corporation* v. *Jubilee Coffee Roasting Company* [1958] 1 Lloyd's Rep 197.
[45] *R* v. *Disciplinary Committee of the Jockey Club, ex parte Aga Khan* [1993] 1 WLR 909.
[46] *Russell* v. *Duke of Norfolk* [1949] 1 All ER 109; (1949) 65 TLR 225; (1949) 93 SJ 132.

adopted, one essential is that the person concerned should have a reasonable opportunity of presenting his case.'

In *Wiseman* v. *Borneman*[47] the House of Lords said:

> 'Natural justice requires that the procedure before any tribunal which is acting judicially shall be fair
> in all the circumstances, and I would be sorry to see this fundamental principle degenerate into hard-
> and-fast rules.'

The process must in essence be fair and allow each side a reasonable opportunity of presenting its case. This was explored in *Macob* v. *Morrison* where there were challenges on the
grounds of breach of the rules of natural justice:

> 'The present case shows how easy it is to mount a challenge based on an alleged breach of natural
> justice. I formed the strong provisional view that the challenge is hopeless. But the fact is that the
> challenge has been made, and a dispute therefore exists between the parties in relation to it. Thus on
> Mr Furst's argument, the party who is unsuccessful before the adjudicator has to do no more than
> assert a breach of the rules of natural justice, or allege that the adjudicator acted partially, and he will
> be able to say that there has been no "decision".
> ... In that event, it would be necessary to ascertain the correct meaning from the scheme of the Act
> and the Scheme, and the background against which it was passed. Adopting that purposive
> approach to the construction of the word "decision", I am in no doubt that it should not be qualified
> in the way suggested by Mr Furst. The plain purpose of the statutory scheme is as I have earlier
> described. Mr Furst would not accept that his construction would drive a coach and horses through
> the scheme. On any view, it would substantially undermine it, and enable a party who was dis-
> satisfied with the decision of an adjudicator to keep the successful party out of his money for longer
> than envisaged by the scheme.'

The robust view adopted by Dyson J in *Macob* v. *Morrison* did not immediately meet with
favour in the Scottish Courts. In *Homer Burgess* v. *Chirex*[48] Lord Macfadyen said:

> 'In coming to that conclusion I also derive support from the views expressed by Dyson J in *The Project
> Consultancy Group* v. *The Trustees of the Gray Trust*. The respect in which the adjudicator's decision in
> that case was beyond the proper scope of his jurisdiction was somewhat different, but the passage
> which I have quoted above from paragraph 6 of Dyson J's judgment figures an example which is
> close to the circumstances of the present case. I would add, however, that I wish to reserve my
> opinion as to the soundness of the distinction which Dyson J drew between the effect of an assertion
> that the decision of the adjudicator was one which he was not empowered to make, and the effect of
> an assertion that the decision of the adjudicator was invalid on some other ground such as breach of
> the rules of natural justice. Although that point does not bear directly on the matter which I have to
> decide, I have some difficulty in reconciling Dyson J's distinction with what was said in *Anisminic*,
> for example by Lord Reid at 171C. Moreover, although I have expressed my views in terms of the
> effect of an error in law on the part of the adjudicator as to the scope of his jurisdiction, since that is
> what is alleged to have happened in the present case, I do not wish to be taken to have decided that
> an error of fact on the part of an adjudicator which resulted in his purporting to make a decision on a
> matter outside his jurisdiction would not have the same effect. I reserve my opinion on that matter
> too. For the purposes of the present case, it is sufficient that I am of opinion that if, as the defenders
> allege, the adjudicator erred in his construction of the term "plant" in section 105(2)(c) and conse-
> quently purported to issue a decision on matters which fell outside his jurisdiction, that is a relevant
> defence to the present action.'

---

[47] *Wiseman* v. *Borneman* [1971] AC 297; [1969] 3 WLR 706; [1969] 3 All ER 275.
[48] *Homer Burgess Limited* v. *Chirex (Annan) Limited* [2000] BLR 124.

This case was really dealing with a straightforward excess of jurisdiction and should not be representative of cases dealing with procedural unfairness.

Another category of procedural unfairness is bias. It does not have to be actual bias, favouring one party over another; it only needs to be a perception or danger of bias having occurred. This was explored in *Discain* v. *Opecprime*[49] where the judge said:

'It does seem to me there was a very serious risk of bias, and there were clear failures to consult with one party on important submissions which were made by the other party. It is said on behalf of the claimant that this is irrelevant because the matter goes only to jurisdiction (that is accepted) and I am looking at the matter of jurisdiction anyway. If there has been a breach of the rules of natural justice (and I find that there has) it is submitted that it does not make any difference because the question of jurisdiction in any event is under review.

I find it distasteful and I cannot bring myself to enforce an adjudication which has been arrived at in that way. I do not wish to criticise the adjudicator, I fully understand his difficulties, but he should have made sure that the other party was involved in the discussions regarding his jurisdiction, which he failed to do. I wish to stress that I am not criticising the adjudicator; I do understand that adjudicators have great difficulties in operating this statutory scheme, and I am not in any way detracting from the decision in *Macob*. It would be quite wrong for parties to search around for breaches of the rules of natural justice. It is a question of fact and degree in each case, and in this case there is an issue to be tried whether the adjudicator overstretched the rules.'

In *Woods Hardwick* v. *Chiltern*[50] the court expressed a straightforward view on partiality and procedural unfairness. This gives with clarity some of the grounds on which an adjudicator's decision will not be enforced:

'39. I am also conscious that the adjudicator attempted to act in an impartial manner and showed no conscious bias or hostility to Chiltern. However, the statutory requirement to act impartially requires the adjudicator to act in a way that does not lead to a perception of partiality by one party which might objectively be held by that party. In this case, the adjudicator led the parties to believe that there would be no need for a hearing. Such a view would only be tenable if the adjudicator took other steps, by way of written communications for example, to inform both parties of any relevant additional information he subsequently obtained to enable them to comment upon it. Having left the parties with the impression that he did, he acted in a manner which could readily be perceived to be partial in approaching one side without informing the other, in seeking much additional information from third parties and in then making adverse findings against the party left in ignorance of these steps. These difficulties were then compounded by the adjudicator voluntarily providing a witness statement which seeks to put forward Woods Hardwick's case in favour of enforcement and which elaborates on the reasons for making adverse findings against Chiltern.

6.3. The Effect of the Findings on the Enforcement Proceedings

40. The adjudicator, in order to make a valid and enforceable decision, must act in conformity with the rules of the Scheme. There will be occasions when an adjudicator's departure from those rules is insignificant and not such as to preclude enforcement. Where, however, the departures are significant, the decision is one taken outside the framework of the Scheme and is not one which a court will ordinarily enforce. The consequence is that I must dismiss the summary judgment application. The claim based on the Chapel Street adjudication will be dealt with and tried at the same time as the action brought by Chiltern against Woods Hardwick that is concerned with this development. Further directions in this action will be given at the first case management conference held in that action. Meanwhile, the costs of the Chapel Street enforcement proceedings will be costs in the case. There will be permission to either party to apply for further directions if necessary.'

---

[49] *Discain Project Services Ltd* v. *Opecprime Development Ltd* (9 August 2000).
[50] *Woods Hardwick Ltd* v. *Chiltern Air Conditioning* [2001] BLR 23.

There is a difficulty when mediation is combined with adjudication. This was explored in *Glencot* v. *Ben Barrett*[51]. This case provides a full discourse on bias. The adjudicator in this case was not found to be actually biased:

'20. It is accepted that the adjudicator has to conduct the proceedings in accordance with the rules of natural justice or as fairly as the limitations imposed by Parliament permit: see *Discain Project Services Ltd* v. *Opecprime Development Ltd* [2000] BLR 402. Secondly, in my judgment it was not intended by Parliament the words should have anything other than their usual meaning, and, in so far as the parties are to be treated as having adopted the wording as a matter of their contract, then contractually they are to bear the same meaning. Thirdly, and similarly, if the concept of impartiality or lack of bias were to be restated or redefined in the light of new legislation, such as the Human Rights Act, then the latest meaning, as it were, applies, obviously in law and as a matter of contract. Otherwise in relation to each of these reasons the words might bear one meaning in general and another meaning by contract or under the HGCRA, which to my mind would be verging on the absurd and is not correct.
21. Accordingly is Mr Talbot's conduct such as to be regarded not "impartial"? It must be emphasised that there is no question of actual bias. It is a case of "apparent bias". Mr Talbot's personal impartiality is presumed. In *Gough* Lord Woolf reiterated that "bias operates in such an insidious manner that the person alleged to be biased may be quite unconscious of its effect". Hence there is a need for an objective test. The views of the person involved are either irrelevant or not determinative. The test is whether the "circumstances would lead a fair-minded and informed observer to conclude that there was a real possibility or a real danger, the two being the same, that the tribunal was biased".
22. In my judgment the defendant has shown that it has real prospects of success in establishing that Mr Talbot was no longer impartial as a result of what took place over most of 29 September. First, Mr Talbot's own appreciation of those events led him to write his letter of 2 October and, in paragraph 7, to ask the parties to decide whether they wished him to continue. I am sure that he was right.'

This case does not rule out mediation during the currency of an adjudication. Had the mediation succeeded the matter would not have been before the courts. The problem arises when matters that are confidential in the mediation are revealed to the mediator/adjudicator and then the adjudication continues and one or both parties end up dissatisfied with the result.

In Discain (No.2)[52] the distinction between procedural error and breach of the rules of natural justice was made:

'The reference in that paragraph to a "procedural error" has been read as meaning that breaches of natural justice are to be regarded as "procedural errors" and to be disregarded when considering whether decisions of adjudicators should be enforced. One can test that proposition by thinking the unthinkable, going to an extreme and asking what would be the approach if it were shown that an adjudicator refused to read the written submissions of one party because they were typed with single rather than double spacing. It would never happen. But if it did, his decision would not be enforced. So there must be some breaches of natural justice that would persuade the court not to enforce the decision of an adjudicator. How is the line to be drawn?
In an article in the *Construction Law Journal* (2000) 16 Const LJ 102, Mr Ian Duncan Wallace QC, who has achieved the remarkable position of being a controversialist in the dry area of construction law, also considered the decision of Dyson J in *Macob*. He said:

---

[51] *Glencot Development & Design Co Ltd* v. *Ben Barrett & Son (Contractors) Limited* (13 February 2001).
[52] *Discain Project Services Limited* v. *Opecprime Developments Limited* (11 April 2001).

"It is respectfully submitted that the words 'a procedural error which invalidates the decision' in this passage go too far if they mean, as Dyson J (probably obiter) states, that enforcement of an award arrived at in breach of the principles of natural justice cannot be resisted on those grounds in summary proceedings at all.

On the other hand, with all respect, it is a startling proposition that an adjudicator's decision, if arrived at in serious breach of a principle of natural justice, must as a matter of law nevertheless be enforced in circumstances where payment under an invalid decision could easily turn out to be irretrievable and precipitate the insolvency of the party affected (particularly where, as here, there had not even been a decision by the adjudicator on the merits, but only a procedural one shutting out consideration of any defence or cross-claim). Even given the inherent and obvious pro-producer and anti-customer and anti- paymaster bias of the HGCRA's statutory adjudication proposals, it is submitted that, in the absence of express wording, Parliament can only have intended adjudicators' decisions validly arrived at on the merits or law of a properly referred dispute to be binding on the parties for the comparatively lengthy period which could be involved before final judgment or award and almost inconceivable that Parliament intended to accord to adjudicators' decisions or conduct an immunity and enforceability not accorded by the law to arbitrators and their awards or even to the judiciary and their judgments."

There is much to be said in support of what Mr Ian Duncan Wallace there writes.'

There is a simple rule that where the adjudicator discovers something or has a view of his own it must be put to the parties and they must be allowed to address the point or points. This is old law in the sense that it has been applied in arbitration. It also applies in adjudication despite the time restraints. This is partly what the facility to extend time is for in the legislation. This was explored fully in *Balfour Beatty* v. *Lambeth*[53] which expounds among other matters the duty to act fairly:

'Thus in *Fox* v. *Wellfair Limited* [1981] 2 Lloyd's Reports 514 at page 529, Dunn LJ said:

"If the expert arbitrator, as he may be entitled to do, forms a view of the facts different from that given in the evidence which might produce a contrary result to that which emerges from the evidence, then he should bring that view to the attention of the parties. This is especially so where there is only one party and the arbitrator is in effect putting the alternative case for the party not present at the arbitration.

Similarly if an arbitrator as a result of a view of the premises reaches a conclusion contrary to or inconsistent with the evidence given at the hearing, then before incorporating that conclusion in his award he should bring it to the attention of the parties so that they may have an opportunity of dealing with it."

In *Interbulk Limited* v. *Aiden Shipping Co Limited (The "Vimeira")* [1984] 2 Lloyd's Reports 66 at page 75, Robert Goff LJ said:

"In truth, we are simply talking about fairness. It is not fair to decide a case against a party on an issue which has never been raised in the case without drawing the point to his attention so that he may have an opportunity of dealing with it, either by calling further evidence or by addressing argument on the facts or the law to the tribunal."

Ackner LJ also said at page 76:

"Where there is a breach of natural justice as a general proposition it is not for the courts to speculate what would have been the result if the principles of fairness had been applied. I adopt, with respect, the words of Mr Justice Megarry in *John* v. *Rees* [1969] 2 All ER 275 at p. 309 where he said: 'As everybody who has anything to do with the law well knows, the path of the law is strewn with examples of open and shut cases which, somehow, were not; of unanswerable charges which, in the event, were completely answered; of inexplicable conduct which was fully explained; of

---

[53] *Balfour Beatty Construction Limited* v. *The Mayor & Burgesses of the London Borough of Lambeth* (12 April 2002).

fixed and unalterable determinations that, by discussion, suffered a change.' But, in this case, speculation does not arise. If the arbitrators had informed the parties of what they had in mind, the consequences would have been obvious. Firstly, the charterers would have sought to persuade the arbitrators that it was common ground on the evidence that there was adequate room to turn the vessel and that, therefore, the arbitrators should decide the dispute according to the evidence. If they failed so to persuade the arbitrators, they would have sought, and would have been entitled to, an adjournment. Having obtained an adjournment, the charterers would have called the evidence which in fact was called at the sub-arbitration and would have satisfied the arbitrators that the turning area was adequate.''

Mr. Acton Davis also referred to a passage in *Gbangbola* v. *Smith & Sherriff* [1988] 3 All ER 730 at page 740B where I said:

'''A tribunal does not act fairly and impartially if it does not give a party an opportunity of dealing with arguments which have not been advanced by either party''.'

The following passage appears in the judgment in *RSL* v. *Stansell*[54]:

'32. It is elementary that the rules of natural justice require that a party to a dispute resolution procedure should know what is the case against him and should have an opportunity to meet it. In paragraph 17 of The Scheme for Construction Contracts it is provided in terms that "The adjudicator . . . shall make available to [the parties] any information to be taken into account in reaching his decision". At one point in her oral submissions Miss Hannaford seemed to come close to relying upon the absence of an equivalent provision in clause 38A of the sub-contract as an indication that there was no similar requirement in relation to an adjudication governed by terms similar to those of the sub-contract. If and insofar as Mr Hinchcliffe, or anyone else, may have thought that the effect of clause 38A.5.7 was that an adjudicator could, subject only to giving the parties to the relevant adjudication advance notice that he was going to seek technical or legal advice, obtain that advice and keep it to himself, not sharing the substance of it with the parties and affording them an opportunity to address it, it seems to me that he or she has fallen into fundamental error. It is absolutely essential, in my judgment, for an adjudicator, if he is to observe the rules of natural justice, to give the parties to the adjudication the chance to comment upon any material, from whatever source, including the knowledge or experience of the adjudicator himself, to which the adjudicator is minded to attribute significance in reaching his decision. Thus in the present case it was plain, in my judgment, that Mr Hinchcliffe should not have had any regard to the final report of Mr Adie without giving both RSL and Stansell the chance to consider the contents of that report and to comment upon it. If he needed an extension of the time allowed for his decision to enable him to provide the necessary chance, then he should have explained that to the parties in seeking their consent to an extension. *If he had explained that he needed an extension in order to afford the parties an opportunity to comment upon Mr Adie's final report, and the parties, with knowledge of the significance of the request for an extension, had not agreed to one, the likelihood is that they would be taken to have waived the right to object that they had not had the opportunity which they had refused* [emphasis added]. However, it is absurd to suggest that a party requested, for no definite reason, to agree to an extension of time for the making by an adjudicator of a decision, who refuses it should be precluded from resisting the enforcement of the decision on the ground that he had no opportunity to address an issue of importance in the ultimate result because there was no time, if the deadline for the making of the decision without an extension was to be met, to give him the opportunity which he was denied.'

We are pleased to note the inclusion by Judge Richard Seymour QC of the sentence that we have emphasised in the above extract. This gives some relief to the adjudicator who has to walk a narrow line between complying with the approach that the courts are taking to the rules of natural justice and complying with the time limits set by Parliament.

---

[54] *RSL (South West) Ltd* v. *Stansell Ltd* (16 June 2003).

We are however a bit concerned about the words: 'It is absolutely essential, in my judgment, for an adjudicator, if he is to observe the rules of natural justice, to give the parties to the adjudication the chance to comment upon any material, from whatever source, including the knowledge or experience of the adjudicator himself, to which the adjudicator is minded to attribute significance in reaching his decision', and particularly, bearing in mind the technical aspects of most issues and the technical background of most adjudicators, about the reference to the adjudicator acquainting the parties of any knowledge or experience he has himself to which he is minded to attach significance.

Not all challenges on the grounds of breach of rules of natural justice or lack of impartiality or procedural unfairness will or have succeeded. It is however clear that the court is looking more and more closely at the actions of the adjudicator and this area is becoming a more and more difficult one.

Human rights

The question of whether or not the Human Rights Act 1998 applies to adjudication is academic in the sense that even if it does not apply to the adjudication process itself, it will apply to the courts at the enforcement stage. In *Elanay* v. *The Vestry*[55] the court stated the following in respect of human rights:

'The question is whether the European Convention on Human Rights Article 6 applies to proceedings before an adjudicator. In the first place, the proceedings before an adjudicator are not in public, whereas the procedure under Article 6 has to be in public. I can see that the problems arise over whether one refers to a decision as a final decision or whether one has to consider whether Article 6 applies to a decision that is not a final decision, but it seems to me that if Article 6 does apply to proceedings before an adjudicator it is manifest that a coach and horses is driven through the whole of the Housing Grants, Construction and Regeneration Act. Maybe it is, of course because the Convention is something which though not at the moment binding by way of statute soon will be and the courts have to take account of it.

In my judgment, Article 6 of the European Convention on Human Rights does not apply to an adjudicator's award or to proceedings before an adjudicator and that is because, although they are the decision or determination of a question of civil rights, they are not in any sense a final determination. When I say that, I am not talking about first instance or appeals, but merely that the determination is itself provisional in the sense that the matter can be re-opened.'

In *Austin Hall* v. *Buckland*[56] the courts made a clear statement concerning the applicability of the human rights legislation and natural justice:

'I should make it plain that although in my view Article 6 of the Convention on Human Rights does not apply to the acts and decisions of adjudicators in construction contracts, adjudicators are required expressly by section 108(2)(e) of the Act of 1996 to act impartially. Where the Statutory Scheme under the 1996 Act applies, paragraph 12 of the Scheme sets out the duty to act impartially in more detailed terms, and paragraph 17 requires that

"The adjudicator shall consider any relevant information submitted to him by any of the parties to the dispute and shall make available to them any information to be taken into account in reaching his decision."

[55] *Elanay Contracts Limited* v. *The Vestry* (30 August 2000).
[56] *Austin Hall Building Limited* v. *Buckland Securities Limited* [2000] BLR 272.

In practice, I would think that adjudications are governed by rules of natural justice that are not very far different from Article 6 of the Convention except for the requirement of a public hearing and public pronouncement of the decision. The time limits that are under attack in this application are also subject to the rules of natural justice, but there is no question of an Act of Parliament being attacked in the courts as being in breach of the rules of natural justice. In our democracy, Parliament is still regarded in the Courts as supreme. The adjudicator was constrained by Act of Parliament to impose the time limits that he did, so he cannot be criticised for breaching the rules of natural justice. I agree with the statement of Judge Humphrey Lloyd QC in *Glencot Development & Design Co Ltd* v. *Ben Barrett & Son (Contractors) Limited* Unreported 13 February, 2001 that, "It is accepted that the adjudicator has to conduct the proceedings in accordance with the rules of natural justice or as fairly as the limitations imposed by Parliament permit".'

The influence of the human rights legislation on the courts at the time of enforcement is unavoidable. We suspect that the applicability or otherwise of the Human Rights Act in the adjudication process itself will be raised again.

Final determination

If the parties so choose they can agree that the adjudicator's decision finally determines the dispute. This agreement ought to be recorded in writing, although an oral agreement would still bind the parties. The parties can agree to amend or modify the adjudicator's decision and reach a new agreement using that decision as a basis, or to reach an entirely new agreement ignoring the adjudicator's decision. If they decide to do this, the agreement should be recorded in writing.

If the adjudicator's decision is unacceptable to one of the parties and a new agreement cannot be reached, the method stipulated in the contract as being the final means of resolving the dispute will apply.

This may be arbitration or litigation. Some of the standard forms now offer this choice where previously the only option was arbitration. If there is no arbitration provision in the contract there is nothing to prevent the parties making ad hoc arrangements to still proceed to arbitration.

The final determination of the dispute is a new process. It does not rely on the adjudicator's decision at all. The parties need not seek to uphold or challenge that decision in these new proceedings. The decision may have some evidential value to support a fact or to show why a party has paid a sum of money that it is now seeking to retrieve. The adjudicator's decision remains binding on the parties until the dispute has been considered afresh and the award is made in an arbitration or judgment given in court.

Appendix 1
Housing Grants, Construction and Regeneration Act 1996

Part II

Construction contracts

Introductory provisions

104.–(1) In this Part a "construction contract" means an agreement with a person for any of the following– Construction
contracts.

   (a) the carrying out of construction operations;

   (b) arranging for the carrying out of construction operations by others, whether under sub-contract to him or otherwise;

   (c) providing his own labour, or the labour of others, for the carrying out of construction operations.

(2) References in this Part to a construction contract include an agreement–

   (a) to do architectural, design, or surveying work, or

   (b) to provide advice on building, engineering, interior or exterior decoration or on the laying-out of landscape,

in relation to construction operations.

(3) References in this Part to a construction contract do not include a contract of employment (within the meaning of the Employment Rights Act 1996).

(4) The Secretary of State may by order add to, amend or repeal any of the provisions of subsection (1), (2) or (3) as to the agreements which are construction contracts for the purposes of this Part or are to be taken or not to be taken as included in references to such contracts.

No such order shall be made unless a draft of it has been laid before and approved by a resolution of each House of Parliament.

(5) Where an agreement relates to construction operations and other matters, this Part applies to it only so far as it relates to construction operations.

An agreement relates to construction operations so far as it makes provision of any kind within subsection (1) or (2).

(6) This Part applies only to construction contracts which–

   (a) are entered into after the commencement of this Part, and

   (b) relate to the carrying out of construction operations in England, Wales or Scotland.

(7) This Part applies whether or not the law of England and Wales or Scotland is otherwise the applicable law in relation to the contract.

105.–(1) In this Part "construction operations" means, subject as follows, operations of any of the following descriptions–

 (a) construction, alteration, repair, maintenance, extension, demolition or dismantling of buildings, or structures forming, or to form, part of the land (whether permanent or not);

 (b) construction, alteration, repair, maintenance, extension, demolition or dismantling of any works forming, or to form, part of the land, including (without prejudice to the foregoing) walls, roadworks, power-lines, telecommunication apparatus, aircraft runways, docks and harbours, railways, inland waterways, pipe-lines, reservoirs, water-mains, wells, sewers, industrial plant and installations for purposes of land drainage, coast protection or defence;

 (c) installation in any building or structure of fittings forming part of the land, including (without prejudice to the foregoing) systems of heating, lighting, air-conditioning, ventilation, power supply, drainage, sanitation, water supply or fire protection, or security or communications systems;

 (d) external or internal cleaning of buildings and structures, so far as carried out in the course of their construction, alteration, repair, extension or restoration;

 (e) operations which form an integral part of, or are preparatory to, or are for rendering complete, such operations as are previously described in this subsection, including site clearance, earth-moving, excavation, tunnelling and boring, laying of foundations, erection, maintenance or dismantling of scaffolding, site restoration, landscaping and the provision of roadways and other access works;

 (f) painting or decorating the internal or external surfaces of any building or structure.

(2) The following operations are not construction operations within the meaning of this Part–

 (a) drilling for, or extraction of, oil or natural gas;

 (b) extraction (whether by underground or surface working) of minerals; tunnelling or boring, or construction of underground works, for this purpose;

 (c) assembly, installation or demolition of plant or machinery, or erection or demolition of steelwork for the purposes of supporting or providing access to plant or machinery, on a site where the primary activity is–

  (i) nuclear processing, power generation, or water or effluent treatment, or

  (ii) the production, transmission, processing or bulk storage (other than warehousing) of chemicals, pharmaceuticals, oil, gas, steel or food and drink;

 (d) manufacture or delivery to site of–

  (i) building or engineering components or equipment,

  (ii) materials, plant or machinery, or

  (iii) components for systems of heating, lighting, air-conditioning, ventilation, power supply, drainage, sanitation, water supply or fire protection, or for security or communications systems,

 except under a contract which also provides for their installation;

 (e) the making, installation and repair of artistic works, being sculptures, murals and other works which are wholly artistic in nature.

(3) The Secretary of State may by order add to, amend or repeal any of the provisions of subsection (1) or (2) as to the operations and work to be treated as construction operations for the purposes of this Part.

(4) No such order shall be made unless a draft of it has been laid before and approved by a resolution of each House of Parliament.

106.–(1) This Part does not apply–
    (a) to a construction contract with a residential occupier (see below), or
    (b) to any other description of construction contract excluded from the operation of this Part by order of the Secretary of State.

Provisions not applicable to contract with residential occupier.

(2) A construction contract with a residential occupier means a construction contract which principally relates to operations on a dwelling which one of the parties to the contract occupies, or intends to occupy, as his residence.

In this subsection "dwelling" means a dwelling-house or a flat; and for this purpose–

"dwelling-house" does not include a building containing a flat; and

"flat" means separate and self-contained premises constructed or adapted for use for residential purposes and forming part of a building from some other part of which the premises are divided horizontally.

(3) The Secretary of State may by order amend subsection (2).

(4) No order under this section shall be made unless a draft of it has been laid before and approved by a resolution of each House of Parliament.

107.–(1) The provisions of this Part apply only where the construction contract is in writing, and any other agreement between the parties as to any matter is effective for the purposes of this Part only if in writing.

The expressions "agreement", "agree" and "agreed" shall be construed accordingly.

Provisions applicable only to agreements in writing.

(2) There is an agreement in writing–
    (a) if the agreement is made in writing (whether or not it is signed by the parties),
    (b) if the agreement is made by exchange of communications in writing, or
    (c) if the agreement is evidenced in writing.

(3) Where parties agree otherwise than in writing by reference to terms which are in writing, they make an agreement in writing.

(4) An agreement is evidenced in writing if an agreement made otherwise than in writing is recorded by one of the parties, or by a third party, with the authority of the parties to the agreement.

(5) An exchange of written submissions in adjudication proceedings, or in arbitral or legal proceedings in which the existence of an agreement otherwise than in writing is alleged by one party against another party and not denied by the other party in his response constitutes as between those parties an agreement in writing to the effect alleged.

(6) References in this Part to anything being written or in writing include its being recorded by any means.

Adjudication

108.–(1) A party to a construction contract has the right to refer a dispute arising under the contract for adjudication under a procedure complying with this section.

For this purpose "dispute" includes any difference.

Right to refer disputes to adjudication.

　　(2) The contract shall–
　　　　(a) enable a party to give notice at any time of his intention to refer a dispute to adjudication;
　　　　(b) provide a timetable with the object of securing the appointment of the adjudicator and referral of the dispute to him within 7 days of such notice;
　　　　(c) require the adjudicator to reach a decision within 28 days of referral or such longer period as is agreed by the parties after the dispute has been referred;
　　　　(d) allow the adjudicator to extend the period of 28 days by up to 14 days, with the consent of the party by whom the dispute was referred;
　　　　(e) impose a duty on the adjudicator to act impartially; and
　　　　(f) enable the adjudicator to take the initiative in ascertaining the facts and the law.

　　(3) The contract shall provide that the decision of the adjudicator is binding until the dispute is finally determined by legal proceedings, by arbitration (if the contract provides for arbitration or the parties otherwise agree to arbitration) or by agreement.
　　The parties may agree to accept the decision of the adjudicator as finally determining the dispute.

　　(4) The contract shall also provide that the adjudicator is not liable for anything done or omitted in the discharge or purported discharge of his functions as adjudicator unless the act or omission is in bad faith, and that any employee or agent of the adjudicator is similarly protected from liability.

　　(5) If the contract does not comply with the requirements of subsections (1) to (4), the adjudication provisions of the Scheme for Construction Contracts apply.

　　(6) For England and Wales, the Scheme may apply the provisions of the Arbitration Act 1996 with such adaptations and modifications as appear to the Minister making the scheme to be appropriate.
　　For Scotland, the Scheme may include provision conferring powers on courts in relation to adjudication and provision relating to the enforcement of the adjudicator's decision.

Payment

Entitlement to stage payments.　　109.–(1) A party to a construction contract is entitled to payment by instalments, stage payments or other periodic payments for any work under the contract unless–
　　　　(a) it is specified in the contract that the duration of the work is to be less than 45 days, or
　　　　(b) it is agreed between the parties that the duration of the work is estimated to be less than 45 days.

　　(2) The parties are free to agree the amounts of the payments and the intervals at which, or circumstances in which, they become due.

　　(3) In the absence of such agreement, the relevant provisions of the Scheme for Construction Contracts apply.

　　(4) References in the following sections to a payment under the contract include a payment by virtue of this section.

Dates for payment.　　110.–(1) Every construction contract shall–
　　　　(a) provide an adequate mechanism for determining what payments become due under the contract, and when, and

(b) provide for a final date for payment in relation to any sum which becomes due.

The parties are free to agree how long the period is to be between the date on which a sum becomes due and the final date for payment.

(2) Every construction contract shall provide for the giving of notice by a party not later than five days after the date on which a payment becomes due from him under the contract, or would have become due if–

(a) the other party had carried out his obligations under the contract, and

(b) no set-off or abatement was permitted by reference to any sum claimed to be due under one or more other contracts,

specifying the amount (if any) of the payment made or proposed to be made, and the basis on which that amount was calculated.

(3) If or to the extent that a contract does not contain such provision as is mentioned in subsection (1) or (2), the relevant provisions of the Scheme for Construction Contracts apply.

111.–(1) A party to a construction contract may not withhold payment after the final date for payment of a sum due under the contract unless he has given an effective notice of intention to withhold payment.

Notice of intention to withhold payment.

The notice mentioned in section 110(2) may suffice as a notice of intention to withhold payment if it complies with the requirements of this section.

(2) To be effective such a notice must specify–

(a) the amount proposed to be withheld and the ground for withholding payment, or

(b) if there is more than one ground, each ground and the amount attributable to it,

and must be given not later than the prescribed period before the final date for payment.

(3) The parties are free to agree what that prescribed period is to be.

In the absence of such agreement, the period shall be that provided by the Scheme for Construction Contracts.

(4) Where an effective notice of intention to withhold payment is given, but on the matter being referred to adjudication it is decided that the whole or part of the amount should be paid, the decision shall be construed as requiring payment not later than–

(a) seven days from the date of the decision, or

(b) the date which apart from the notice would have been the final date for payment,

whichever is the later.

112.–(1) Where a sum due under a construction contract is not paid in full by the final date for payment and no effective notice to withhold payment has been given, the person to whom the sum is due has the right (without prejudice to any other right or remedy) to suspend performance of his obligations under the contract to the party by whom payment ought to have been made ("the party in default").

Right to suspend performance for non-payment.

(2) The right may not be exercised without first giving to the party in default at least seven days' notice of intention to suspend performance, stating the ground or grounds on which it is intended to suspend performance.

(3) The right to suspend performance ceases when the party in default makes payment in full of the amount due.

(4) Any period during which performance is suspended in pursuance of the right conferred by this section shall be disregarded in computing for the purposes of any contractual time limit the time taken, by the party exercising the right or by a third party, to complete any work directly or indirectly affected by the exercise of the right.
   Where the contractual time limit is set by reference to a date rather than a period, the date shall be adjusted accordingly.

Prohibition of conditional payment provisions.

113.-(1) A provision making payment under a construction contract conditional on the payer receiving payment from a third person is ineffective, unless that third person, or any other person payment by whom is under the contract (directly or indirectly) a condition of payment by that third person, is insolvent.

(2) For the purposes of this section a company becomes insolvent–
   (a) on the making of an administration order against it under Part II of the Insolvency Act 1986,
   (b) on the appointment of an administrative receiver or a receiver or manager of its property under Chapter I of Part III of that Act, or the appointment of a receiver under Chapter II of that Part,
   (c) on the passing of a resolution for voluntary winding-up without a declaration of solvency under section 89 of that Act, or
   (d) on the making of a winding-up order under Part IV or V of that Act.

(3) For the purposes of the section a partnership becomes insolvent–
   (a) on the making of a winding-up order against it under any provision of the Insolvency Act 1986 as applied by an order under section 420 of that Act, or
   (b) when sequestration is awarded on the estate of the partnership under section 12 of the Bankruptcy (Scotland) Act 1985 or the partnership grants a trust deed for its creditors.

(4) For the purposes of this section an individual becomes insolvent–
   (a) on the making of a bankruptcy order against him under Part IX of the Insolvency Act 1986, or
   (b) on the sequestration of his estate under the Bankruptcy (Scotland) Act 1985 or when he grants a trust deed for his creditors.

(5) A company, partnership or individual shall also be treated as insolvent on the occurrence of any event corresponding to those specified in subsection (2), (3) or (4) under the law of Northern Ireland or of a country outside the United Kingdom.

(6) Where a provision is rendered ineffective by subsection (1), the parties are free to agree other terms for payment.
   In the absence of such agreement, the relevant provisions of the Scheme for Construction Contracts apply.

Supplementary provisions

The Scheme for Construction Contracts.

114.-(1) The Minister shall by regulations make a scheme ("the Scheme for Construction Contracts") containing provision about the matters referred to in the preceding provisions of this Part.

(2) Before making any regulations under this section the Minister shall consult such persons as he thinks fit.

(3) In this section ''the Minister'' means–

    (a) for England and Wales, the Secretary of State, and

    (b) for Scotland, the Lord Advocate.

(4) Where any provisions of the Scheme for Construction Contracts apply by virtue of this Part in default of contractual provision agreed by the parties, they have effect as implied terms of the contract concerned.

(5) Regulations under this section shall not be made unless a draft of them has been approved by resolution of each House of Parliament.

115.–(1) The parties are free to agree on the manner of service of any notice or other document required or authorised to be served in pursuance of the construction contract or for any of the purposes of this Part.   **Service of notices, &c.**

(2) If or to the extent that there is no such agreement the following provisions apply.

(3) A notice or other document may be served on a person by any effective means.

(4) If a notice or other document is addressed, pre-paid and delivered by post–

    (a) to the addressee's last known principal residence or, if he is or has been carrying on a trade, profession or business, his last known principal business address, or

    (b) where the addressee is a body corporate, to the body's registered or principal office,

it shall be treated as effectively served.

(5) This section does not apply to the service of documents for the purposes of legal proceedings, for which provision is made by rules of court.

(6) References in this Part to a notice or other document include any form of communication in writing and references to service shall be construed accordingly.

116.–(1) For the purposes of this Part periods of time shall be reckoned as follows.   **Reckoning periods of time.**

(2) Where an act is required to be done within a specified period after or from a specified date, the period begins immediately after that date.

(3) Where the period would include Christmas Day, Good Friday or a day which under the Banking and Financial Dealings Act 1971 is a bank holiday in England and Wales or, as the case may be, in Scotland, that day shall be excluded.

117.–(1) This Part applies to a construction contract entered into by or on behalf of the Crown otherwise than by or on behalf of Her Majesty in her private capacity.   **Crown application.**

(2) This Part applies to a construction contract entered into on behalf of the Duchy of Cornwall notwithstanding any Crown interest.

(3) Where a construction contract is entered into by or on behalf of Her Majesty in right of the Duchy of Lancaster, Her Majesty shall be represented, for the purposes of any adjudication or other proceedings arising out of the contract by virtue of this Part, by the Chancellor of the Duchy or such person as he may appoint.

(4) Where a construction contract is entered into on behalf of the Duchy of Cornwall, the Duke of Cornwall or the possessor for the time being of the Duchy shall be represented, for the purposes of any adjudication or other proceedings arising out of the contract by virtue of this Part, by such person as he may appoint.

Appendix 2
The Scheme for Construction Contracts (England and Wales) Regulations 1998

Statutory Instrument 1998 No. 649

The Secretary of State, in exercise of the powers conferred on him by sections 108(6), 114 and 146(1) and (2) of the Housing Grants, Construction and Regeneration Act 1996, and of all other powers enabling him in that behalf, having consulted such persons as he thinks fit, and draft Regulations having been approved by both Houses of Parliament, hereby makes the following Regulations:

Citation, commencement, extent and interpretation

1.–(1) These Regulations may be cited as the Scheme for Construction Contracts (England and Wales) Regulations 1998 and shall come into force at the end of the period of 8 weeks beginning with the day on which they are made (the ''commencement date'').

(2) These Regulations shall extend only to England and Wales.

(3) In these Regulations, ''the Act'' means the Housing Grants, Construction and Regeneration Act 1996.

The Scheme for Construction Contracts

2. Where a construction contract does not comply with the requirements of section 108(1) to (4) of the Act, the adjudication provisions in Part I of the Schedule to these Regulations shall apply.

3. Where–
    (a) the parties to a construction contract are unable to reach agreement for the purposes mentioned respectively in sections 109, 111 and 113 of the Act, or
    (b) a construction contract does not make provision as required by section 110 of the Act,
the relevant provisions in Part II of the Schedule to these Regulations shall apply.

4. The provisions in the Schedule to these Regulations shall be the Scheme for Construction Contracts for the purposes of section 114 of the Act.

SCHEDULE   Regulations 2, 3 and 4

THE SCHEME FOR CONSTRUCTION CONTRACTS
PART I–ADJUDICATION

Notice of Intention to seek Adjudication

1.–(1) Any party to a construction contract (the ''referring party'') may give written notice (the ''notice of adjudication'') of his intention to refer any dispute arising under the contract, to adjudication.

(2) The notice of adjudication shall be given to every other party to the contract.

(3) The notice of adjudication shall set out briefly–
   (a) the nature and a brief description of the dispute and of the parties involved,
   (b) details of where and when the dispute has arisen,
   (c) the nature of the redress which is sought, and
   (d) the names and addresses of the parties to the contract (including, where appropriate, the addresses which the parties have specified for the giving of notices).

2.–(1) Following the giving of a notice of adjudication and subject to any agreement between the parties to the dispute as to who shall act as adjudicator–
   (a) the referring party shall request the person (if any) specified in the contract to act as adjudicator, or
   (b) if no person is named in the contract or the person named has already indicated that he is unwilling or unable to act, and the contract provides for a specified nominating body to select a person, the referring party shall request the nominating body named in the contract to select a person to act as adjudicator, or
   (c) where neither paragraph (a) nor (b) above applies, or where the person referred to in (a) has already indicated that he is unwilling or unable to act and (b) does not apply, the referring party shall request an adjudicator nominating body to select a person to act as adjudicator.

(2) A person requested to act as adjudicator in accordance with the provisions of paragraph (1) shall indicate whether or not he is willing to act within two days of receiving the request.

(3) In this paragraph, and in paragraphs 5 and 6 below, an ''adjudicator nominating body'' shall mean a body (not being a natural person and not being a party to the dispute) which holds itself out publicly as a body which will select an adjudicator when requested to do so by a referring party.

3. The request referred to in paragraphs 2, 5 and 6 shall be accompanied by a copy of the notice of adjudication.

4. Any person requested or selected to act as adjudicator in accordance with paragraphs 2, 5 or 6 shall be a natural person acting in his personal capacity. A person requested or selected to act as an adjudicator shall not be an employee of any of the parties to the dispute and shall declare any interest, financial or otherwise, in any matter relating to the dispute.

5.–(1) The nominating body referred to in paragraphs 2(1)(b) and 6(1)(b) or the adjudicator nominating body referred to in paragraphs 2(1)(c), 5(2)(b) and 6(1)(c) must communicate the selection of an adjudicator to the referring party within five days of receiving a request to do so.

(2) Where the nominating body or the adjudicator nominating body fails to comply with paragraph (1), the referring party may–
   (a) agree with the other party to the dispute to request a specified person to act as adjudicator, or
   (b) request any other adjudicator nominating body to select a person to act as adjudicator.

(3) The person requested to act as adjudicator in accordance with the provisions of paragraphs (1) or (2) shall indicate whether or not he is willing to act within two days of receiving the request.

6.–(1) Where an adjudicator who is named in the contract indicates to the parties that he is unable or unwilling to act, or where he fails to respond in accordance with paragraph 2(2), the referring party may–
   (a) request another person (if any) specified in the contract to act as adjudicator, or
   (b) request the nominating body (if any) referred to in the contract to select a person to act as adjudicator, or

(c) request any other adjudicator nominating body to select a person to act as adjudicator.

(2) The person requested to act in accordance with the provisions of paragraph (1) shall indicate whether or not he is willing to act within two days of receiving the request.

7.–(1) Where an adjudicator has been selected in accordance with paragraphs 2, 5 or 6, the referring party shall, not later than seven days from the date of the notice of adjudication, refer the dispute in writing (the "referral notice") to the adjudicator.

(2) A referral notice shall be accompanied by copies of, or relevant extracts from, the construction contract and such other documents as the referring party intends to rely upon.

(3) The referring party shall, at the same time as he sends to the adjudicator the documents referred to in paragraphs (1) and (2), send copies of those documents to every other party to the dispute.

8.–(1) The adjudicator may, with the consent of all the parties to those disputes, adjudicate at the same time on more than one dispute under the same contract.

(2) The adjudicator may, with the consent of all the parties to those disputes, adjudicate at the same time on related disputes under different contracts, whether or not one or more of those parties is a party to those disputes.

(3) All the parties in paragraphs (1) and (2) respectively may agree to extend the period within which the adjudicator may reach a decision in relation to all or any of these disputes.

(4) Where an adjudicator ceases to act because a dispute is to be adjudicated on by another person in terms of this paragraph, that adjudicator's fees and expenses shall be determined in accordance with paragraph 25.

9.–(1) An adjudicator may resign at any time on giving notice in writing to the parties to the dispute.

(2) An adjudicator must resign where the dispute is the same or substantially the same as one which has previously been referred to adjudication, and a decision has been taken in that adjudication.

(3) Where an adjudicator ceases to act under paragraph 9(1)–
   (a) the referring party may serve a fresh notice under paragraph 1 and shall request an adjudicator to act in accordance with paragraphs 2 to 7; and
   (b) if requested by the new adjudicator and insofar as it is reasonably practicable, the parties shall supply him with copies of all documents which they had made available to the previous adjudicator.

(4) Where an adjudicator resigns in the circumstances referred to in paragraph (2), or where a dispute varies significantly from the dispute referred to him in the referral notice and for that reason he is not competent to decide it, the adjudicator shall be entitled to the payment of such reasonable amount as he may determine by way of fees and expenses reasonably incurred by him. The parties shall be jointly and severally liable for any sum which remains outstanding following the making of any determination on how the payment shall be apportioned.

10. Where any party to the dispute objects to the appointment of a particular person as adjudicator, that objection shall not invalidate the adjudicator's appointment nor any decision he may reach in accordance with paragraph 20.

11.–(1) The parties to a dispute may at any time agree to revoke the appointment of the adjudicator. The adjudicator shall be entitled to the payment of such reasonable amount as he may determine by way of fees and expenses incurred by him. The parties shall be jointly and severally liable for any sum which remains outstanding following the making of any determination on how the payment shall be apportioned.

(2) Where the revocation of the appointment of the adjudicator is due to the default or misconduct of the adjudicator, the parties shall not be liable to pay the adjudicator's fees and expenses.

Powers of the adjudicator

12. The adjudicator shall–
- (a) act impartially in carrying out his duties and shall do so in accordance with any relevant terms of the contract and shall reach his decision in accordance with the applicable law in relation to the contract; and
- (b) avoid incurring unnecessary expense.

13. The adjudicator may take the initiative in ascertaining the facts and the law necessary to determine the dispute, and shall decide on the procedure to be followed in the adjudication. In particular he may–
- (a) request any party to the contract to supply him with such documents as he may reasonably require including, if he so directs, any written statement from any party to the contract supporting or supplementing the referral notice and any other documents given under paragraph 7(2),
- (b) decide the language or languages to be used in the adjudication and whether a translation of any document is to be provided and if so by whom,
- (c) meet and question any of the parties to the contract and their representatives,
- (d) subject to obtaining any necessary consent from a third party or parties, make such site visits and inspections as he considers appropriate, whether accompanied by the parties or not,
- (e) subject to obtaining any necessary consent from a third party or parties, carry out any tests or experiments,
- (f) obtain and consider such representations and submissions as he requires, and, provided he has notified the parties of his intention, appoint experts, assessors or legal advisers,
- (g) give directions as to the timetable for the adjudication, any deadlines, or limits as to the length of written documents or oral representations to be complied with, and
- (h) issue other directions relating to the conduct of the adjudication.

14. The parties shall comply with any request or direction of the adjudicator in relation to the adjudication.

15. If, without showing sufficient cause, a party fails to comply with any request, direction or timetable of the adjudicator made in accordance with his powers, fails to produce any document or written statement requested by the adjudicator, or in any other way fails to comply with a requirement under these provisions relating to the adjudication, the adjudicator may–
- (a) continue the adjudication in the absence of that party or of the document or written statement requested,
- (b) draw such inferences from that failure to comply as circumstances may, in the adjudicator's opinion, be justified, and
- (c) make a decision on the basis of the information before him attaching such weight as he thinks fit to any evidence submitted to him outside any period he may have requested or directed.

16.–(1) Subject to any agreement between the parties to the contrary, and to the terms of paragraph (2) below, any party to the dispute may be assisted by, or represented by, such advisers or representatives (whether legally qualified or not) as he considers appropriate.

(2) Where the adjudicator is considering oral evidence or representations, a party to the dispute may not be represented by more than one person, unless the adjudicator gives directions to the contrary.

17. The adjudicator shall consider any relevant information submitted to him by any of the parties to the dispute and shall make available to them any information to be taken into account in reaching his decision.

18. The adjudicator and any party to the dispute shall not disclose to any other person any information or document provided to him in connection with the adjudication which the party supplying it has indicated is to be treated as confidential, except to the extent that it is necessary for the purposes of, or in connection with, the adjudication.

19.–(1) The adjudicator shall reach his decision not later than–
    (a) twenty eight days after the date of the referral notice mentioned in paragraph 7(1), or
    (b) forty two days after the date of the referral notice if the referring party so consents, or
    (c) such period exceeding twenty eight days after the referral notice as the parties to the dispute may, after the giving of that notice, agree.

(2) Where the adjudicator fails, for any reason, to reach his decision in accordance with paragraph (1)
    (a) any of the parties to the dispute may serve a fresh notice under paragraph 1 and shall request an adjudicator to act in accordance with paragraphs 2 to 7; and
    (b) if requested by the new adjudicator and insofar as it is reasonably practicable, the parties shall supply him with copies of all documents which they had made available to the previous adjudicator.

(3) As soon as possible after he has reached a decision, the adjudicator shall deliver a copy of that decision to each of the parties to the contract.

Adjudicator's decision

20. The adjudicator shall decide the matters in dispute. He may take into account any other matters which the parties to the dispute agree should be within the scope of the adjudication or which are matters under the contract which he considers are necessarily connected with the dispute. In particular, he may–
    (a) open up, revise and review any decision taken or any certificate given by any person referred to in the contract unless the contract states that the decision or certificate is final and conclusive,
    (b) decide that any of the parties to the dispute is liable to make a payment under the contract (whether in sterling or some other currency) and, subject to section 111(4) of the Act, when that payment is due and the final date for payment,
    (c) having regard to any term of the contract relating to the payment of interest decide the circumstances in which, and the rates at which, and the periods for which simple or compound rates of interest shall be paid.

21. In the absence of any directions by the adjudicator relating to the time for performance of his decision, the parties shall be required to comply with any decision of the adjudicator immediately on delivery of the decision to the parties in accordance with this paragraph.

22. If requested by one of the parties to the dispute, the adjudicator shall provide reasons for his decision.

Effects of the decision

23.–(1) In his decision, the adjudicator may, if he thinks fit, order any of the parties to comply peremptorily with his decision or any part of it.

(2) The decision of the adjudicator shall be binding on the parties, and they shall comply with it until the dispute is finally determined by legal proceedings, by arbitration (if the contract provides for arbitration or the parties otherwise agree to arbitration) or by agreement between the parties.

24. Section 42 of the Arbitration Act 1996 shall apply to this Scheme subject to the following modifications–
  (a) in subsection (2) for the word "tribunal" wherever it appears there shall be substituted the word "adjudicator",
  (b) in subparagraph (b) of subsection (2) for the words "arbitral proceedings" there shall be substituted the word "adjudication",
  (c) subparagraph (c) of subsection (2) shall be deleted, and
  (d) subsection (3) shall be deleted.

25. The adjudicator shall be entitled to the payment of such reasonable amount as he may determine by way of fees and expenses reasonably incurred by him. The parties shall be jointly and severally liable for any sum which remains outstanding following the making of any determination on how the payment shall be apportioned.

26. The adjudicator shall not be liable for anything done or omitted in the discharge or purported discharge of his functions as adjudicator unless the act or omission is in bad faith, and any employee or agent of the adjudicator shall be similarly protected from liability.

PART II–PAYMENT

Entitlement to and amount of stage payments

1. Where the parties to a relevant construction contract fail to agree–
  (a) the amount of any instalment or stage or periodic payment for any work under the contract, or
  (b) the intervals at which, or circumstances in which, such payments become due under that contract, or
  (c) both of the matters mentioned in sub-paragraphs (a) and (b) above,
the relevant provisions of paragraphs 2 to 4 below shall apply.

2.–(1) The amount of any payment by way of instalments or stage or periodic payments in respect of a relevant period shall be the difference between the amount determined in accordance with sub-paragraph (2) and the amount determined in accordance with sub-paragraph (3).

(2) The aggregate of the following amounts–
  (a) an amount equal to the value of any work performed in accordance with the relevant construction contract during the period from the commencement of the contract to the end of the relevant period (excluding any amount calculated in accordance with sub-paragraph (b)),
  (b) where the contract provides for payment for materials, an amount equal to the value of any materials manufactured on site or brought onto site for the purposes of the works during the period from the commencement of the contract to the end of the relevant period, and
  (c) any other amount or sum which the contract specifies shall be payable during or in respect of the period from the commencement of the contract to the end of the relevant period.

(3) The aggregate of any sums which have been paid or are due for payment by way of instalments,

stage or periodic payments during the period from the commencement of the contract to the end of the relevant period.

(4) An amount calculated in accordance with this paragraph shall not exceed the difference between–
  (a) the contract price, and
  (b) the aggregate of the instalments or stage or periodic payments which have become due.

Dates for payment

3. Where the parties to a construction contract fail to provide an adequate mechanism for determining either what payments become due under the contract, or when they become due for payment, or both, the relevant provisions of paragraphs 4 to 7 shall apply.

4. Any payment of a kind mentioned in paragraph 2 above shall become due on whichever of the following dates occurs later–
  (a) the expiry of 7 days following the relevant period mentioned in paragraph 2(1) above, or
  (b) the making of a claim by the payee.

5. The final payment payable under a relevant construction contract, namely the payment of an amount equal to the difference (if any) between–
  (a) the contract price, and
  (b) the aggregate of any instalment or stage or periodic payments which have become due under the contract,
shall become due on the expiry of–
  (a) 30 days following completion of the work, or
  (b) the making of a claim by the payee,
whichever is the later.

6. Payment of the contract price under a construction contract (not being a relevant construction contract) shall become due on
  (a) the expiry of 30 days following the completion of the work, or
  (b) the making of a claim by the payee,
whichever is the later.

7. Any other payment under a construction contract shall become due
  (a) on the expiry of 7 days following the completion of the work to which the payment relates, or
  (b) the making of a claim by the payee,
whichever is the later.

Final date for payment

8.–(1) Where the parties to a construction contract fail to provide a final date for payment in relation to any sum which becomes due under a construction contract, the provisions of this paragraph shall apply.

(2) The final date for the making of any payment of a kind mentioned in paragraphs 2, 5, 6 or 7, shall be 17 days from the date that payment becomes due.

Notice specifying amount of payment

9. A party to a construction contract shall, not later than 5 days after the date on which any payment–
  (a) becomes due from him, or
  (b) would have become due, if–

(i) the other party had carried out his obligations under the contract, and

(ii) no set-off or abatement was permitted by reference to any sum claimed to be due under one or more other contracts,

give notice to the other party to the contract specifying the amount (if any) of the payment he has made or proposes to make, specifying to what the payment relates and the basis on which that amount is calculated.

Notice of intention to withhold payment

10. Any notice of intention to withhold payment mentioned in section 111 of the Act shall be given not later than the prescribed period, which is to say not later than 7 days before the final date for payment determined either in accordance with the construction contract, or where no such provision is made in the contract, in accordance with paragraph 8 above.

Prohibition of conditional payment provisions

11. Where a provision making payment under a construction contract conditional on the payer receiving payment from a third person is ineffective as mentioned in section 113 of the Act, and the parties have not agreed other terms for payment, the relevant provisions of–

(a) paragraphs 2, 4, 5, 7, 8, 9 and 10 shall apply in the case of a relevant construction contract, and

(b) paragraphs 6, 7, 8, 9 and 10 shall apply in the case of any other construction contract.

Interpretation

12. In this Part of the Scheme for Construction Contracts–

''claim by the payee'' means a written notice given by the party carrying out work under a construction contract to the other party specifying the amount of any payment or payments which he considers to be due and the basis on which it is, or they are calculated;

''contract price'' means the entire sum payable under the construction contract in respect of the work;

''relevant construction contract'' means any construction contract other than one–

(a) which specifies that the duration of the work is to be less than 45 days, or

(b) in respect of which the parties agree that the duration of the work is estimated to be less than 45 days;

''relevant period'' means a period which is specified in, or is calculated by reference to the construction contract or where no such period is so specified or is so calculable, a period of 28 days;

''value of work'' means an amount determined in accordance with the construction contract under which the work is performed or where the contract contains no such provision, the cost of any work performed in accordance with that contract together with an amount equal to any overhead or profit included in the contract price;

''work'' means any of the work or services mentioned in section 104 of the Act.

EXPLANATORY NOTE

(This note is not part of the Order)

Part II of the Housing Grants, Construction and Regeneration Act 1996 makes provision in relation to construction contracts. Section 114 empowers the Secretary of State to make the Scheme for Construction Contracts. Where a construction contract does not comply with the requirements of sections 108 to 111 (adjudication of disputes and payment provisions), and section 113 (prohibition of conditional payment provisions), the relevant provisions of the Scheme for Construction Contracts have effect.

The Scheme which is contained in the Schedule to these Regulations is in two parts. Part I provides for the selection and appointment of an adjudicator, gives powers to the adjudicator to gather and consider information, and makes provisions in respect of his decisions. Part II makes provision with respect to payments under a construction contract where either the contract fails to make provision or the parties fail to agree–

    (a) the method for calculating the amount of any instalment, stage or periodic payment,

    (b) the due date and the final date for payments to be made, and

    (c) prescribes the period within which a notice of intention to withhold payment must be given.

Appendix 3
The Construction Contracts (England and Wales) Exclusion Order 1998

Statutory Instrument 1998 No. 648

The Secretary of State, in exercise of the powers conferred on him by sections 106(1)(b) and 146(1) of the Housing Grants, Construction and Regeneration Act 1996 and of all other powers enabling him in that behalf, hereby makes the following Order, a draft of which has been laid before and approved by resolution of, each House of Parliament:

Citation, commencement and extent

1.–(1) This Order may be cited as the Construction Contracts Exclusion Order 1998 and shall come into force at the end of the period of 8 weeks beginning with the day on which it is made ("the commencement date").

(2) This Order shall extend to England and Wales only.

Interpretation

2. In this Order, "Part II" means Part II of the Housing Grants, Construction and Regeneration Act 1996.

Agreements under statute

3. A construction contract is excluded from the operation of Part II if it is–
(a) an agreement under section 38 (power of highway authorities to adopt by agreement) or section 278 (agreements as to execution of works) of the Highways Act 1980;
(b) an agreement under section 106 (planning obligations), 106A (modification or discharge of planning obligations) or 299A (Crown planning obligations) of the Town and Country Planning Act 1990;
(c) an agreement under section 104 of the Water Industry Act 1991 (agreements to adopt sewer, drain or sewage disposal works); or
(d) an externally financed, development agreement within the meaning of section 1 of the National Health Service (Private Finance) Act 1997 (powers of NHS Trusts to enter into agreements).

Private finance initiative

4..–(1) A construction contract is excluded from the operation of Part II if it is a contract entered into under the private finance initiative, within the meaning given below.

(2) A contract is entered into under the private finance initiative if all the following conditions are fulfilled–
(a) it contains a statement that it is entered into under that initiative or, as the case may be, under a project applying similar principles;

(b) the consideration due under the contract is determined at least in part by reference to one or more of the following–
    (i) the standards attained in the performance of a service, the provision of which is the principal purpose or one of the principal purposes for which the building or structure is constructed;
    (ii) the extent, rate or intensity of use of all or any part of the building or structure in question; or
    (iii) the right to operate any facility in connection with the building or structure in question; and
(c) one of the parties to the contract is–
    (i) a Minister of the Crown;
    (ii) a department in respect of which appropriation accounts are required to be prepared under the Exchequer and Audit Departments Act 1866;
    (iii) any other authority or body whose accounts are required to be examined and certified by or are open to the inspection of the Comptroller and Auditor General by virtue of an agreement entered into before the commencement date or by virtue of any enactment;
    (iv) any authority or body listed in Schedule 4 to the National Audit Act 1983 (nationalised industries and other public authorities);
    (v) a body whose accounts are subject to audit by auditors appointed by the Audit Commission;
    (vi) the governing body or trustees of a voluntary school within the meaning of section 31 of the Education Act 1996 (county schools and voluntary schools), or
    (vii) a company wholly owned by any of the bodies described in paragraphs (i) to (v).

Finance agreements

5.–(1) A construction contract is excluded from the operation of Part II if it is a finance agreement, within the meaning given below.

(2) A contract is a finance agreement if it is any one of the following–
(a) any contract of insurance;
(b) any contract under which the principal obligations include the formation or dissolution of a company, unincorporated association or partnership;
(c) any contract under which the principal obligations include the creation or transfer of securities or any right or interest in securities;
(d) any contract under which the principal obligations include the lending of money;
(e) any contract under which the principal obligations include an undertaking by a person to be responsible as surety for the debt or default of another person, including a fidelity bond, advance payment bond, retention bond or performance bond.

Development agreements

6.–(1) A construction contract is excluded from the operation of Part II if it is a development agreement, within the meaning given below.

(2) A contract is a development agreement if it includes provision for the grant or disposal of a relevant interest in the land on which take place the principal construction operations to which the contract relates.

(3) In paragraph (2) above, a relevant interest in land means–
(a) a freehold; or
(b) a leasehold for a period which is to expire no earlier than 12 months after the completion of the construction operations under the contract.

EXPLANATORY NOTE

(This note is not part of the Order)

Part II of the Housing Grants, Construction and Regeneration Act 1996 makes provision in relation to the terms of construction contracts. Section 106 confers power on the Secretary of State to exclude descriptions of contracts from the operation of Part II. This Order excludes contracts of four descriptions.

Article 3 excludes agreements made under specified statutory provisions dealing with highway works, planning obligations, sewage works and externally financed NHS Trust agreements. Article 4 excludes agreements entered into by specified public bodies under the private finance initiative (or a project applying similar principles). Article 5 excludes agreements which primarily relate to the financing of works. Article 6 excludes development agreements, which contain provision for the disposal of an interest in land.

Appendix 4
The Scheme for Construction Contracts (Scotland) Regulations 1998

SI 1998 NO. 687 (S. 34)

CONSTRUCTION CONTRACTS

THE SCHEME FOR CONSTRUCTION CONTRACTS
(SCOTLAND) REGULATIONS 1998

Made – – – – – –	*6th March 1998*
Coming into force – – – –	*1st May 1998*

The Lord Advocate, in exercise of the powers conferred on him by sections 108(6), 114 and 146 of the Housing Grants, Construction and Regeneration Act 1996 and of all other powers enabling him in that behalf, having consulted such persons as he thinks fit, hereby makes the following Regulations, a draft of which has been laid before and has been approved by resolutions of each House of Parliament:

Citation, commencement and extent

1.–(1) These Regulations may be cited as the Scheme for Construction Contracts (Scotland) Regulations 1998 and shall come into force at the end of the period of 8 weeks beginning with the day on which they are made.

(2) These Regulations extend to Scotland only.

Interpretation

2. In these Regulations, ''the Act'' means the Housing Grants, Construction and Regeneration Act 1996.

The Scheme for Construction Contracts (Scotland)

3. Where a construction contract does not comply with the requirements of subsections (1) to (4) of section 108 of the Act, the adjudication provisions in Part I of the Schedule to these Regulations shall apply.

4. Where—

(a) the parties to a construction contract are unable to reach agreement for the purposes mentioned respectively in sections 109, 111 and 113 of the Act; or

(b) a construction contract does not make provision as required by section 110 of the Act,

the relevant provisions in Part II of the Schedule to these Regulations shall apply.

5. The provisions in the Schedule to these Regulations shall be the Scheme for Construction Contracts (Scotland) for the purposes of section 114 of the Act.

Edinburgh	*Hardie*
6 March 1998	*Lord Advocate*

SCHEDULE Regulations 3 to 5

THE SCHEME FOR CONSTRUCTION CONTRACTS
(SCOTLAND)

PART I

ADJUDICATION

Notice of intention to seek adjudication

1.-(1) Any party to a construction contract ("the referring party") may give written notice ("the notice of adjudication") of his intention to refer any dispute arising under the contract to adjudication.

(2) The notice of adjudication shall be given to every other party to the contract.

(3) The notice of adjudication shall set out briefly–

(a) the nature and a brief description of the dispute and of the parties involved;
(b) details of where and when the dispute has arisen;
(c) the nature of the redress which is sought; and
(d) the names and addresses of the parties to the contract (including, where appropriate, the addresses which the parties have specified for the giving of notices.)

2.-(1) Following the giving of a notice of adjudication and subject to any agreement between the parties to the disputes as to who shall act as adjudicator–

(a) the referring party shall request the person (if any) specified in the contract to act as adjudicator;
(b) if no person is named in the contract or the person named has already indicated that he is unwilling or unable to act, and the contract provides for a specified nominating body to select a person, the referring party shall request the nominating body named in the contract to select a person to act as adjudicator; or
(c) where neither head (a) nor (b) above applies, or where the person referred to in (a) has already indicated that he is unwilling or unable to act and (b) does not apply, the referring party shall request an adjudicator nominating body to select a person to act as adjudicator.

(2) A person requested to act as adjudicator in accordance with the provisions of sub-paragraph (1) shall indicate whether or not he is willing to act within two days of receiving the request.

(3) In this paragraph, and in paragraphs 5 and 6 below, "an adjudicator nominating body" shall mean a body (not being a natural person and not being a party to the dispute) which holds itself out publicly as a body which will select an adjudicator when requested to do so by a referring party.

3. The request referred to in paragraphs 2, 5 and 6 shall be accompanied by a copy of the notice of adjudication.

4. Any person requested or selected to act as adjudicator in accordance with paragraphs 2, 5 or 6 shall be a natural person acting in his personal capacity. A person requested or selected to act as an adjudicator shall not be an employee of any of the parties to the dispute and shall declare any interest, financial or otherwise, in any matter relating to the dispute.

5.-(1) The nominating body referred to in paragraphs 2(1)(b) and 6(1)(b) or the adjudicator nominating body referred to in paragraphs 2(1)(c), 5(2)(b) and 6(1)(c) must communicate the selection of an adjudicator to the referring party within five days of receiving a request to do so.

(2) Where the nominating body or the adjudicator nominating body fails to comply with sub-paragraph (1), the referring party may–

(a) agree with the other party to the dispute to request a specified person to act as adjudicator; or
(b) request any other adjudicator nominating body to select a person to act as adjudicator.

(3) The person requested to act as adjudicator in accordance with the provisions of sub-paragraph (1) or (2) shall indicate whether or not he is willing to act within two days of receiving the request.

6.–(1) Where an adjudicator who is named in the contract indicates to the parties that he is unable or unwilling to act, or where he fails to respond in accordance with paragraph 2(2), the referring party may–

(a) request another person (if any) specified in the contract to act as adjudicator;
(b) request the nominating body (if any) referred to in the contract to select a person to act as adjudicator; or
(c) request any other adjudicator nominating body to select a person to act as adjudicator.

(2) The person requested to act in accordance with the provisions of sub-paragraph (1) shall indicate whether or not he is willing to act within two days of receiving the request.

7.–(1) Where an adjudicator has been selected in accordance with paragraphs 2, 5 or 6, the referring party shall, not later than seven days from the date of the notice of adjudication, refer the dispute in writing ("the referral notice") to the adjudicator.

(2) A referral notice shall be accompanied by copies of, or relevant extracts from, the construction contract and such other documents as the referring party intends to rely upon.

(3) The referring party shall, at the same time as he sends to the adjudicator the documents referred to in sub-paragraphs (1) and (2), send copies of those documents to every other party to the dispute.

8.–(1) The adjudicator may, with the consent of all the parties to those disputes, adjudicate at the same time on more than one dispute under the same contract.

(2) The adjudicator may, with the consent of all the parties to those disputes, adjudicate at the same time on related disputes under different contracts, whether or not one or more of those parties is a party to those disputes.

(3) All the parties in sub-paragraphs (1) and (2) respectively may agree to extend the period within which the adjudicator may reach a decision in relation to all or any of these disputes.

(4) Where an adjudicator ceases to act because a dispute is to be adjudicated on by another person in terms of this paragraph, that adjudicator's fees and expenses shall be determined and payable in accordance with paragraph 25.

9.–(1) An adjudicator may resign at any time on giving notice in writing to the parties to the dispute.

(2) An adjudicator must resign where the dispute is the same or substantially the same as one which has previously been referred to adjudication, and a decision has been taken in that adjudication.

(3) Where an adjudicator ceases to act under sub-paragraph (1)–

(a) the referring party may serve a fresh notice under paragraph 1 and shall request an adjudicator to act in accordance with paragraphs 2 to 7; and
(b) if requested by the new adjudicator and in so far as it is reasonably practicable, the parties shall supply him with copies of all documents which they had made available to the previous adjudicator.

(4) Where an adjudicator resigns in the circumstances mentioned in sub-paragraph (2), or where a dispute varies significantly from the dispute referred to him and for that reason he is not competent to decide it, that adjudicator's fees and expenses shall be determined and payable in accordance with paragraph 25.

10. Where any party to the dispute objects to the appointment of a particular person as adjudicator, that objection shall not invalidate the adjudicator's appointment nor any decision he may reach in accordance with paragraph 20.

11.–(1) The parties to a dispute may at any time agree to revoke the appointment of the adjudicator and in such circumstances the fees and expenses of that adjudicator shall, subject to sub-paragraph (2), be determined and payable in accordance with paragraph 25.

(2) Where the revocation of the appointment of the adjudicator is due to the default or misconduct of the adjudicator, the parties shall not be liable to pay the adjudicator's fees and expenses.

Powers of the adjudicator

12. The adjudicator shall–

(a) act impartially in carrying out his duties and shall do so in accordance with any relevant terms of the contract and shall reach his decision in accordance with the applicable law in relation to the contract; and

(b) avoid incurring unnecessary expense.

13. The adjudicator may take the initiative in ascertaining the facts and the law necessary to determine the dispute, and shall decide on the procedure to be followed in the adjudication. In particular, he may–

(a) request any party to the contract to supply him with such documents as he may reasonably require including, if he so directs, any written statement from any party to the contract supporting or supplementing the referral notice and any other documents given under paragraph 7(2);

(b) decide the language or languages to be used in the adjudication and whether a translation of any document is to be provided and, if so, by whom;

(c) meet and question any of the parties to the contract and their representatives;

(d) subject to obtaining any necessary consent from a third party or parties, make such site visits and inspections as he considers appropriate, whether accompanied by the parties or not;

(e) subject to obtaining any necessary consent from a third party or parties, carry out any tests or experiments;

(f) obtain and consider such representations and submissions as he requires, and provided he has notified the parties of his intention, appoint experts, assessors or legal advisers;

(g) give directions as to the timetable for the adjudication, any deadlines, or limits as to the length of written documents or oral representations to be complied with; and

(h) issue other directions relating to the conduct of the adjudication.

14. The parties shall comply with any request or direction of the adjudicator in relation to the adjudication.

15. If, without showing sufficient cause, a party fails to comply with any request, direction or timetable of the adjudicator made in accordance with his powers, fails to produce any document or written statement requested by the adjudicator, or in any other way fails to comply with a requirement under these provisions relating to the adjudication, the adjudicator may–

(a) continue the adjudication in the absence of that party or of the document or written statement requested;

(b) draw such inferences from that failure to comply as may, in the adjudicator's opinion, be justified in the circumstances; and

(c) make a decision on the basis of the information before him, attaching such weight as he thinks fit to any evidence submitted to him outside any period he may have requested or directed.

16.–(1) Subject to any agreement between the parties to the contrary and to the terms of sub-paragraph (2), any party to the dispute may be assisted by, or represented by, such advisers or representatives (whether legally qualified or not) as he considers appropriate.

(2) Where the adjudicator is considering oral evidence or representations, a party to the dispute may not be represented by more than one person, unless the adjudicator gives directions to the contrary.

17. The adjudicator shall consider any relevant information submitted to him by any of the parties to the dispute and shall make available to them any information to be taken into account in reaching his decision.

18. The adjudicator and any party to the dispute shall not disclose to any other person any information or document provided to him in connection with the adjudication which the party supplying it has indicated is to be treated as confidential, except to the extent that it is necessary for the purposes of, or in connection with, the adjudication.

19.–(1) The adjudicator shall reach his decision not later than–

  (a) twenty eight days after the date of the referral notice mentioned in paragraph 7(1);
  (b) forty two days after the date of the referral notice if the referring party so consents; or
  (c) such period exceeding twenty eight days after the referral notice as the parties to the dispute may, after the giving of that notice, agree.

(2) Where the adjudicator fails, for any reason, to reach his decision in accordance with sub-paragraph (1)–

  (a) any of the parties to the dispute may serve a fresh notice under paragraph 1 and shall request an adjudicator to act in accordance with paragraphs 2 to 7; and
  (b) if requested by the new adjudicator and in so far as it is reasonably practicable, the parties shall supply him with copies of all documents which they had made available to the previous adjudicator.

(3) As soon as possible after he has rerached a decision, the adjudicator shall deliver a copy of that decision to each of the parties to the contract.

Adjudicator's decision

20.–(1) The adjudicator shall decide the matters in dispute and may make a decision on different aspects of the dispute at different times.

(2) The adjudicator may take into account any other matters which the parties to the dispute agree should be within the scope of the adjudication or which are matters under the contract which he considers are necessarily connected with the dispute and, in particular, he may–

  (a) open up, review and revise any decision taken or any certificate given by any person referred to in the contract, unless the contract states that the decision or certificate is final and conclusive;
  (b) decide that any of the parties to the dispute is liable to make a payment under the contract (whether in sterling or some other currency) and, subject to section 111(4) of the Act, when that payment is due and the final date for payment;
  (c) having regard to any term of the contract relating to the payment of interest, decide the circumstances in which, the rates at which, and the periods for which simple or compound rates of interest shall be paid.

21. In the absence of any directions by the adjudicator relating to the time for performance of his decision, the parties shall be required to comply with any decision of the adjudicator immediately on delivery of the decision to the parties in accordance with paragraph 19(3).

22. If requested by one of the parties to the dispute, the adjudicator shall provide reasons for his decision.

Effects of the decision

23.–(1) In his decision, the adjudicator may, if he thinks fit, order any of the parties to comply peremptorily with his decision or any part of it.

(2) The decision of the adjudicator shall be binding on the parties, and they shall comply with it, until the dispute is finally determined by legal proceedings, by arbitration (if the contract provides for arbitration or the parties otherwise agree to arbitration) or by agreement between the parties.

24. Where a party or the adjudicator wishes to register the decision for execution in the Books of Council and Session, any other party shall, on being requested to do so, forthwith consent to such registration by subscribing the decision before a witness.

25.–(1) The adjudicator shall be entitled to the payment of such reasonable amount as he may determine by way of fees and expenses incurred by him and the parties shall be jointly and severally liable to pay that amount to the adjudicator.

(2) Without prejudice to the right of the adjudicator to effect recovery from any party in accordance with sub-paragraph (1), the adjudicator may by direction determine the apportionment between the parties of liability for his fees and expenses.

26. The adjudicator shall not be liable for anything done or omitted in the discharge or purported discharge of his functions as adjudicator unless the act or omission is in bad faith, and any employee or agent of the adjudicator shall be similarly protected from liability.

PART II

PAYMENT

Entitlement to and amount of stage payments

1. Where the parties to a relevant construction contract fail to agree–

(a) the amount of any instalment or stage or periodic payment for any work under the contract;
(b) the intervals at which, or circumstances in which, such payments become due under that contract; or
(c) both of the matters mentioned in sub-paragraphs (a) and (b),

the relevant provisions of paragraphs 2 to 4 shall apply.

2.–(1) The amount of any payment by way of instalments or stage or periodic payments in respect of a relevant period shall be the difference between the amount determined in accordance with sub-paragraph (2) and the amount determined in accordance with sub-paragraph (3).

(2) The aggregate of the following amounts:–

(a) an amount equal to the value of any work performed in accordance with the relevant construction contract during the period from the commencement of the contract to the end of the relevant period (excluding any amount calculated in accordance with head (b));
(b) where the contract provides for payment for materials, an amount equal to the value of any materials manufactured on site or brought onto site for the purposes of the works during the period from the commencement of the contract to the end of the relevant period; and
(c) any other amount or sum which the contract specifies shall be payable during or in respect of the period from the commencement of the contract to the end of the relevant period.

(3) The aggregate of any sums which have been paid or are due for payment by way of instalments, stage or periodic payments during the period from the commencement of the contract to the end of the relevant period.

(4) An amount calculated in accordance with this paragraph shall not exceed the difference between–

(a) the contract price; and
(b) the aggregate of the instalments or stage or periodic payments which have become due.

Dates for payment

3. Where the parties to a construction contract fail to provide an adequate mechanism for determining either what payments become due under the contract, or when they become due for payment, or both, the relevant provisions of paragraphs 4 to 7 shall apply.

4. Any payment of a kind mentioned in paragraph 2 above shall become due on whichever of the following dates occurs later:–

  (a) the expiry of 7 days following the relevant period mentioned in paragraph 2(1); or
  (b) the making of a claim by the payee.

5. The final payment payable under a relevant construction contract, namely the payment of an amount equal to the difference (if any) between–

  (a) the contract price; and
  (b) the aggregate of any instalment or stage or periodic payments which have become due under the contract,

shall become due on–

   (i) the expiry of 30 days following completion of the work; or
  (ii) the making of a claim by the payee,

whichever is the later.

6. Payment of the contract price under a construction contract (not being a relevant construction contract) shall become due on–

  (a) the expiry of 30 days following the completion of the work; or
  (b) the making of a claim by the payee,

whichever is the later.

7. Any other payment under a construction contract shall become due on–

  (a) the expiry of 7 days following the completion of the work to which the payment relates; or
  (b) the making of a claim by the payee,

whichever is the later.

Final date for payment

8.–(1) Where the parties to a construction contract fail to provide a final date for payment in relation to any sum which becomes due under a construction contract, the provisions of this paragraph shall apply.

(2) The final date for the making of any payment of a kind mentioned in paragraph 2, 5, 6 or 7 shall be 17 days from the date that payment becomes due.

Notice specifying amount of payment

9. A party to a construction contract shall, not later than 5 days after the date on which any payment–

  (a) becomes due from him; or
  (b) would have become due, if
       (i) the other party had carried out his obligations under the contract; and
      (ii) no set-off or abatement was permitted by reference to any sum claimed to be due under one or more other contracts,

give notice to the other party to the contract specifying the amount (if any) of the payment he has made or proposes to make, specifying to what the payment relates and the basis on which that amount is calculated.

Notice of intention to withhold payment

10. Any notice of intention to withhold payment mentioned in section 111 of the Act shall be given not later than the prescribed period, which is to say not later than 7 days before the final date for payment determined either in accordance with the construction contract or, where no such provision is made in the contract, in accordance with paragraph 8.

Prohibition of conditional payment provisions

11. Where a provision making payment under a construction contract conditional on the payer receiving payment from a third person is ineffective as mentioned in section 113 of the Act and the parties have not agreed other terms for payment, the relevant provisions of–

(a) paragraphs 2, 4, 5, and 7 to 10 shall apply in the case of a relevant construction contract; and
(b) paragraphs 6 to 10 shall apply in the case of any other construction contract.

Interpretation

12. In this Part–

"claim by the payee" means a written notice given by the party carrying out work under a construction contract to the other party specifying the amount of any payment or payments which he considers to be due, specifying to what the payment relates (or payments relate) and the basis on which it is, or they are, calculated;

"contract price" means the entire sum payable under the construction contract in respect of the work;

"relevant construction contract" means any construction contract other than one–

(a) which specifies that the duration of the work is to be less than 45 days; or
(b) in respect of which the parties agree that the duration of the work is estimated to be less than 45 days;

"relevant period" means a period which is specified in, or is calculated by reference to, the construction contract or, where no such period is so specified or is so calculable, a period of 28 days;

"value of work" means an amount determined in accordance with the construction contract under which the work is performed or, where the contract contains no such provision, the cost of any work performed in accordance with that contract together with an amount equal to any overhead or profit included in the contract price;

"work" means any of the work or services mentioned in section 104 of the Act.

EXPLANATORY NOTE

(This note is not part of the Regulations)

Part II of the Housing Grants, Construction and Regeneration Act 1996 makes provision in relation to construction contracts. Section 114 empowers the Lord Advocate to make the Scheme for Construction Contracts (as regards Scotland). Where a construction contract does not comply with the requirements of sections 108 to 111 (adjudication of disputes and payment provisions), and section 113 (prohibition of conditional payment provisions), the relevant provisions of the Scheme for Construction Contracts have effect.

The Scheme which is contained in the Schedule to these Regulations is in two parts. Part I provides for the selection and appointment of an adjudicator, gives powers to the adjudicator to gather and consider information, and makes provisions in respect of his decisions. Part II makes provision with respect to payments under a construction contract where either the contract fails to make provision or the parties fail to agree–

(a) the method for calculating the amount of any instalment, stage or periodic payment;
(b) the due date and the final date for payments to be made; and
(c) the prescribed period within which a notice of intention to withhold payment must be given.

Appendix 5
The Construction Contracts (Scotland) Exclusion Order 1998

SI 1998 NO. 686 (S. 33)

CONSTRUCTION CONTRACTS

THE CONSTRUCTION CONTRACTS (SCOTLAND)
EXCLUSION ORDER 1998

Made – – – – – –	*6th March 1998*
Coming into force – – – –	*1st May 1998*

The Secretary of State, in exercise of the powers conferred on him by sections 106(1)(b) and 146(1) of the Housing Grants, Construction and Regeneration Act 1996 and of all other powers enabling him in that behalf, hereby makes the following Order, a draft of which has been laid before, and approved by resolution of, each House of Parliament:

Citation, commencement and extent

1.–(1) This Order may be cited as the Construction Contracts (Scotland) Exclusion Order 1998 and shall come into force at the end of the period of 8 weeks beginning with the day on which it is made.

(2) This Order shall extend to Scotland only.

Interpretation

2. In this Order, "Part II" means Part II of the Housing Grants, Construction and Regeneration Act 1996.

Agreements under statute

3. A construction contract is excluded from the operation of Part II if it is–

(a) an agreement under section 48 (contributions towards expenditure on constructing or improving roads) of the Roads (Scotland) Act 1984;

(b) an agreement under section 75 (agreements regulating development or use of land) or 246 (agreements relating to Crown land) of the Town and Country Planning (Scotland) Act 1997;

(c) an agreement under section 8 (agreements as to provision of sewers etc. for new premises) of the Sewerage (Scotland) Act 1968; or

(d) an externally financed development agreement within the meaning of section 1 (powers of NHS Trusts to enter into agreements) of the National Health Service (Private Finance) Act 1997.

Private finance initiative

4.–(1) A construction contract is excluded from the operation of Part II if it is a contract entered into under the private finance initiative, within the meaning given below.

(2) A contract is entered into under the private finance initiative if all the following conditions are fulfilled–

    (a) it contains a statement that it is entered into under that initiative or, as the case may be, under a project applying similar principles;

    (b) the consideration due under the contract is determined at least in part by reference to one or more of the following:–

        (i) the standards attained in the performance of a service, the provision of which is the principal purpose or one of the principal purposes for which the building or structure is constructed;

        (ii) the extent, rate or intensity of use of all or any part of the building or structure in question; or

        (iii) the right to operate any facility in connection with the building or structure in question; and

    (c) one of the parties to the contract is–

        (i) a Minister of the Crown;

        (ii) a department in respect of which appropriation accounts are required to be prepared under the Exchequer and Audit Departments Act 1866;

        (iii) any other authority or body whose accounts are required to be examined and certified by or are open to the inspection of the Comptroller and Auditor General by virtue of an agreement entered into before the date on which this Order comes into force, or by virtue of any enactment;

        (iv) any authority or body listed in Schedule 4 (nationalised industries and other public authorities) to the National Audit Act 1983;

        (v) a body whose accounts are subject to audit by auditors appointed by the Accounts Commission for Scotland;

        (vi) a water and sewerage authority established under section 62 (new water and sewerage authorities) of the Local Government etc. (Scotland) Act 1994;

        (vii) the board of management of a self-governing school within the meaning of section 1(3) (duty of Secretary of State to maintain self-governing schools) of the Self-Governing Schools etc. (Scotland) Act 1989; or

        (viii) a company wholly owned by any of the bodies described in heads (i) to (v) above.

Finance agreements

5.–(1) A construction contract is excluded from the operation of Part II if it is a finance agreement, within the meaning given below.

    (2) A contract is a finance agreement if it is any one of the following:–

    (a) any contract of insurance;

    (b) any contract under which the principal obligations include the formation or dissolution of a company, unincorporated association or partnership;

    (c) any contract under which the principal obligations include the creation or transfer of securities or any right or interest in securities;

    (d) any contract under which the principal obligations include the lending of money;

    (e) any contract under which the principal obligations include an undertaking by a person to be responsible as surety for the debt or default of another person, including a fidelity bond, advance payment bond, retention bond or performance bond.

Development agreements

6.–(1) A construction contract is excluded from the operation of Part II if it is a development agreement, within the meaning given below.

(2) A contract is a development agreement if it includes provision for the grant or disposal of a relevant interest in the land on which take place the principal construction operations to which the contract relates.

(3) In paragraph (2) above, a relevant interest in land means–

(a) ownership; or
(b) a tenant's interest under a lease for a period which is to expire no earlier than 12 months after the completion of the construction operations under the contract.

Calum MacDonald

St Andrew's House, Edinburgh
6th March 1998   Parliamentary Under Secretary of State, Scottish Office

EXPLANATORY NOTE

(This note is not part of the Order)

Part II of the Housing Grants, Construction and Regeneration Act 1996 makes provision in relation to the terms of construction contracts. Section 106 confers power on the Secretary of State to exclude descriptions of contracts from the operation of Part II. This Order excludes, as regards Scotland, contracts of four descriptions.

Article 3 excludes agreements made under specified statutory provisions dealing with works relating to roads, planning obligations, sewerage works and externally financed NHS Trust agreements. Article 4 excludes agreements entered into by specified public bodies under the private finance initiative (or a project applying similar principles). Article 5 excludes agreements which primarily relate to the financing of works. Article 6 excludes development agreements, which contain provision for the disposal of an interest in land.

Appendix 6
The Construction Contracts (Northern Ireland) Order 1997

SI 1997 NO. 274 (N.I. 1)

NORTHERN IRELAND

THE CONSTRUCTION CONTRACTS
(NORTHERN IRELAND) ORDER 1997

Made – – – – – – *12th February 1997*
Laid before Parliament – – – – *27th February 1997*
Coming into operation on a day to be appointed under Article 1(2)

ARRANGEMENT OF ORDER

Introductory

15. Crown application.
16. Orders and regulations.

At the Court at Buckingham Palace, the 12th day of February 1997
Present,
The Queen's Most Excellent Majesty in Council

Whereas this Order is made only for purposes corresponding to those of Part II of the Housing Grants, Construction and Regeneration Act 1996:

Now, therefore, Her Majesty, in exercise of the powers conferred by paragraph 1 of Schedule 1 to the Northern Ireland Act 1974 (as modified by section 149 of the said Act of 1996) and of all other powers enabling Her in that behalf, is pleased, by and with the advice of Her Privy Council, to order, and it is hereby ordered, as follows:–

Introductory

Title and commencement

1.–(1) This Order may be cited as the Construction Contracts (Northern Ireland) Order 1997.

(2) This Order shall come into operation on such day as the Head of the Department may by order appoint.

Interpretation

2.–(1) The Interpretation Act (Northern Ireland) 1954 shall apply to Article 1 and the following provisions of this Order as it applies to a Measure of the Northern Ireland Assembly.

(2) In this Order–

"the Department" means the Department of the Environment;

"the Scheme" means the Scheme for Construction Contracts in Northern Ireland, made under Article 13(1).

Construction contracts.

3.–(1) In this Order a "construction contract" means an agreement with a person for any of the following–

(a) the carrying out of construction operations;

(b) arranging for the carrying out of construction operations by others, whether under sub-contract to him or otherwise;

(c) providing his own labour, or the labour of others, for the carrying out of construction operations.

(2) References in this Order to a construction contract include an agreement–

(a) to do architectural, design, or surveying work, or

(b) to provide advice on building, engineering, interior or exterior decoration or on the laying-out of landscape,

in relation to construction operations.

(3) References in this Order to a construction contract do not include a contract of employment (within the meaning of the Employment Rights (Northern Ireland) Order 1996).

(4) The Department may by order add to, amend or repeal any of the provisions of paragraph (1), (2) or (3) as to the agreements which are construction contracts for the purposes of this Order or are to be taken or not to be taken as included in references to such contracts.

(5) Where an agreement relates to construction operations and other matters, this Order applies to it only so far as it relates to construction operations.

An agreement relates to construction operations so far as it makes provision of any kind within paragraph (1) or (2).

(6) This Order applies only to construction contracts which–

(a) are entered into after the coming into operation of this Order, and
(b) relate to the carrying out of construction operations in Northern Ireland.

(7) This Order applies whether or not the law of Northern Ireland is otherwise the applicable law in relation to the contract.

Meaning of ''construction operations''

4.–(1) In this Order ''construction operations'' means, subject as follows, operations of any of the following descriptions–

(a) construction, alteration, repair, maintenance, extension, demolition or dismantling of buildings, or structures forming, or to form, part of the land (whether permanent or not);
(b) construction, alteration, repair, maintenance, extension, demolition or dismantling of any works forming, or to form, part of the land, including (without prejudice to the foregoing) walls, roadworks, power-lines, telecommunication apparatus, aircraft runways, docks and harbours, railways, inland waterways, pipe-lines, reservoirs, water-mains, wells, sewers, industrial plant and installations for purposes of land drainage, coast protection or defence;
(c) installation in any building or structure of fittings forming part of the land, including (without prejudice to the foregoing) systems of heating, lighting, air-conditioning, ventilation, power supply, drainage, sanitation, water supply or fire protection, or security or communications systems;
(d) external or internal cleaning of buildings and structures, so far as carried out in the course of their construction, alteration, repair, extension or restoration;
(e) operations which form an integral part of, or are preparatory to, or are for rendering complete, such operations as are previously described in this paragraph, including site clearance, earthmoving, excavation, tunnelling and boring, laying of foundations, erection, maintenance or dismantling of scaffolding, site restoration, landscaping and the provision of roadways and other access works;
(f) painting or decorating the internal or external surfaces of any building or structure.

(2) The following operations are not construction operations within the meaning of this Order–

(a) drilling for, or extraction of, oil or natural gas;
(b) extraction (whether by underground or surface working) of minerals; tunnelling or boring, or construction of underground works, for this purpose;
(c) assembly, installation or demolition of plant or machinery, or erection or demolition of steelwork for the purposes of supporting or providing access to plant or machinery, on a site where the primary activity is–
    (i) nuclear processing, power generation, or water or effluent treatment, or
    (ii) the production, transmission, processing or bulk storage (other than warehousing) of chemicals, pharmaceuticals, oil, gas, steel or food and drink;

(d) manufacture or delivery to site of–
    (i) building or engineering components or equipment,
    (ii) materials, plant or machinery, or

    (iii) components for systems of heating, lighting, air-conditioning, ventilation, power supply, drainage, sanitation, water supply or fire protection, or for security or communications systems,

except under a contract which also provides for their installation;

(e) the making, installation and repair of artistic works, being sculptures, murals and other works which are wholly artistic in nature.

(3) The Department may by order add to, amend or repeal any of the provisions of paragraph (1) or (2) as to the operations and work to be treated as construction operations for the purposes of this Order.

Provisions not applicable to contract with residential occupier

5.–(1) This Order does not apply–

(a) to a construction contract with a residential occupier (see below), or
(b) to any other description of construction contract excluded from the operation of this Order by order of the Department.

(2) A construction contract with a residential occupier means a construction contract which principally relates to operations on a dwelling which one of the parties to the contract occupies, or intends to occupy, as his residence.

In this paragraph "dwelling" means a dwelling-house or a flat; and for this purpose–

"dwelling-house" does not include a building containing a flat; and

"flat" means separate and self-contained premises constructed or adapted for use for residential purposes and forming part of a building from some other part of which the premises are divided horizontally.

(3) The Department may by order amend paragraph (2).

Provisions applicable only to agreements in writing

6.–(1) The provisions of this Order apply only where the construction contract is in writing, and any other agreement between the parties as to any matter is effective for the purposes of this Order only if in writing.

The expression "agreement" shall be construed accordingly.

(2) There is an agreement in writing–

(a) if the agreement is made in writing (whether or not it is signed by the parties),
(b) if the agreement is made by exchange of communications in writing, or
(c) if the agreement is evidenced in writing.

(3) Where parties agree otherwise than in writing by reference to terms which are in writing, they make an agreement in writing.

(4) An agreement is evidenced in writing if an agreement made otherwise than in writing is recorded by one of the parties, or by a third party, with the authority of the parties to the agreement.

(5) An exchange of written submissions in adjudication proceedings, or in arbitral or legal proceedings in which the existence of an agreement otherwise than in writing is alleged by one party against another party and not denied by the other party in his response constitutes as between the parties an agreement in writing to the effect alleged.

(6) References in this Order to anything being written or in writing include its being recorded by any means.

Adjudication

Right to refer disputes to adjudication

7.–(1) A party to a construction contract has the right to refer a dispute arising under the contract for adjudication under a procedure complying with this Article.

For this purpose ''dispute'' includes any difference.

(2) The contract shall–

    (a) enable a party to give notice at any time of his intention to refer a dispute to adjudication;

    (b) provide a timetable with the object of securing the appointment of the adjudicator and referral of the dispute to him within 7 days of such notice;

    (c) require the adjudicator to reach a decision within 28 days of referral or such longer period as is agreed by the parties after the dispute has been referred;

    (d) allow the adjudicator to extend the period of 28 days by up to 14 days, with the consent of the party by whom the dispute was referred;

    (e) impose a duty on the adjudicator to act impartially; and

    (f) enable the adjudicator to take the initiative in ascertaining the facts and the law.

(3) The contract shall provide that the decision of the adjudicator is binding until the dispute is finally determined by legal proceedings, by arbitration (if the contract provides for arbitration or the parties otherwise agree to arbitration) or by agreement.

The parties may agree to accept the decision of the adjudicator as finally determining the dispute.

(4) The contract shall also provide that the adjudicator is not liable for anything done or omitted in the discharge or purported discharge of his functions as adjudicator unless the act or omission is in bad faith, and that any employee or agent of the adjudicator is similarly protected from liability.

(5) If the contract does not comply with the requirements of paragraphs (1) to (4), the adjudication provisions of the Scheme apply.

(6) The Scheme may apply the provisions of the Arbitration Act 1996 with such adaptations and modifications as appear to the Department to be appropriate.

Payment

Entitlement to stage payments

8.–(1) A party to a construction contract is entitled to payment by instalments, stage payments or other periodic payments for any work under the contract unless–

    (a) it is specified in the contract that the duration of the work is to be less than 45 days, or

    (b) it is agreed between the parties that the duration of the work is estimated to be less than 45 days.

(2) The parties are free to agree the amounts of the payments and the intervals at which, or circumstances in which, they become due.

(3) In the absence of such agreement, the relevant provisions of the Scheme apply.

(4) References in the following Articles to a payment under the contract include a payment by virtue of this Article.

Dates for payment

9.–(1) Every construction contract shall–

    (a) provide an adequate mechanism for determining what payments become due under the contract, and when, and

    (b) provide for a final date for payment in relation to any sum which becomes due.

The parties are free to agree how long the period is to be between the date on which a sum becomes due and the final date for payment.

(2) Every construction contract shall provide for the giving of notice by a party not later than 5 days after the date on which a payment becomes due from him under the contract, or would have become due if–

> (a) the other party had carried out his obligations under the contract, and
> (b) no set-off or abatement was permitted by reference to any sum claimed to be due under one or more other contracts,

specifying the amount (if any) of the payment made or proposed to be made, and the basis on which that amount was calculated.

(3) If or to the extent that a contract does not contain such provision as is mentioned in paragraph (1) or (2), the revelevant provisions of the Scheme apply.

Notice of intention to withhold payment

10.–(1) A party to a construction contract may not withhold payment after the final date for payment of a sum due under the contract unless he has given an effective notice of intention to withhold payment.

The notice mentioned in Article 9(2) may suffice as a notice of intention to withhold payment if it complies with the requirements of this Article.

(2) To be effective such a notice must specify–

> (a) the amount proposed to be withheld and the ground for withholding payment, or
> (b) if there is more than one ground, each ground and the amount attributable to it,

and must be given not later than the prescribed period before the final date for payment.

(3) The parties are free to agree what that prescribed period is to be.

In the absence of such agreement, the period shall be that provided by the Scheme.

(4) Where an effective notice of intention to withhold payment is given, but on the matter being referred to adjudication it is decided that the whole or part of the amount should be paid, the decision shall be construed as requiring payment not later than–

> (a) 7 days from the date of the decision, or
> (b) the date which apart from the notice would have been the final date for payment,

whichever is the later.

Right to suspend performance for non-payment

11.–(1) Where a sum due under a construction contract is not paid in full by the final date for payment and no effective notice to withhold payment has been given, the person to whom the sum is due has the right (without prejudice to any other right or remedy) to suspend performance of his obligations under the contract to the party by whom payment ought to have been made ("the party in default").

(2) The right may not be exercised without first giving to the party in default at least 7 days' notice of intention to suspend performance, stating the ground or grounds on which it is intended to suspend performance.

(3) The right to suspend performance ceases when the party in default makes payment in full of the amount due.

(4) Any period during which performance is suspended in pursuance of the right conferred by this Article shall be disregarded in computing for the purposes of any contractual time limit the time taken, by the party exercising the right or by a third party, to complete any work directly or indirectly affected by the exercise of the right.

Where the contractual time limit is set by reference to a date rather than a period, the date shall be adjusted accordingly.

Prohibition of conditional payment provisions

12.–(1) A provision making payment under a construction contract conditional on the payer receiving payment from a third person is ineffective, unless that third person, or any other person payment by whom is under the contract (directly or indirectly) a condition of payment by that third person, is insolvent.

(2) For the purposes of this Article a company becomes insolvent–

    (a) on the making of an administration order against it under Part III of the Insolvency (Northern Ireland) Order 1989,

    (b) on the appointment of an administrative receiver or a receiver or manager of its property under Part IV of that Order,

    (c) on the passing of a resolution for voluntary winding-up without a declaration of solvency under Article 75 of that Order, or

    (d) on the making of a winding-up order under Part V or Part VI of that Order.

(3) For the purposes of this Article a partnership becomes insolvent on the making of a winding-up order against it under any provision of the Insolvency (Northern Ireland) Order 1989 as applied by an order under Article 364 of that Order.

(4) For the purposes of this Article an individual becomes insolvent on the making of a bankruptcy order against him under Part IX of the Insolvency (Northern Ireland) Order 1989.

(5) A company, partnership or individual shall also be treated as insolvent on the occurrence of any event corresponding to those specified in paragraph (2), (3) or (4) under the law of England and Wales, or of Scotland, or of a country outside the United Kingdom.

(6) Where a provision is rendered ineffective by paragraph (1), the parties are free to agree other terms for payment.

In the absence of such agreement, the relevant provisions of the Scheme apply.

Supplementary provisions

The Scheme for Construction Contracts in Northern Ireland

13.–(1) The Department shall by regulations make a scheme (''the Scheme for Construction Contracts in Northern Ireland'') containing provision about the matters referred to in the preceding provisions of this Order.

Before making any regulations under this Article the Department shall consult such persons as it thinks fit.

(3) Where any provisions of the Scheme apply by virtue of this Order in default of contractual provision agreed by the parties, they have effect as implied terms of the contract concerned.

Service of notices, &c.

14.–(1) The parties are free to agree on the manner of service of any notice or other document required or authorised to be served in pursuance of the construction contract or for any of the purposes of this Order.

(2) If or to the extent that there is no such agreement the following provisions apply.

(3) A notice or other document may be served on a person by any effective means.

(4) Section 24 of the Interpretation Act (Northern Ireland) 1954 (service of documents), as it applies to the service by post of such a notice or other document, shall have effect with the omission of the word ''registering'' in subsection (1).

(5) This Article does not apply to the service of documents for the purposes of legal proceedings, for which provision is made by rules of court.

(6) References in this Order to a notice or other document include any form of communication in writing and references to service shall be construed accordingly.

Crown application

15. This Order applies to a construction contract entered into by or on behalf of the Crown otherwise than by or on behalf of Her Majesty in her private capacity.

Orders and regulations

16.–(1) Orders and regulations under this Order may contain such incidental, supplementary or transitional provisions and savings as the Department considers appropriate.

(2) Subject to paragraph (3), orders and regulations made under this Order shall be subject to affirmative resolution.

(3) Paragraph (2) does not apply to an order made under Article 1(2).

N.H. Nicholls
Clerk of the Privy Council

EXPLANATORY NOTE

(This note is not part of the Order)

This Order is made only for purposes corresponding to those of Part II of the Housing Grants, Construction and Regeneration Act 1996.

The Order gives any party to a construction contract in Northern Ireland a right to refer disputes to a quick, impartial and investigative adjudication procedure. If no adequate procedure is agreed in the contract, one set out in a scheme made by the Department of the Environment will apply (*Article 7*).

The Order also–

(a) gives a contractor a statutory right to insist on payment by instalments where the work is to last 45 days or more (*Article 8*);

(b) requires construction contracts to provide a mechanism for determining the amount of payments and set a final date for payment of sums due (*Article 9*);

(c) obliges a payer to give a contractor notice with reasons, before the final date of payment, if he intends to withhold payment (*Article 10*); and

(d) makes unenforceable contract provisions which have the effect of making payment dependent upon receipt of payment from a third party, unless payment is delayed because a third party has become insolvent (*Article 12*).

JCT Standard Form of Building Contract 1998

Article 5

If any dispute or difference arises under this Contract either Party may refer it to adjudication in accordance with clause 41A.

The Conditions

1·3   Unless the context otherwise requires or the Articles or the Conditions or an item in or entry in the Appendix specifically otherwise provides, the following words and phrases in the Articles of Agreement, the Conditions and the Appendix shall have the meanings given below or as ascribed in the article, clause or Appendix item to which reference is made:

Part 4: Settlement of disputes – adjudication – arbitration – legal proceedings [uu]

41A   **Adjudication [uu·1]**

41A·1   Clause 41A applies where, pursuant to article 5, either Party refers any dispute or difference arising under this Contract to adjudication.

41A·2   The Adjudicator to decide the dispute or differences shall be either an individual agreed by the Parties or, on the application of either Party, an individual to be nominated as the Adjudicator by the person named in the Appendix ('the nominator'). Provided that [vv]

41A·2 ·1   no Adjudicator shall be agreed or nominated under clause 41A·2 or clause 41A·3 who will not execute the Standard Agreement for the appointment of an Adjudicator issued by the JCT (the 'JCT Adjudication Agreement' [ww]) with the Parties, [vv] and

41A·2 ·2   where either Party has given notice of his intention to refer a dispute or difference to adjudication then

– any agreement by the Parties on the appointment of an adjudicator must be reached with the object of securing the appointment of, and the referral of the dispute or difference to, the Adjudicator within 7 days of the date of the notice of intention to refer (*see clause* **41A·4·1**);

– any application to the nominator must be made with the object of securing the appointment of, and the referral of the dispute or difference to, the Adjudicator within 7 days of the date of the notice of intention to refer.

Upon agreement by the Parties on the appointment of the Adjudicator or upon receipt by the Parties from the nominator of the name of the nominated

Adjudicator the Parties shall thereupon execute with the Adjudicator the JCT Adjudication Agreement.

Death of Adjudicator – inability to adjudicate

41A·3 If the Adjudicator dies or becomes ill or is unavailable for some other cause and is thus unable to adjudicate on a dispute or difference referred to him, then either the Parties may agree upon an individual to replace the Adjudicator or either Party may apply to the nominator for the nomination of an adjudicator to adjudicate that dispute or difference; and the Parties shall execute the JCT Adjudication Agreement with the agreed or nominated Adjudicator.

Dispute or difference – notice of intention to refer to adjudication – referral

41A·4 .1 When pursuant to article 5 a Party requires a dispute or difference to be referred to adjudication then that Party shall give notice to the other Party of his intention to refer the dispute or difference, briefly identified in the notice, to adjudication. If an Adjudicator is agreed or appointed within 7 days of the notice then the Party giving the notice shall refer the dispute or difference to the Adjudicator ('the referral') within 7 days of the notice. If an Adjudicator is not agreed or appointed within 7 days of the notice the referral shall be made immediately on such agreement or appointment. The said Party shall include with that referral particulars of the dispute or difference together with a summary of the contentions on which he relies, a statement of the relief or remedy which is sought and any material he wishes the Adjudicator to consider. The referral and its accompanying documentation shall be copied simultaneously to the other Party.

41A·4 ·2 The referral by a Party with its accompanying documentation to the Adjudicator and the copies thereof to be provided to the other Party shall be given by actual delivery or by FAX or by special delivery or recorded delivery. If given by FAX then, for record purposes, the referral and its accompanying documentation must forthwith be sent by first class post or given by actual delivery. If sent by special delivery or recorded delivery the referral and its accompanying documentation shall, subject to proof to the contrary, be deemed to have been received 48 hours after the date of posting subject to the exclusion of Sundays and any Public Holiday.

Footnotes

[uu] It is open to the Employer and the Contractor to resolve disputes by the process of Mediation: see Practice Note 28 'Mediation on a Building Contractor or Sub-Contract Dispute'.

[uu·1] The time periods generally specified in this clause are those defined by statute. Where the nature of the dispute or the work concerned may have any significant effect upon the progress or cost of the Works such as works relating to the primary structural elements the Adjudicator should consider an accelerated time table for the adjudication procedures: see JCT Practice Note 2 (Series 2): Adjudication under JCT Forms.

[vv] The nominators named in the Appendix have agreed with the JCT that they will comply with the requirements of clause 41A on the nomination of an adjudicator including the requirement in clause **41A·2·2** for the nomination to be made with the object of securing the appointment of, and the referral of the dispute or difference to, the Adjudicator within 7 days of the date of the notice of intention to refer; and will only nominate adjudicators who will enter into the 'JCT Adjudication Agreement'.

[ww] The JCT Adjudication Agreement is available from the retailers of JCT Forms.
A version of this Agreement is also available for use if the Parties have named an Adjudicator in their contract.

41A·5 .1 The Adjudicator shall immediately upon receipt of the referral and its accompanying documentation confirm the date of that receipt to the Parties.

41A.5 .2 The Party not making the referral may, by the same means stated in clause 41A·4·2, send to the Adjudicator within 7 days of the date of the referral, with a copy to the other Party, a written statement of the contentions on which he relies and any material he wishes the Adjudicator to consider.

41A·5 .3 The Adjudicator shall within 28 days of the referral under clause 41A·4·1 and acting as an Adjudicator for the purposes of S.108 of the Housing Grants, Construction and Regeneration Act 1996 and not as an expert or an arbitrator reach his decision and forthwith send that decision in writing to the Parties. Provided that the Party who has made the referral may consent to allowing the Adjudicator to extend the period of 28 days by up to 14 days; and that by agreement between the Parties after the referral has been made a longer period than 28 days may be notified jointly by the Parties to the Adjudicator within which to reach his decision.

41A·5 ·4 The Adjudicator shall not be obliged to give reasons for his decision.

41A·5 .5 In reaching his decision the Adjudicator shall act impartially and set his own procedure; and at his absolute discretion may take the initiative in ascertaining the facts and the law as he considers necessary in respect of the referral which may include the following:

·5 ·1 using his own knowledge and/or experience;

·5 ·2 subject to clause 30·9, opening up, reviewing and revising any certificate, opinion, decision, requirement or notice issued, given or made under this Contract as if no such certificate, opinion, decision, requirement or notice had been issued, given or made;

·5 ·3 requiring from the Parties further information than that contained in the notice of referral and its accompanying documentation or in any written statement provided by the Parties including the results of any tests that have been made or of any opening up;

·5 ·4 requiring the Parties to carry out tests or additional tests or to open up work or further open up work;

·5 ·5 visiting the site of the Works or any workshop where work is being or has been prepared for this Contract;

·5 ·6 obtaining such information as he considers necessary from any employee or representative of the Parties provided that before obtaining information from an employee of a Party he has given prior notice to that Party;

·5 ·7 obtaining from others such information and advice as he considers necessary on technical and on legal matters subject to giving prior notice to the Parties together with a statement or estimate of the cost involved:

·5 ·8 having regard to any term of this Contract relating to the payment of interest, deciding the circumstances in which or the period for which a simple rate of interest shall be paid.

41A·5 ·6 Any failure by either Party to enter into the JCT Adjudication Agreement or to comply with any requirement of the Adjudicator under clause 41A·5·5 or with any provision in or requirement under clause 41A shall not invalidate the decision of the Adjudicator.

41A·5 ·7 The Parties shall meet their own costs of the adjudication except that the Adjudicator may direct as to who should pay the cost of any test or opening up if required pursuant to clause 41A·5·5·4.

41A·5 ·8 Where any dispute or difference arises under clause 8·4·4 as to whether an instruction issued thereunder is reasonable in all the circumstances the following provisions shall apply:

·8 ·1 The Adjudicator to decide such dispute or difference shall (where practicable) be an individual with appropriate expertise and experience in the specialist area or discipline relevant to the instruction or issue in dispute.

·8 ·2 Where the Adjudicator does not have the appropriate expertise and experience referred to in clause 41A·5·8·1 above the Adjudicator shall appoint an independent expert with such relevant expertise and experience to advise and report in writing on whether or not any instruction issued under clause 8·4·4 is reasonable in all the circumstances.

·8 ·3 Where an expert has been appointed by the Adjudicator pursuant to clause 41A·5·8·2 above the Parties shall be jointly and severally responsible for the expert's fees and expenses but, in his decision, the Adjudicator shall direct as to who should pay the fees and expenses of such expert or the proportion in which such fees and expenses are to be shared between the Parties.

·8 ·4 Notwithstanding the provisions of clause 41A·5·4 above, where an independent expert has been appointed by the Adjudicator pursuant to clause 41A·5·8·2 above, copies of the Adjudicator's instructions to the expert and any written advice or reports received from such expert shall be supplied to the Parties as soon as practicable.

Adjudicator's fee and reasonable expenses – payment

41A·6 ·1 The Adjudicator in his decision shall state how payment of his fee and reasonable expenses is to be apportioned as between the Parties. In default of such statement the Parties shall bear the cost of the Adjudicator's fee and reasonable expenses in equal proportions.

41A·6 ·2 The Parties shall be jointly and severally liable to the Adjudicator for his fee and for all expenses reasonably incurred by the Adjudicator pursuant to the adjudication.

Effect of Adjudicator's decision

41A·7 ·1 The decision of the Adjudicator shall be binding on the Parties until the dispute or difference is finally determined by arbitration or by legal proceedings **[xx]** or by an agreement in writing between the Parties made after the decision of the Adjudicator has been given.

41A·7 ·2 The Parties shall, without prejudice to their other rights under this Contract, comply with the decision of the Adjudicator; and the Employer and the Contractor shall ensure that the decision of the Adjudicator is given effect.

41A·7 ·3 If either Party does not comply with the decision of the Adjudicator the other Party shall be entitled to take legal proceedings to secure such compliance pending any final determination of the referred dispute or difference pursuant to clause 41A·7·1.

41A·8 The Adjudicator shall not be liable for anything done or omitted in the discharge or purported discharge of his functions as Adjudicator unless the act or omission is in bad faith and this protection from liability shall similarly extend to any employee or agent of the Adjudicator.   **Immunity**

Footnote [xx] The arbitration or legal proceedings are *not* an appeal against the decision of the Adjudicator but are a consideration of the dispute or difference as if no decision had been made by an Adjudicator.

JCT Adjudication Agreement

This Agreement

is made on the _____ day of _____ 19 _____

BETWEEN ('the Contracting Parties')

Insert names and
addresses of the
Contracting Parties

(1)

_____

_____

_____

_____

(2)

_____

_____

_____

_____

and ('the Adjudicator')

Insert name and
address of
Adjudicator

_____

_____

_____

_____

JCT Adjudication Agreement

Whereas

the Contracting Parties have entered into a
*Contract/Sub-Contract/Agreement (the 'contract') for

Brief description of
the works/the sub-
contract works

_____

_____

_____

_____

_____

on the terms of

Insert the title of the
JCT Contract/Sub-
Contract/Agreement
and any amendments
thereto incorporated
therein

_____

_____

_____

in which the provisions on adjudication ('the Adjudication Provisions') are set out in

clause _____

And Whereas

a dispute or difference has arisen under the contract which the Contracting Parties wish
to be referred to adjudication in accordance with the said Adjudication Provisions.

*Delete as appropriate.

© RIBA Publications 1998

JCT Adjudication Agreement

Now it is agreed that

Appointment and acceptance

1　The Contracting Parties hereby appoint the Adjudicator and the Adjudicator hereby accepts such appointment in respect of the dispute briefly identified in the attached notice.

Adjudication Provisions

2　The Adjudicator shall observe the Adjudication Provisions as if they were set out in full in this Agreement.

Adjudicator's fee and reasonable expenses

3　The Contracting Parties will be jointly and severally liable to the Adjudicator for his fee as stated in the Schedule hereto for conducting the adjudication and for all expenses reasonably incurred by the Adjudicator as referred to in the Adjudication Provisions.

Unavailability of Adjudicator to act on the referral

4　If the Adjudicator becomes ill or becomes unavailable for some other cause and is thus unable to complete the adjudication he shall immediately give notice to the Contracting Parties to such effect.

Termination

5　·1　The Contracting Parties jointly may terminate the Adjudication Agreement at any time on written notice to the Adjudicator. Following such termination the Contracting Parties shall, subject to clause 5·2, pay the Adjudicator his fee or any balance thereof and his expenses reasonably incurred prior to the termination.

　　·2　Where the decision of the Contracting Parties to terminate the Adjudication Agreement under clause 5·1 is because of a failure by the Adjudicator to give his decision on the dispute or difference within the time-scales in the Adjudication Provisions or at all, the Adjudicator shall not be entitled to recover from the Contracting Parties his fee and expenses.

　　　　　　　　　　　　　　　　　　　Adj1/3

JCT Adjudication Agreement

As Witness
the hands of the Contracting Parties and the Adjudicator

Signed by or on behalf of: the Contracting Parties

(1) _____

in the presence of _____

(2) _____

in the presence of _____

Signed by: the Adjudicator _____

in the presence of _____

Schedule

Fee The lump sum fee is £_____

or

The hourly rate is £_____

© RIBA Publications 1998

Construction Confederation Contract DOM/1

38   Settlement of disputes

38A Adjudication

Application of clause 38A

.1   Clause 38A applies, where pursuant to Article 3, either Party refers any dispute or difference arising under this Sub-Contract to adjudication.

Identity of Adjudicator

.2   The Adjudicator to decide the dispute or difference shall be either an individual agreed by the Parties or, on the application of either party, an individual to be nominated as the Adjudicator by the person named in the Appendix part 8 ("the Nominator") provided that:–

.1   no Adjudicator shall be agreed or nominated under clause 38.A.2.2 or clause 38A.3 who will not execute the Standard Agreement with the Parties and

.2   where either Party has given notice of his intention to refer a dispute to adjudication then

– any agreement by the Parties on the appointment of an Adjudicator must be reached with the object of securing the appointment and of the referral of the dispute or difference to the Adjudicator within 7 days of the date of the notice of intention to refer, (see clause 38A.4.1);

– any application to the nominator must be made with the object of securing the appointment of, and the referral of the dispute or difference to, the Adjudicator within 7 days of the date of the notice of intention to refer:

.3   upon agreement by the Parties on the appointment of the Adjudicator or upon receipt by the Parties from the nominator of the name of the nominated Adjudicator the Parties shall thereupon execute with the Adjudicator the JCT Adjudication Agreement.

Death of Adjudicator – inability to adjudicate

.3   If the Adjudicator dies or becomes ill or is unavailable for some other cause and is thus unable to adjudicate on a dispute or difference referred to him, the Parties may either agree upon a person to replace the Adjudicator **or** either Party may apply to the nominator for the nomination of an adjudicator to adjudicate that dispute or difference; ;and the Parties shall execute the JCT Adjudication Agreement with the agreed or nominated Adjudicator.

Dispute or difference – notice of intention to Adjudication – referral

.4 .1 When pursuant to Article 3 a Party requires a dispute or difference to be referred to adjudication then that Party shall give notice to the other Party of his intention to refer the dispute or difference, briefly identified in the notice, to adjudication. Within 7 days from the date of such notice or the execution of the JCT Adjudication Agreement by the Adjudicator if later the Party giving the notice of intention shall refer the dispute or difference to the Adjudicator for his decision (''the referral''); and shall include with that referral particulars of the dispute or difference together with a summary of the contentions on which he relies, a statement of the relief or remedy which is sought and any material he wishes the Adjudicator to consider. The referral and its accompanying documentation shall be copied simultaneously to the other Party.

.2 The referral by a Party with its accompanying documentation to the Adjudicator and the copies thereof to be provided to the other Party shall be given by actual delivery or by FAX or by registered post or recorded delivery. If given by FAX then, for record purposes, the referral and its accompanying documentation must forthwith be sent by first class post or given by actual delivery. If sent by registered post or recorded delivery the referral and its accompanying documentation shall, subject to proof to the contrary, he deemed to have been received 48 hours after the date of posting subject to the exclusion of Sundays and any Public Holiday.

Conduct of the Adjudicator

38A .5 .1 The Adjudicator shall immediately upon receipt of the referral and its accompanying documentation confirm that receipt to the Parties.

.2 The Party not making the referral may, by the same means stated in clause 38A.4.2, send to the Adjudicator within 7 days of the date of the referral with a copy to the other Party, a written statement of the contentions on which he relies and any material he wishes the Adjudicator to consider.

.3 The Adjudicator shall within 28 days of his receipt of the referral and its accompanying documentation under clause 38A.4.1 and acting as an Adjudicator for the purposes of s.108 of the Housing Grants, Construction and Regeneration Act 1996 and not as an expert or an arbitrator reach his decision and forthwith send that decision in writing to the Parties. Provided that the Party who has made the referral may consent to allowing the Adjudicator to extend the period of 28 days by up to 14 days; and that by agreement between the Parties after the referral has been made a longer period than 28 days may be notified jointly by the Parties to the Adjudicator within which to reach his decision.

.4 The Adjudicator shall not be obliged to give reasons for his decision.

.5 In reaching his decision the Adjudicator shall act impartially, set his own procedure and at his absolute discretion may take the initiative in ascertaining the facts and the law as he considers necessary in respect of the referral which may include the following:

.1 using his own knowledge and/or experience;
.2 opening up, reviewing and revising any certificate, direction, opinion, decision, requirement or notice issued given or made under the Sub-Contract as if no such certificate, direction, opinion, decision, requirement or notice had been given or made;
.3 requiring from the Parties further information than that contained in the notice of referral and its accompanying documentation or in any written statement pro-

vided by the Parties including the results of any tests that have been made or of any opening up;

   .4   requiring the Parties to carry out tests or additional tests or to open up work or further open up work;

   .5   visiting the site of the Works or any workshop where work is being or has been prepared for this Sub-Contract;

   .6   obtaining such information as he considers necessary from any employee or representative of the Parties provided that before obtaining information from an employee of a Party he has given prior notice to that Party;

   .7   obtaining from others such information and advice as he considers necessary on technical and on legal matters subject to giving prior notice to the Parties together with a statement or estimate of the cost involved;

   .8   having regard to any term of the Sub-Contract relating to the payment of interest deciding the circumstances in which or the period for which a simple rate of interest shall be paid.

   .6   Any failure by either Party to enter into the JCT Adjudication Agreement or to comply with any requirement of the Adjudicator under clause 38A.5.5. or with any provision in or requirement under clause 38A shall not invalidate the decision of the Adjudicator.

   .7   The Parties shall meet their own costs of the Adjudication except that the Adjudicator may direct as to who should pay the cost of any test or opening up if required pursuant to clause 38A.5.5.4.

Adjudicator's fee and reasonable expenses – payment

   .6   .1   The Adjudicator in his decision shall state how payment of his fee and reasonable expenses is to be apportioned as between the Parties. In default of such statement the Parties shall bear the cost of the Adjudicator's fee and reasonable expenses in equal proportions.

   .2   The Parties shall be jointly and severally liable to the Adjudicator for his fee and for all expenses incurred by the Adjudicator pursuant to the Adjudication.

Effect of Adjudicator's decision

   .7   .1   The decision of the Adjudicator shall be binding on the Parties until the dispute or difference is finally determined by arbitration or by legal proceedings or by an agreement in writing between the Parties made after the decision of the Adjudicator has been given. [ff]

   .2   The Parties shall, without prejudice to their other rights under the Contract, comply with the decisions of the Adjudicator; and the Contractor and the Sub-Contractor shall ensure that the decisions of the Adjudicator are given effect.

   .3   If either Party does not comply with the decision of the Adjudicator the other Party shall be entitled, to take proceedings in the Courts to secure such compliance pending any final determination of the referred dispute or difference pursuant to Clause 38A.7.1.

[ff]   The arbitration or legal proceedings are *not* an appeal against the decision of the Adjudicator but are a consideration of the dispute or difference as if no decision had been made by an Adjudicator.

Immunity

.8 The Adjudicator shall not be liable for anything done or omitted in the discharge or purported discharge of his functions as Adjudicator unless the act or omission is in bad faith and this protection from liability shall similarly extend to any employee or agent of the Adjudicator.

ICE Conditions of Contract 7th Edition

AVOIDANCE AND SETTLEMENT OF DISPUTES

Avoidance of disputes 66

(1)   In order to overcome where possible the causes of disputes and in those cases where disputes are likely still to arise to facilitate their clear definition and early resolution (whether by agreement or otherwise) the following procedure shall apply for the avoidance and settlement of disputes.

Matters of dissatisfaction

(2)   If at any time

(a)   the Contractor is dissatisfied with any act or instruction of the Engineer's Representative or any other person responsible to the Engineer or

(b)   the Employer or the Contractor is dissatisfied with any decision opinion instruction direction certificate or valuation of the Engineer or with any other matter arising under or in connection with the Contract or the carrying out of the Works

the matter of dissatisfaction shall be referred to the Engineer who shall notify his written decision to the Employer and the Contractor within one month of the reference to him.

Disputes

(3)   The Employer and the Contractor agree that no matter shall constitute nor be said to give rise to a dispute unless and until in respect of that matter

(a)   the time for the giving of a decision by the Engineer on a matter of dissatisfaction under Clause 66(2) has expired or the decision given is unacceptable or has not been implemented and in consequence the Employer or the Contractor has served on the other and on the Engineer a notice in writing (hereinafter called the Notice of Dispute)

(b)   an adjudicator has given a decision on a dispute under Clause 66(6) and the Employer or the Contractor is not giving effect to the decision, and in consequence the other has served on him and the Engineer a Notice of Dispute

and the dispute shall be that stated in the Notice of Dispute. For the purposes of all matters arising under or in connection with the Contract or the carrying out of the Works the word "dispute" shall be construed accordingly and shall include any difference.

(4)   (a)   Notwithstanding the existence of a dispute following the service of a Notice under Clause 66(3) and unless the Contract has already been determined or abandoned the Employer and the Contractor shall continue to perform their obligations.

(b)   The Employer and the Contractor shall give effect forthwith to every decision of

(i)   the Engineer on a matter of dissatisfaction given under Clause 66(2) and

(ii)   the adjudicator on a dispute given under Clause 66(6)

unless and until that decision is revised by agreement of the Employer and Contractor or pursuant to Clause 66.

(5)   (a)   The Employer or the Contractor may at any time before service of a Notice to Refer to arbitration under Clause 66(9) by notice in writing seek the agreement of the other for the dispute to be considered under "The Institution of Civil Engineers' Conciliation Procedure 1999" or any amendment or modification thereof being in force at the date of such notice.   **Conciliation**

(b)   If the other party agrees to this procedure any recommendations of the conciliator shall be deemed to have been accepted as finally determining the dispute by agreement so that the matter is no longer in dispute unless a Notice of Adjudication under Clause 66(6) or a Notice to Refer to arbitration under Clause 66(9) has been served in respect of that dispute not later than one month after receipt of the recommendation by the dissenting party.

(6)   (a)   The Employer and the Contractor each has the right to refer a dispute as to a matter under the Contract for adjudication and either party may give notice in writing (hereinafter called the Notice of Adjudication) to the other at any time of his intention so to do. The adjudication shall be conducted under "The Institution of Civil Engineers' Adjudication Procedure 1997" or any amendment or modification thereof being in force at the time of the said Notice.   **Adjudication**

(b)   Unless the adjudicator has already been appointed he is to be appointed by a timetable with the object of securing his appointment and referral of the dispute to him within 7 days of such notice.

(c)   The adjudicator shall reach a decision within 28 days of referral or such longer period as is agreed by the parties after the dispute has been referred.

(d)   The adjudicator may extend the period of 28 days by up to 14 days with the consent of the party by whom the dispute was referred.

(e)   The adjudicator shall act impartially.

(f)   The adjudicator may take the initiative in ascertaining the facts and the law.

(7)   The decision of the adjudicator shall be binding until the dispute is finally determined by legal proceedings or by arbitration (if the contract provides for arbitration or the parties otherwise agree to arbitration) or by agreement.

(8)   The adjudicator is not liable for anything done or omitted in the discharge or purported discharge of his functions as adjudicator unless the act or omission is in bad faith and any employee or agent of the adjudicator is similarly not liable.

(9)   (a)   All disputes arising under or in connection with the Contract or the carrying out of the Works other than failure to give effect to a decision of an adjudicator shall be finally determined by reference to arbitration. The party seeking arbitration shall serve on the other party a notice in writing (called the Notice to Refer) to refer the dispute to arbitration.   **Arbitration**

(b)   Where an adjudicator has given a decision under Clause 66(6) in respect of the particular dispute the Notice to Refer must be served within three months of the giving of the decision otherwise it shall be final as well as binding.

Appointment of arbitrator President or Vice-President to act

(10) (a)   The arbitrator shall be a person appointed by agreement of the parties.

(b)   If the parties fail to appoint an arbitrator within one month of either party serving on the other party a notice in writing (hereinafter called the Notice to Concur) to concur in the appointment of an arbitrator the dispute shall be referred to a person to be appointed on the application of either party by the President for the time being of the Institution of Civil Engineers.

(c)   If an arbitrator declines the appointment or after appointment is removed by order of a competent court or is incapable of acting or dies and the parties do not within one month of the vacancy arising fill the vacancy then either party may apply to the President for the time being of the Institution of Civil Engineers to appoint another arbitrator to fill the vacancy.

(d)   In any case where the President for the time being of the Institution of Civil Engineers is not able to exercise the functions conferred on him by this Clause the said functions shall be exercised on his behalf by a Vice-President for the time being of the said Institution.

Arbitration – procedure and powers

(11) (a)   Any reference to arbitration under this Clause shall be deemed to be a submission to arbitration within the meaning of the Arbitration Act 1996 or any statutory re-enactment or amendment thereof for the time being in force. The reference shall be conducted in accordance with the procedure set out in the Appendix to the Form of Tender or any amendment or modification thereof being in force at the time of the appointment of the arbitrator. Such arbitrator shall have full power to open up review and revise any decision opinion instruction direction certificate or valuation of the Engineer or an adjudicator.

(b)   Neither party shall be limited in the arbitration to the evidence or arguments put to the Engineer or to any adjudicator pursuant to Clause 66(2) or 66(6) respectively.

(c)   The award of the arbitrator shall be binding on all parties.

(d)   Unless the parties otherwise agree in writing any reference to arbitration may proceed notwithstanding that the Works are not then complete or alleged to be complete.

Witnesses

(12) (a)   No decision opinion instruction direction certificate or valuation given by the Engineer shall disqualify him from being called as a witness and giving evidence before a conciliator adjudicator or arbitrator on any matter whatsoever relevant to the dispute.

(b)   All matters and information placed before a conciliator pursuant to a reference under sub-clause (5) of this Clause shall be deemed to be submitted to him without prejudice and the conciliator shall not be called as witness by the parties or anyone claiming through them in connection with any adjudication arbitration or other legal proceedings arising out of or connected with any matter so referred to him.

Appendix 10
New Engineering and Construction Contract

NEC MAIN CONTRACT

REFERENCE: NEC/ECC/Y(UK)2/APRIL 1998

Clause 90 is deleted and replaced by the following

90   Avoidance and settlement of disputes

90.1   The Parties and the *Project Manager* follow this procedure for the avoidance and settlement of disputes.

90.2   If a Party is dissatisfied with an action, a failure to take action by the *Project Manager*, he notifies his dissatisfaction to the other party no later than

- four weeks after he became aware of the action,
- four weeks after he became aware that the action had not been taken.

Within two weeks of such notification of dissatisfaction, the *Contractor* and the *Project Manager* attend a meeting to discuss and seek to resolve the matter.

90.3   If either Party is dissatisfied with any other matter, he notifies his dissatisfaction to the *Project Manager* and to the other Party no later than four weeks after he became aware of the matter. Within two weeks of such notification of dissatisfaction, the Parties and the *Project Manager* attend a meeting to discuss and seek to resolve the matter.

90.4   The Parties agree that no matter shall be a dispute unless a notice of dissatisfaction has been given and the matter has not been resolved within four weeks. The word dispute (which includes a difference) has that meaning.

90.5   Either Party may give notice to the other Party at any time of his intention to refer a dispute to adjudication. The notifying Party refers the dispute to the *Adjudicator* within seven days of the notice.

90.6   The Party referring the dispute to the *Adjudicator* includes with his submission information to be considered by the *Adjudicator*. Any further information from a Party to be considered by the *Adjudicator* is provided within fourteen days of referral.

90.7   Unless and until the *Adjudicator* has given his decision on the dispute, the Parties and the *Project Manager* proceed as if the action, failure to take action or other matters were not disputed.

90.8   The *Adjudicator* acts impartially. The *Adjudicator* may take the initiative in ascertaining the facts and the law.

90.9   The *Adjudicator* reaches a decision within twenty eight days of referral or such longer period as is agreed by the Parties after the dispute has been referred. The *Adjudicator* may extend the period of twenty eight days by up to fourteen days with the consent of the notifying Party.

90.10 The *Adjudicator* provides his reasons to the Parties and to the *Project Manager* with his decision.

90.11 The decision of the *Adjudicator* is binding until the dispute is finally determined by the *tribunal* or by agreement.

90.12 The *Adjudicator* is not liable for anything done or omitted in the discharge or purported discharge of his functions as adjudicator unless the act or omission is in bad faith and any employee or agent of the *Adjudicator* is similarly protected from liability.

Clause 91 is amended as follows:—

Side heading "**The adjudication**" is replaced with "**Combining procedures**"

Clause 91.1 is deleted and replaced by the following:—

91.1 If a matter causing dissatisfaction under or in connection with a subcontract is also a matter causing dissatisfaction under or in connection with this contract, the subcontractor may attend the meeting between the Parties and the *Project Manager* to discuss and seek to resolve the matter.

Clause 91.2   line 4   "settles" is replaced with "gives his decision on"

Clause 92 is amended as follows:—

Clause 92.1   line 1   "settles" is replaced with "gives his decision on"

Clause 92.2   line 6   "settle" is replaced with "decide on"

Clause 92.2   line 7   "had not been settled" is replaced with "a decision had not been given"

Contract Data Part 1 – Optional statements

The fifth optional statement is deleted and replaced by the following:—

"**If the period for payment is not twenty one days**
• The period within which payments are made is ___ days"

NEC SUB-CONTRACT

REFERENCE: NEC/ECS/Y(UK)2/APRIL 1998

Clause 90 is deleted and replaced by the following:

90   Avoidance and settlement of disputes

90.1 The Parties follow this procedure for the avoidance and settlement of disputes.

90.2 If a Party is dissatisfied with an action, a failure to take action, or any other matter, he notifies his dissatisfaction to the other Party no later than

• four weeks after he became aware of the action,
• four weeks after he became aware that the action had not been taken or
• four weeks after he became aware of the matter he is dissatisfied with.

90.3 Within two weeks of such notification of dissatisfaction, the Parties attend a meeting to discuss and seek to resolve the matter.

90.4 The Parties agree that no matter shall be a dispute unless a notice of dissatisfaction has been given and the matter has not been resolved within four weeks. The word dispute (which includes a difference) has that meaning.

90.5 Either Party may given notice to the other Party at any time of his intention to refer a dispute to adjudication. The notifying Party refers the dispute to the *Adjudicator* within seven days of the notice.

90.6 The Party referring the dispute to the *Adjudicator* includes with his submission information to be considered by the *Adjudicator*. Any further information from a Party to be considered by the *Adjudicator* is provided within fourteen days of referral.

90.7 Unless and until the *Adjudicator* has given his decision on the dispute, the Parties proceed as if the action, failure to take action or other matters were not disputed.

90.8 The *Adjudicator* acts impartially. The *Adjudicator* may take the initiative in ascertaining the facts and the law.

90.9 The *Adjudicator* reaches a decision within twenty eight days of referral or such longer period as is agreed by the Parties after the dispute has been referred. The *Adjudicator* may extend the period of twenty eight days by up to fourteen days with the consent of the notifying Party.

90.10 The *Adjudicator* provides his reasons to the Parties with his decision.

90.11 The decision of the *Adjudicator* is binding until the dispute is finally determined by the *tribunal* or by agreement.

90.12 The *Adjudicator* is not liable for anything done or omitted in the discharge or purported discharged of his functions as adjudicator unless the act or omission is in bad faith and any employee or agent of the *Adjudicator* is similarly protected from liability.

Clause 91 is amended as follows:—

Side heading "**The adjudication**" is replaced with "**Combining procedures**"

Clause 91.1 is deleted and replaced by the following:—

91.1 If a matter causing dissatisfaction under or in connection with a subcontract is also a matter causing dissatisfaction under or in connection with this contract, the subcontractor may attend the meeting between the Parties to discuss and seek to resolve the matter.

Clause 91.2 line 4 "settles" is replaced with "gives his decision on"

Clause 91.3 is renumbered to clause 91.4 and a new clause 91.3 inserted as follows:—

91.3 If the main contract provides for the *Subcontractor* to attend a meeting to discuss and seek to resolve a matter of dissatisfaction under or in connection with the main contract which is also a matter causing dissatisfaction under or in connection with this subcontract, the *Subcontractor* attends the meeting.

renumbered clause 91.4 line 10 "settles" is replaced with "gives his decision on".

Clause 92 is amended as follows:—

Clause 92.1 line 1 "settles" is replaced with "gives his decision on"

Clause 92.2 line 6 "settle" is replaced with "decide on"

Clause 92.2 line 7 "had not been settled" is replaced with "a decision had not been given"

Subcontract Data Part 1 – Optional statements

The fifth optional statement is deleted and replaced by the following:—

"**If the period for payment is not twenty one days**
● The period within which payments are made is … days"

Appendix 11
Civil Engineering Contractors Association Sub-contract for use with ICE 6th Edition Main Contract

Disputes

18. (1) If any dispute or difference shall arise between the Contractor and the Sub-Contractor in connection with or arising out of the Sub-Contract, or the carrying out of the Sub-Contract Works (excluding a dispute concerning VAT but including a dispute as to any act or omission of the Engineer) whether arising during the progress of the Sub-Contract Works or after their completion it shall be settled in accordance with the following provisions.

(2) (a) Where the Sub-Contractor seeks to make a submission that payment is due of any amount exceeding the amount determined by the Contractor as due to the Sub-Contractor, or that any act, decision, opinion, instruction or direction of the Contractor or any other matter arising under the Sub-Contract is unsatisfactory, the Sub-Contractor shall so notify the Contractor in writing, stating the grounds for such submission in sufficient detail for the Contractor to understand and consider the Sub-Contractor's submission.

(b) Where in the opinion of the Contractor such a submission gives rise to a matter of dissatisfaction under the Main Contract, the Contractor shall so notify the Sub-Contractor in writing as soon as possible. In that event, the Contractor shall pursue the matter of dissatisfaction under the Main Contract promptly and shall keep the Sub-Contractor fully informed in writing of progress. The Sub-Contractor shall promptly provide such information and attend such meetings in connection with the matter of dissatisfaction as the Contractor may request. The Contractor and the Sub-Contractor agree that no such submission shall constitute nor be said to give rise to a dispute under the Sub-Contract unless and until the Contractor has had the time and opportunity to refer the matter of dissatisfaction to the Engineer under the Main Contract and either the Engineer has given his decision or the time for the giving of a decision by the Engineer has expired.

(3) (a) The Contractor or the Sub-Contractor may at any time before service of a Notice to Refer to arbitration under sub-clause 18(7) by notice in writing seek the agreement of the other for the dispute to be considered under the Institution of Civil Engineers' Conciliation Procedure (1994) or any amendment or modification thereof being in force at the date of such notice.

(b) If the other party agrees to this procedure any recommendation of the conciliator shall be deemed to have been accepted as finally determining the dispute by agreement so that the matter is no longer in dispute unless a Notice of Adjudication under sub-clause 18(4) or a Notice to Refer to arbitration under sub-clause 18(7) is served within 28 days of receipt by the dissenting party of the conciliator's recommendation.

(4) (a) The Contractor and the Sub-Contractor each has the right to refer any dispute under the Sub-Contract for adjudication and either party may at any time give notice in writing (hereinafter called the Notice of Adjudication) to the other of his intention to refer the dispute to adjudication. The Notice of Adjudication and the appointment of the adjudicator shall,

save as provided under sub-clause 18(10)(b), be as provided at paragraphs 2 and 3 of the Institution of Civil Engineers' Adjudication Procedure (1997). Any dispute referred to adjudication shall be conducted in accordance with the Institution of Civil Engineers' Adjudication Procedure (1997) or any amendment or modification thereof being in force at the time of the appointment of the adjudicator.

(b)   Unless the adjudicator has already been appointed he is to be appointed by a timetable with the object of securing his appointment and referral of the dispute to him within 7 days of such notice.

(c)   The adjudicator shall reach a decision within 28 days of referral or such longer period as is agreed by the parties after the dispute has been referred.

(d)   The adjudicator may extend the period of 28 days by up to 14 days with the consent of the party by whom the dispute was referred.

(e)   The adjudicator shall act impartially.

(f)   The adjudicator may take the initiative in ascertaining the facts and the law.

(5)   The decision of the adjudicator shall be binding until the dispute is finally determined by legal proceedings or by arbitration (if the Sub-Contract provides for arbitration or the parties otherwise agree to arbitration).

(6)   The adjudicator shall not be liable for anything done or omitted in the discharge or purported discharge of his functions as adjudicator unless the act or omission is in bad faith and any employee or agent of the adjudicator shall similarly not be liable.

(7)   (a)   All disputes arising under or in connection with the Sub-Contract, other than failure to give effect to a decision of an adjudicator, shall be finally determined by reference to arbitration. The party seeking arbitration shall serve on the other party a notice in writing (called the Notice to Refer) to refer the dispute to arbitration.

(b)   Where an adjudicator has given a decision under sub-clause 18(4) in respect of the particular dispute the Notice to Refer must be served within three months of the giving of the decision, otherwise it shall be final as well as binding.

(c)   The date upon which the Notice to Refer is served shall be regarded as the date upon which the arbitral proceedings are commenced.

(8)   (a)   The arbitrator shall be a person appointed by agreement of the parties.

(b)   If the parties fail to appoint an arbitrator within 28 days of either party serving on the other party a notice in writing (hereinafter called the Notice to Concur) to concur in the appointment of an arbitrator the dispute shall be referred to a person to be appointed on the application of either party by the President for the time being of the Institution of Civil Engineers.

(c)   If an arbitrator declines the appointment or after appointment is removed by order of a competent court or is incapable of acting or dies and the parties do not within one month of the vacancy arising fill the vacancy then either party may apply to the President for the time being of the Institution of Civil Engineers to appoint another arbitrator to fill the vacancy.

(d)   In any case where the President for the time being of the Institution of Civil Engineers is not able to exercise the functions conferred on him by this Clause the said functions shall be exercised on his behalf by a Vice-President for the time being of the said Institution.

(9)   (a)   Any reference to arbitration under this Clause shall be deemed to be a submission to arbitration within the meaning of the Arbitration Act 1996 or any statutory re-enactment or amendment thereof for the time being in force. The reference shall be conducted in accordance with the procedure set out in the Second Schedule or any amendment or modification thereof being in force at the time of the appointment of the arbitrator. In the event of any inconsistency between the procedure set out in the Second Schedule and this Clause, this Clause shall prevail.

(b)   Neither party shall be limited in the arbitration to evidence or argument put to any adjudicator pursuant to sub-clause 18(4).

(c)   The award of the arbitrator shall be binding on the parties.

(d)   Unless the parties otherwise agree in writing any reference to arbitration may proceed notwithstanding that the Sub-Contract Works are not then complete or alleged to be complete.

(e)   The arbitrator shall have full power to open up, review and revise any decision, opinion, instruction, direction or valuation of the Contractor or an adjudicator.

(10) (a)   If, when a dispute in connection with the Main Contract (hereinafter called a Main Contract Dispute) is referred to a conciliator or an adjudicator under the Main Contract, and the Contractor is of the opinion that the Main Contract Dispute has any connection with the Sub-Contract Works then the Contractor may by notice in writing require that the Sub-Contractor shall as soon as is practicable provide such information and attend such meetings in connection with the Main Contract Dispute as the Contractor may request.

(b)   If a Main Contract Dispute has been referred to conciliation or adjudication under the Main Contract and the Contractor is of the opinion that the Main Contract Dispute has any connection with a dispute which is to be (but has not yet been) referred for conciliation or adjudication under this Sub-Contract (hereinafter called a Connected Dispute), the Contractor may by notice in writing require that the Connected Dispute be referred to the conciliator or adjudicator to whom the Main Contract Dispute has been referred.

(c)   If the Contractor is of the opinion that a Main Contract Dispute has any connection with a dispute in connection with the Sub-Contract (hereinafter called a Related Sub-Contract Dispute) and the Main Contract Dispute is referred to an arbitrator under the Main Contract, the Contractor may by notice in writing require that the Sub-Contractor provide such information and attend such meetings in connection with the Main Contract Dispute as the Contractor may request. The Contractor may also by notice in writing require that any Related Sub-Contract Dispute be dealt with jointly with the Main Contract Dispute and in like manner. In connection with any Related Sub-Contract Dispute the Sub-Contractor shall be bound in like manner as the Contractor by any award by an arbitrator in relation to the Main Contract Dispute.

(d)   If a dispute arises under or in connection with the Sub-Contract (hereinafter called a Sub-Contract Dispute) and the Contractor is of the opinion that the Sub-Contract Dispute raises a matter or has any connection with a matter which the Contractor wishes to refer to arbitration under the Main Contract, the Contractor may by notice in writing require that the Sub-Contract Dispute be finally determined jointly with any arbitration to be commenced in accordance with the Main Contract. In connection with the Sub-Contract Dispute, the Sub-Contractor shall be bound in like manner as the Contractor by any award by an arbitrator concerning the matter referred to arbitration under the Main Contract.

(11) All matters and information placed before a conciliator pursuant to a reference under sub-clause 18(3) shall be deemed to be submitted to him without prejudice and the conciliator shall not be called as witness by the parties or anyone claiming through them in connection with any adjudication, arbitration or other legal proceedings arising out of or connected with any matter so referred to him.

Appendix 12
GC/Works/1 With Quantities 1998

59 Adjudication

(1) The Employer or the Contractor may at any time notify the other of intention to refer a dispute, difference or question arising under, out of, or relating to, the Contract to adjudication. Within 7 Days of such notice, the dispute, may by further notice be referred to the adjudicator specified in the Abstract of Particulars.

(2) The notice of referral shall set out the principal facts and arguments relating to the dispute. Copies of all relevant documents in the possession of the party giving the notice of referral shall be enclosed with the notice. A copy of the notice and enclosures shall at the same time be sent by the party giving the notice to the PM, the QS and the other party.

(3) (a) If the person named as adjudicator in the Abstract of Particulars is unable to act, or is not or ceases to be independent of the Employer, the Contractor, the PM and the QS, he shall be substituted as provided in the Abstract of Particulars.

  (b) It shall be a condition precedent to the appointment of an adjudicator that he shall notify both parties that he will comply with this Condition and its time limits.

  (c) The adjudicator, unless already appointed, shall be appointed within 7 Days of the giving of a notice of intention to refer a dispute to adjudication under paragraph (1). The Employer and the Contractor shall jointly proceed to use all reasonable endeavours to complete the appointment of the adjudicator and named substitute adjudicator. If either or both such joint appointments has not been completed within 28 Days of the acceptance of the tender, either the Employer or the Contractor alone may proceed to complete such appointments. If it becomes necessary to substitute as adjudicator a person not named as adjudicator or substitute adjudicator in the Abstract of Particulars, the Employer and Contractor shall jointly proceed to use all reasonable endeavours to appoint the substitute adjudicator. If such joint appointment has not been made within 28 Days of the selection of the substitute adjudicator, either the Employer or Contractor alone may proceed to make such appointment. For all such appointments, the form of adjudicator's appointment prescribed by the Contract shall be used, so far as is reasonably practicable. A copy of each such appointment shall be supplied too each party. No such appointment shall be amended or replaced without the consent of both parties.

(4) The PM, the QS and the other party may submit representations to the adjudicator not later than 7 Days from the receipt of the notice of referral.

(5) The adjudicator shall notify his decision to the PM, the QS, the Employer and the Contractor not earlier than 10 and not later than 28 Days from receipt of the notice of referral, or such longer period as is agreed by the Employer and the Contractor after the dispute has been referred. The adjudicator may extend the period of 28 Days by up to 14 Days, with the consent of the party by whom the dispute was referred. The adjudicator's decision shall nevertheless be valid if issued after the time allowed. The adjudicator's decision shall state how the cost of the adjudicator's fee or salary (including overheads) shall be apportioned between the parties, and whether one party is to bear the whole or part of the reasonable legal and other costs and expenses of the other, relating to the adjudication.

(6) The adjudicator may take the initiative in ascertaining the facts and the law, and the Employer and the Contractor shall enable him to do so. In coming to a decision the adjudicator shall have regard to how far the parties have complied with any procedures in the Contract relevant to the matter in dispute and to what extent each of them has acted promptly, reasonably and in good faith. The adjudicator shall act independently and impartially, as an expert adjudicator and not as an arbitrator. The adjudicator shall have all the powers of an arbitrator acting in accordance with Condition 60 (Arbitration and choice of law), and the fullest possible powers to assess and award damages and legal and other costs and expenses; and, in addition to, and notwithstanding the terms of, Condition 47 (Finance charges), to award interest. In particular, without limitation, the adjudicator may award simple or compound interest from such dates, at such rates and with such rests as he considers meet the justice of the case–

    (a) on the whole or part of any amount awarded by him, in respect of any period up to the date of the award;

    (b) on the whole or part of any amount claimed in the adjudication proceedings and outstanding at the commencement of the adjudication proceedings but paid before the award was made, in respect of any period up to the date of payment;

and may award such interest from the date of the award (or any later date) until payment, on the outstanding amount of any award (including any award of interest and any award of damages and legal and other costs and expenses).

(7) Subject to the proviso to Condition 60(1) (Arbitration and choice of law), the decision of the adjudicator is binding until the dispute is finally determined by legal proceedings, by arbitration (if the Contract provides for arbitration, or the parties otherwise agree to arbitration), or by agreement: and the parties do not agree to accept the decision of the adjudicator as finally determining the dispute.

(8) In addition to his other powers, the adjudicator shall have power to vary or overrule any decision previously made under the Contract by the Employer; the PM or the QS, other than decisions in respect of the following matters–

    (a) decisions by or on behalf of the Employer under Condition 26 (Site admittance);

    (b) decisions by or on behalf of the Employer under Condition 27 (Passes) (if applicable);

    (c) provided that the circumstances mentioned in Condition 56(1)(a) or (b) (Determination by Employer) have arisen, and have not been waived by the Employer, decisions of the Employer to give notice under Condition 56(1)(a), or to give notice of determination under Condition 56(1);

    (d) decisions or deemed decisions of the Employer to determine the Contract under Condition 56(8) (Determination by Employer);

    (e) provided that the circumstances mentioned in Condition 58A(1) (Determination following suspension of Works) have arisen, and have not been waived by the Employer, decisions of the Employer to give notice of determination under Condition 58A(1); and

    (f) decisions of the Employer under Condition 61 (Assignment).

In relation to decisions in respect of those matters, the Contractors's only remedy against the Employer shall be financial compensation.

(9) Notwithstanding Condition 60 (Arbitration and choice of law), the Employer and the Contractor shall comply forthwith with any decision of the adjudicator; and shall submit to summary judgment and enforcement in respect of all such decisions.

(10) If requested by one of the parties to the dispute, the adjudicator shall provide reasons for his decision. Such requests shall only be made within 14 Days of the decision being notified to the requesting party.

(11) The adjudicator is not liable for anything done or omitted in the discharge or purported discharge of his functions as adjudicator, unless the act or omission is in bad faith. Any employee or agent of the adjudicator is similarly protected from liability.

MODEL FORM 8

ADJUDICATOR'S APPOINTMENT
(CONDITION 59)

THIS AGREEMENT is made the   day of   —

BETWEEN:

(1)

    of

    ('the Employer', which term shall include its successors and assignees);

(2)

    [of] OR [whose registered office is at]

    ('the Contractor'); and

(3)

    of

    ('the Adjudicator').

WHEREAS:

(A) The Employer has entered into a contract dated   ('the Contract')
     with the Contractor for the execution of certain Works, and a copy of the Contract has been
     supplied to the Adjudicator.
(B) The Adjudicator has agreed to act as [adjudicator] OR [named substitute adjudicator] in accor-
     dance with the Contract.

NOW THIS DEED WITNESSETH as follows:

1 The Adjudicator shall, as and when required, act as [adjudicator] OR [named substitute adjudi-
   cator] in accordance with the Contract, except when unable so to act because of facts or circumstances beyond his reasonable control.
2 The Adjudicator confirms that he is independent of the Employer, the Contractor, and the Project
   Manager and Quantity Surveyor under the Contract, and undertakes to use reasonable endeavours
   to remain so, and that he shall exercise his task in an impartial manner. He shall promptly inform
   the Employer and the Contractor of any facts or circumstances which may cause him to cease to be
   so independent.
3 The Adjudicator hereby notifies the Employer and the Contractor that he will comply with Condition 59 (*Adjudication*) of the Contract, and its time limits.
4 The Adjudicator shall be entitled to take independent legal and other professional advice as reasonably necessary in connection with the performance of his duties as adjudicator. The reasonable
   net cost to the Adjudicator of such advice shall constitute expenses recoverable by the Adjudicator
   under this Agreement.
5 The Adjudicator shall comply, and shall take all reasonable steps to ensure that any persons
   advising or aiding him shall comply, with the Official Secrets Act 1989 and, where appropriate, with
   the provisions of Section 11 of the Atomic Energy Act 1946. Any information concerning the
   Contract obtained either by the Adjudicator or any person advising or aiding him is confidential,

and shall not be used or disclosed by the Adjudicator or any such person except for the purposes of this Agreement.

6   The Employer and the Contractor shall pay the Adjudicator fees, expenses and other sums (if any) in accordance with the Contract and the Schedule, plus applicable Value Added Tax.

7   The Adjudicator is not liable for anything done or omitted in the discharge or purported discharge of his functions as adjudicator, unless the act or omission is in bad faith. Any employee or agent of the Adjudicator is similarly protected from liability.

8   The proper law of this Agreement shall be the same as that of the Contract. Where the proper law of this Agreement is Scots law, the parties prorogate the non-exclusive jurisdiction of the Scottish courts.

IN WITNESS whereof the Employer, the Contractor and the Adjudicator have executed this Deed in triplicate on the date first stated above.

SCHEDULE

Adjudicator's Fees, Expenses, etc.

NOTE: Where the proper law of the above document is Scots law, the format will be subject to alteration to reflect the requirements of Scots law in relation to the execution of a document.

Appendix 13
Technology and Construction Solicitors' Association 2002 Version 2.0 Procedural Rules for Adjudication

1.   The following rules

(i)   may be incorporated into any contract by reference to the "TeCSA Adjudication Rules" or the "ORSA Adjudication Rules", which expressions shall mean, in relation to any adjudication, the most recent edition hereof as at the date of the written notice requiring that adjudication.

(ii)   meet the requirements of adjudication procedure as set out in section 108 of the Housing Grants, Construction and Regeneration Act 1996; Part I of the Scheme for Construction Contracts shall thus not apply.

DEFINITIONS

2.   In these Rules:

"Contract"	means the agreement which includes the agreement to adjudicate in accordance with these Rules
"Party"	means any party to the Contract
"Chairman"	means the Chairman for the time being of the Technology and Construction Solicitors Association ("TeCSA"), or such other officer thereof as is authorised to deputise for him.
"days"	shall have the same meaning as and be calculated in accordance with Part II of the Housing Grants, Construction and Regeneration Act 1996.

COMMENCEMENT AND APPOINTMENT

3.   These Rules shall apply upon any Party giving written notice to any other Party requiring adjudication, and identifying in general terms the dispute in respect of which adjudication is required.

4.   Where the Parties have agreed upon the identity of an adjudicator who confirms his readiness and willingness to embark upon the Adjudication within 7 days of the notice requiring adjudication, then that person shall be the Adjudicator.

5.   Where the Parties have not so agreed upon an adjudicator, or where such person has not so confirmed his willingness to act, then any Party shall apply to the Chairman of TeCSA for a nomination. The following procedure shall apply:-

(i)   The application shall be in writing, accompanied by a copy of the Contract, a copy of the written notice requiring adjudication, and TeCSA's appointment fee of £100.

(ii)   The Chairman of TeCSA shall endeavour to secure the appointment of an Adjudicator within 7 days from the notice requiring adjudication.

(iii)   Any person so appointed, and not any person named in the Contract whose readiness or willingness is in question, shall be the Adjudicator.

6.   Within 7 days from the date of the Notice referred to in Rule 3:-

(i)   provided he is willing and able to act, any agreed Adjudicator under Rule 4 or nominated Adjudicator under Rule 5(ii) shall give written notice of his acceptance of appointment to all parties; and

(ii)   the referring party shall serve the Referral Notice on the Adjudicator and the Responding Party.

7.   The date of the referral of the dispute shall be the date the Referral Notice is received by the Adjudicator. The Adjudicator shall confirm to the Parties the date of receipt of the Referral Notice.

8.   The Chairman of TeCSA shall have the power by written notice to the Parties to replace the Adjudicator with another nominated person if and when it appears necessary to him to do so. The Chairman of TeCSA shall consider whether to exercise such power if any Party shall represent to him that the Adjudicator is not acting impartially, or that the Adjudicator is physically or mentally incapable of conducting the Adjudication, or that the Adjudicator is failing with necessary dispatch to proceed with the Adjudication or make his decision. In the event of a replacement under this Rule, directions and decisions of the previous Adjudicator shall remain in effect unless reviewed and replaced by the new Adjudicator, and all timescales shall be recalculated from the date of the replacement. Any replacement Adjudicator shall give written notice of acceptance of his appointment.

9.   Where an Adjudicator has already been appointed in relation to another dispute arising out of the Contract, the Chairman of TeCSA may appoint either the same or a different person as Adjudicator.

10.   Notice requiring adjudication may be given at any time and notwithstanding that arbitration or litigation has been commenced in respect of such dispute.

11.   More than one such notice requiring adjudication may be given in respect of disputes arising out of the same contract.

AGREEMENT

12.   An agreement to adjudicate in accordance with these Rules shall be treated as an offer made by each of the Parties to TeCSA and to any Adjudicator to abide by these Rules, which offer may be accepted by conduct by appointing an Adjudicator or embarking upon the Adjudication respectively.

SCOPE OF THE ADJUDICATION

13.   The scope of the Adjudication shall be the matters identified in the notice requiring adjudication, together with

(i)   any further matters which all Parties agree should be within the scope of the Adjudication, and

(ii)   any further matters which the Adjudicator determines must be included in order that the Adjudication may be effective and/or meaningful.

14.   The Adjudicator may decide upon his own substantive jurisdiction, and as to the scope of the Adjudication.

THE PURPOSE OF THE ADJUDICATION AND THE ROLE OF THE ADJUDICATOR

15.   The underlying purpose of the Adjudication is to resolve disputes between the Parties that are within the scope of the Adjudication as rapidly and economically as is reasonably possible.

16.   Unless the Parties agree that any decisions of the Adjudicator shall be final and binding, any decision of the Adjudicator shall be binding until the dispute is finally determined by legal proceedings, by arbitration (if the Contract provides for arbitration or the parties otherwise agree to arbitration) or by agreement.

17.   Wherever possible, any decision of the Adjudicator shall reflect the legal entitlements of the Parties. Where it appears to the Adjudicator impossible to reach a concluded view upon the legal entitlements of the Parties within the practical constraints of a rapid and economical adjudication process, any decision shall represent his fair and reasonable view, in light of the facts and the law insofar as they have been ascertained by the Adjudicator, of how the disputed matter should lie unless and until resolved by litigation or arbitration.

18.   The Adjudicator shall have the like power to open up and review any certificates or other things issued or made pursuant to the Contract as would an arbitrator appointed pursuant to the Contract and/or a court.

19.   The Adjudicator shall act fairly and impartially, but shall not be obliged or empowered to act as though he were an arbitrator.

CONDUCT OF THE ADJUDICATION

20.   The Adjudicator shall establish the procedure and timetable for the Adjudication.

21.   Without prejudice to the generality of Rule 20, the Adjudicator may if he thinks fit:

(i)   Require the delivery of written statements of case,

(ii)   Require any party to produce a bundle of key documents, whether helpful or otherwise to that Party's case, and to draw such inference as may seem proper from any imbalance in such bundle that may become apparent,

(iii)   Require the delivery to him and/or the other parties of copies of any documents other than documents that would be privileged from production to a court,

(iv)   Limit the length of any written or oral submission,

(v)   Require the attendance before him for questioning of any Party or employee or agent of any Party,

(vi)   Make site visits,

(vii)   Make use of hiw own specialist knowledge,

(viii) Obtain advice from specialist consultants, provided that at least one of the Parties so requests or consents,

(ix)   Meet and otherwise communicate with any Party without the presence of other Parties,

(x)   Make directions for the conduct of the Adjudication orally or in writing,

(xi)   Review and revise any of his own previous directions,

(xii)   Conduct the Adjudication inquisitorially, and take the initiative in ascertaining the facts and the law,

(xiii) Reach his decision(s) with or without holding an oral hearing, and with or without having endeavoured to facilitate an agreement between the Parties.

22.   The Adjudicator shall exercise such powers with a view of fairness and impartiality, giving each Party a reasonable opportunity, in light of the timetable, of putting his case and dealing with that of his opponents.

23.   The Adjudicator may not:

(i)   Require any advance payment of or security for his fees,

(ii)   Receive any written submissions from one Party that are not also made available to the others,

(iii) Refuse any Party the right at any hearing or meeting to be represented by any representative of that Party's choosing who is present,

(iv)   Act or continue to act in the face of a conflict of interest,

(v)   Subject to Rule 28, require any Party to pay or make contribution to the legal costs of another Party arising in the Adjudication.

24.   The Adjudicator shall reach a decision within 28 days of referral or such longer period as is agreed by the Parties after the dispute has been referred to him. The Adjudicator shall be entitled to extend the said period of 28 days by up to 14 days with the consent of the Party by whom the dispute was referred.

ADJUDICATOR'S FEES AND EXPENSES

25.   If a Party shall request Adjudication, and it is subsequently established that he is not entitled to do so, that Party shall be solely responsible for the Adjudicator's fees and expenses.

26.   Save as aforesaid, the Parties shall be jointly responsible for the Adjudicator's fees and expenses including those of any specialist consultant appointed under Rule 19(viii). In his decision, the Adjudicator shall have the discretion to make directions with regard to those fees and expenses. If no such directions are made, the Parties shall bear such fees and expenses in equal shares, and if any Party has paid more than such equal share, that Party shall be entitled to contribution from other Parties accordingly.

27.   The Adjudicator's fees shall not exceed the rate of £1,250 per day plus expenses and VAT.

COSTS

28.   If the Parties so agree, the Adjudicator shall have jurisdiction to award costs to the successful party.

29.   Notwithstanding anything to the contrary in any contract between the Parties, the Adjudicator shall have no jurisdiction to require the Party which referred the dispute to adjudication to pay the costs of any other Party solely by reason of having referred the dispute to adjudication.

DECISIONS

30.   The Adjudicator may in any decision direct the payment of such compound or simple interest as may be commercially reasonable.

31.   All decisions shall be in writing. The Adjudicator shall provide written reasons for any decision if any or all of the Parties make a request for written reasons within seven days of the date of the referral of the dispute. If requested by one Party, reasons are to be delivered to all Parties.

32.
(i)   The Adjudicator may, on his own initiative or on the application of a Party, correct his decision so as to remove any clerical mistake or error arising from an accidental slip or omission;

(ii)   Any application for the exercise of the Adjudicator's powers under paragraph (i) shall be made within 5 days of the date that the decision is delivered to the Parties or such shorter period as the Adjudicator may specify in his decision;

(iii) Any correction of a decision shall be made as soon as possible after the date that the application was received by the Adjudicator or, where the correction is made by the Adjudicator on his own initiative as soon as possible after he becomes aware of the need to make a correction.

ENFORCEMENT

33. Every decision of the Adjudicator shall be implemented without delay. The Parties shall be entitled to such reliefs and remedies as are set out in the decision, and shall be entitled to summary enforcement thereof, regardless of whether such decision is or is to be the subject of any challenge or review. No party shall be entitled to raise any right of set-off, counterclaim or abatement in connection with any enforcement proceedings.

IMMUNITY, CONFIDENTIALITY AND NON-COMPELLABILITY

34. Neither TeCSA, nor its Chairman, nor deputy, nor the Adjudicator nor any employee or agent of any of them shall be liable for anything done or not done in the discharge or purported discharge of his functions as Adjudicator, whether in negligence or otherwise, unless the act or omission is in bad faith.

35. The Adjudication and all matters arising in the course thereof are and will be kept confidential by the Parties except insofar as necessary to implement or enforce any decision of the Adjudicator or as may be required for the purpose of any subsequent proceedings.

36. In the event that any Party seeks to challenge or review any decision of the Adjudicator in any subsequent litigation or arbitration, the Adjudicator shall not be joined as a party to, nor shall be summoned or otherwise required to give evidence or provide his notes in such litigation or arbitration.

LAW

37. These Rules shall be governed by English law and under the jurisdiction of the English Courts.

38. No Party shall, save in case of bad faith on the part of the Adjudicator, make any application to the courts whatsoever in relation to the conduct of the Adjudication or the decision of the Adjudicator until such time as the Adjudicator has made his decision, or refused to make a decision, and until the Party making the application has complied with any such decision.

October 2002

These notes do not form part of the TeCSA Adjudication Rules

These rules may be incorporated into contracts, including contracts contained in correspondence, by suitable wording along the following lines:

> ''Any dispute arising under this agreement shall in the first instance be referred to adjudication in accordance with the TeCSA Adjudication Rules. [Current at the date of this Contract] [Version 2.0]''
> N.B. Delete as appropriate.

The Housing Grants, Construction and Regeneration Act 1996 gives parties to a construction contract other than with a residential occupier or an excluded contract entered into after 1st May 1998 a right to refer a dispute arising under the contract to adjudication. If the contract does not incorporate the TeCSA Adjudication Rules or other provisions meeting the compliance criteria set out in the Act, then the terms of Part I of the Scheme for Construction Contracts become applicable.

If the contract does not incorporate the TeCSA Adjudication rules or otherwise comply with the compliance criteria such that there is no agreed adjudication or nominating body, then a party in dispute may yet ask TeCSA to appoint an adjudicator; TeCSA is an ''adjudicator nominating body'' within the meaning of paragraph 2(3) of The Scheme for Construction Contracts (England and Wales) Regulations 1998. An adjudicator so appointed will conduct the adjudication in accordance with the Scheme, or if the parties so agree, the TeCSA Adjudication Rules.

If the contract contains an arbitration clause, then in order to prevent enforcement difficulties arising out of Section 9 of the Arbitration Act 1996, TeCSA recommends that the arbitration clause should contain wording along the following lines:

"Provided always that the enforcement of any decision of an adjudicator is not a matter which may be referred to arbitration."

Applications to the Chairman of TeCSA should be addressed to:

Mr Peter Rees
Technology and Construction Solicitors Association
Kempson House
Camomile Street
London EC3A 7AN
Tel: 020 7283 6000
Fax: 020 7283 6500

TeCSA and its members take no responsibility for loss or damage caused to any user of these Rules or these Notes.

Appendix 14
Centre for Effective Dispute Resolution (CEDR) Rules for Adjudication

Commencing adjudication and nomination of the Adjudicator

1 A party to a contract (the 'Referring Party') may at any time give notice (the 'Notice') in writing to the other Party(ies) of its intention to refer a dispute arising under, out of, or in connection with, the contract to adjudication.

   The Notice shall contain:
   - names, addresses and full contact details of the Parties and of any representatives appointed by the Parties
   - a copy of the relevant provisions of the contract providing for adjudication
   - brief details of the dispute to be referred to adjudication
   - details of remedy sought.

   A copy of the Notice shall be sent by the Referring Party to the Adjudicator, if named in the contract, at the same time as it is sent to the other Party(ies). The Adjudicator shall, within 2 days of receiving the Notice, confirm in writing to the Parties that he or she is available to act.

2 If no Adjudicator is named in the contract, or if the named Adjudicator does not confirm his or her availability to act, then the Referring Party shall immediately apply to the Centre for Effective Dispute Resolution ('CEDR Solve') using CEDR Solve's application form to nominate an Adjudicator. CEDR Solve shall nominate an Adjudicator and communicate the nomination to all the Parties within 5 days of receipt of:
   - the completed application form
   - a copy of the Notice of Adjudication
   - CEDR Solve's nomination fee.

3 The Adjudicator shall, within 24 hours of receipt of the nomination, confirm in writing to the Parties that he or she is available to act, whether in response to receiving the Notice or to a nomination by CEDR Solve. The Adjudicator shall provide to them, at the same time, a copy of the terms on which he or she is prepared to act including information regarding fees and expenses.

Conduct of the adjudication

4 Within 7 days of the giving of the Notice, the Referring Party shall send to the Adjudicator, copied at the same time and by the same method to the other Party(ies), a concise statement of case which shall include:
   - a copy of the Notice
   - a copy of the conditions of the contract and other provisions in the contract on which the Referring Party intends to rely
   - details of the circumstances giving rise to the dispute
   - the reasons for entitlement to the remedy sought
   - the evidence, including relevant documentation, in support of its case.

5   Under these Rules the date of referral is the date on which both the Adjudicator and the other Party(ies) receive the concise statement of case from the Referring Party.

6   The Adjudicator shall reach a decision within 28 days of the date of referral. Subject to the Adjudicator's agreement, this period may be extended by 14 days with the consent of the Referring Party or longer if agreed by all the Parties.

7   The Adjudicator may take the initiative in ascertaining the facts and the law.

8   The Adjudicator shall establish the timetable and procedure for the adjudication which may include the consideration of:
    - the extent, form and time limits applying to any documentary or oral submission of the Parties
    - site visits or inspections
    - meeting the Parties
    - issuing particular directions
    - the appointment of an Expert or Assessor subject to paragraph 13 of these rules.

9   Copies of all documents submitted by a Party to the Adjudicator shall be sent simultaneously and by the same method to the other Party(ies). Similarly, all documents issued by the Adjudicator shall be sent simultaneously to the Parties.

10 The Adjudicator shall not take into consideration any document or statement, whether of a Party or Witness, that has not been made available to the other Party(ies) for comment.

11 Any failure by any Party to respond to any request or direction by the Adjudicator shall not invalidate the adjudication or the Adjudicator's decision.

12 A party may at any time request additional Parties to be joined in the adjudication. Joinder of additional Parties shall be subject to the agreement of the Adjudicator, the existing Parties and additional Parties.

13 The Adjudicator may, at any time, obtain legal or technical advice on any matter provided that the Parties are informed with reasons beforehand. Prior to making the decision, the Adjudicator shall provide the Parties with copies of any written advice so obtained.

Decision of the Adjudicator

14 The Adjudicator shall decide the dispute acting impartially and in good faith. The Adjudicator shall have the power to open up, review and revise any certificate, decision, direction, instruction, notice, requirement or valuation made under the contract to which the dispute relates except where the contract precludes this.

15 The Adjudicator may decide any other matters which the Adjudicator determines should be taken into account in deciding the dispute.

16 The Adjudicator shall reach a decision and communicate the decision in writing to the Parties in accordance with the time limits in paragraph 6.

17 The Adjudicator shall give reasons with the decision unless the Parties agree to the contrary.

18 The Adjudicator may, on his or her own initiative or at the request of a Party made within 5 days of the date that the decision is communicated to the Parties, correct the decision in respect of any typographical or arithmetical error as a result of an accidental slip or omission.

19 The Adjudicator's decision shall be binding unless or until the dispute is finally determined by agreement, court proceedings or by reference to arbitration in accordance with the contract. Unless otherwise agreed by the Parties, the Court, or the Arbitrator(s) shall not be bound by the Adjudicator's decision.

Enforcement

20 The Parties shall implement the Adjudicator's decision without delay and shall be entitled to such relief or remedies as are set out in the decision.

21 Any payment to be made in accordance with the Adjudicator's decision shall be paid in full without the paying Party(ies) having a right of set-off, counterclaim or abatement.

Cost of the parties

22 Each Party shall bear its own costs. The Adjudicator may not decide the Parties' legal and other costs arising out of or in connection with the adjudication unless the Parties otherwise agree.

Fees and expenses of the Adjudicator

23 The Parties shall be jointly and severally responsible for the Adjudicator's fees and expenses including the fees and expenses of any legal or technical adviser instructed under paragraph 13.

24 In the decision, the Adjudicator shall have discretion to apportion liability with regard to the Adjudicator's fees and expenses referred to in paragraph 23.

Resignation of the Adjudicator

25 The Adjudicator shall resign if:
- the dispute has already been referred to Adjudication and a decision has been made
- the Adjudicator is not competent to decide because the nature of the dispute is significantly different to the dispute referred in the Notice, or
- the Adjudicator becomes unable to give a decision in accordance with the timescales set out in paragraph 6.

26 The Adjudicator shall notify the Parties of his or her resignation in writing and the Parties shall be liable for the Adjudicator's fees and expenses up to the date of resignation in accordance with paragraph 23.

Mediation

27 At any time before the issue of the Adjudicator's decision the Parties may agree to refer the dispute to mediation. In that case each Party shall notify the Adjudicator in writing and the adjudication shall be suspended. The time in which the Adjudicator must decide the dispute shall be extended by the period of suspension.

28 If the Parties are unable to agree a Mediator within 7 days from the date they agree to refer the dispute to mediation then any Party may apply to CEDR Solve to nominate the mediator.

29 The Adjudicator shall not act as the Mediator and shall not take part in any such mediation.

30 If the dispute is settled by mediation, the adjudication shall be at an end and the Parties shall promptly settle the Adjudicator's fees and expenses referred to in paragraph 23. If a settlement is not reached within 28 days from the date on which the Parties agree to refer the dispute to mediation, or if at any time a Party abandons the mediation, the adjudication shall recommence on written confirmation to the Parties by the Adjudicator that he or she is able to continue pursuant to a request in writing by any Party.

Other provisions

31 If at any time after the date of referral the Adjudicator is unable or unwilling to act or fails to reach a decision in accordance with the time limits in paragraph 6, a Party may apply to CEDR Solve to mominate a replacement Adjudicator.

32 The Adjudicator shall not be liable for anything done or omitted in the discharge of his or her functions unless the act or omission was in bad faith. The same immunity shall extend to CEDR Solve as the Adjudicator Nominating Body and any employee or agent of the Adjudicator or CEDR Solve.

33 The Adjudicator's decision may not be relied upon by third parties to whom the Adjudicator shall owe no duty of care.

Law

34 These rules shall be governed by English Law and under the jurisdiction of the English Courts.

Appendix 15

Construction Industry Council (CIC) Model Adjudication Procedure: Third Edition

General Principles

1. The object of adjudication is to reach a fair, rapid and inexpensive decision upon a dispute arising under the Contract and this procedure shall be interpreted accordingly.

   Object

2. The Adjudicator shall act impartially.

   Impartiality

3. The Adjudicator may take the initiative in ascertaining the facts and the law. He may use his own knowledge and experience. The adjudication shall be neither an arbitration nor an expert determination.

   The Adjudicator's role

4. The Adjudicator's decision shall be binding until the dispute is finally determined by legal proceedings, by arbitration (if the contract provides for arbitration or the parties otherwise agree to arbitration) or by agreement.

   Decision binding in interim

5. The Parties shall implement the Adjudicator's decision without delay whether or not the dispute is to be referred to legal proceedings or arbitration.

   Implementation of the decision

Application

6. If this procedure is incorporated into the Contract by reference, the reference shall be deemed to be to the edition current at the date of the Notice, unless expressly stated otherwise in the Contract.

   Application

7. If a conflict arises between this procedure and the Contract, unless the Contract provides otherwise, this procedure shall prevail.

   Conflict

Appointment of the Adjudicator

8. Either Party may give notice at any time of its intention to refer a dispute arising under the Contract to adjudication by giving a written Notice to the other Party. The Notice shall include a brief statement of the issues or issues which it is desired to refer and the redress sought. The referring Party shall send a copy of the Notice to any adjudicator named in the Contract.

   Notice of adjudication

9. The object of the procedure in paragraphs 10–14 is to secure the appointment of the Adjudicator and referral of the dispute to him within 7 days of the giving of the Notice.

   Time for appointment and referral

10. If an adjudicator is named in the Contract, he shall within 2 days of receiving the Notice indicate whether or not he is willing to act. If no adjudicator is named, or if

    Appointment

the named adjudicator does not indicate that he is willing to act, the referring Party shall request the body stated in the Contract if any, or if none the Construction Industry Council, to nominate an adjudicator within 5 days of receipt of the request. The request shall be in writing, accompanied by a copy of the Notice and the appropriate fee. Alternatively the Parties may, within 2 days of the giving of the Notice, appoint an adjudicator by agreement.

Adjudicator unable to act

11. If, for any reason, the Adjudicator is unwilling to act, or fails to reach his decision within the time required by this procedure, either Party may request the body stated in the Contract if any, or if none the Construction Industry Council, to nominate a replacement adjudicator. No such request may be made after the Adjudicator has notified the Parties that he has reached his decision.

Adjudicator's terms and conditions

12. Unless the Parties and the Adjudicator otherwise agree, the Adjudicator shall be appointed on the terms and conditions set out in the attached Agreement and shall be entitled to a reasonable fee and expenses.

Objection to appointment

13. If a Party objects to the appointment of a particular person as adjudicator, that objection shall not invalidate the Adjudicator's appointment or any decision he may reach.

Conduct of the Adjudication

Statement of case

14. The referring Party shall send to the Adjudicator within 7 days of the giving of the Notice (or as soon thereafter as the Adjudicator is appointed), and at the same time copy to the other Party, a statement of its case including a copy of the Notice, the Contract, details of the circumstances giving rise to the dispute, the reasons why it is entitled to the redress sought, and the evidence upon which it relies. The statement of case shall be confined to the issues raised in the Notice.

Date of referral

15. The date of referral shall be the date on which the Adjudicator receives this statement of case.

Period for decision

16. The Adjudicator shall reach his decision within 28 days of the date of referral, or such longer period as is agreed by the Parties after the dispute has been referred. The Adjudicator may extend the period of 28 days by up to 14 days with the consent of the referring Party.

Procedure

17. The Adjudicator shall have complete discretion as to how to conduct the adjudication, and shall establish the procedure and timetable, subject to any limitation there may be in the Contract or the Act. He shall not be required to observe any rule of evidence, procedure or otherwise, of any court or tribunal. Without prejudice to the generality of these powers he may:
    (i) request a written response, further argument or counter argument;
    (ii) request the production of documents or the attendance of people whom he considers could assist;
    (iii) visit the site;
    (iv) meet and question the Parties and their representatives;
    (v) meet the parties separately;
    (vi) limit the length or time for submission of any statement, response or argument;
    (vii) proceed with the adjudication and reach a decision even if a Party fails to comply with a request or direction of the Adjudicator;
    (viii) issue such further directions as he considers to be appropriate.

18. The Parties shall comply with any request or direction of the Adjudicator in relation to the adjudication.

Parties to comply

19. The Adjudicator may obtain legal or technical advice, provided that he has notified the Parties of his intention first. He shall provide the Parties with copies of any written advice received.

Obtaining advice

20. The Adjudicator shall decide the matters set out in the Notice, together with any other matters which the Parties and the Adjudicator agree shall be within the scope of the adjudication.

Matters to be determined

21. The Adjudicator shall determine the rights and obligations of the Parties in accordance with the law of the Contract.

Adjudicator to apply the law

22. Any Party may at any time ask that additional parties shall be joined in the adjudication. Joinder of additional parties shall be subject to the agreement of the Adjudicator and the existing and additional parties. An additional party shall have the same rights and obligations as the other Parties, unless otherwise agreed by the Adjudicator and the Parties.

Joining third parties

23. The Adjudicator may resign at any time on giving notice in writing to the Parties.

Resignation

The Decision

24. The Adjudicator shall reach his decision within the time limits in paragraph 16. He shall be required to give reasons unless both Parties agree at any time that he shall not be required to give reasons.

The decision

25. If the adjudicator fails to reach his decision within the time permitted by this procedure, his decision shall nonetheless be effective if reached before the referral of the dispute to any replacement adjudicator under paragraph 11 but not otherwise. If he fails to reach such an effective decision, he shall not be entitled to any fees or expenses (save for the cost of any legal or technical advice subject to the Parties having received such advice).

Late decisions

26. The Adjudicator may open up, review and revise any certificate, decision, direction, instruction, notice, opinion, requirement or valuation made in relation to the Contract.

Power to open up certificates

27. The Adjudicator may in any decision direct the payment of such simple or compound interest from such dates, at such rates and with such rests, as he considers appropriate.

Interest

28. The Adjudicator may, within 5 days of delivery of the decision to the Parties, correct his decision so as to remove any error arising from an accidental error or omission or to clarify or remove any ambiguity.

Correction of errors

29. The Parties shall bear their own costs and expenses incurred in the adjudication.

Costs

30. The Parties shall be jointly and severally liable for the Adjudicator's fees and expenses, including those of any legal or technical adviser appointed under paragraph 19, but the Adjudicator may direct a Party to pay all or part of the fees and expenses. If he makes no such direction, the Parties shall pay them in equal shares. The Party requesting the adjudication shall be liable for the Adjudicator's fees and expenses if the adjudication does not proceed.

Adjudicator's fees and expenses

Enforcement   31. The Parties shall be entitled to the redress set out in the decision and to seek summary enforcement, whether or not the dispute is to be finally determined by legal proceedings or arbitration. No issue decided by the Adjudicator may subsequently be referred for decision by another adjudicator unless so agreed by the Parties.

Subsequent decision by arbitration or court   32. In the event that the dispute is referred to legal proceedings or arbitration, the Adjudicator's decision shall not inhibit the right of the court or arbitrator to determine the Parties' rights or obligations as if no adjudication had taken place.

Miscellaneous Provisions

Adjudicator not to be appointed arbitrator   33. Unless the Parties agree, the Adjudicator shall not be appointed arbitrator in any subsequent arbitration between the Parties under the Contract. No Party may call the Adjudicator as a witness in any legal proceedings or arbitration concerning the subject matter of the adjudication.

Immunity of the Adjudicator   34. The Adjudicator is not liable for anything done or omitted in the discharge or purported discharge of his functions as adjudicator (whether in negligence or otherwise) unless the act or omission is in bad faith, and any employee or agent of the Adjudicator is similarly protected from liability.

Reliance   35. The Adjudicator is appointed to determine the dispute or disputes between the Parties and his decision may not be relied upon by third parties, to whom he shall owe no duty of care.

Proper law   36. This procedure shall be interpreted in accordance with the law of England and Wales.

Definitions

'Act' means the Housing Grants, Construction and Regeneration Act 1996.

'Adjudicator' means the person named as such in the Contract or appointed in accordance with this procedure.

'Contract' means the contract between the Parties which contains the provision for adjudication.

'Notice' means the notice given under paragraph 8.

'Party' means a party to the Contract, and any additional parties joined under paragraph 22 and 'referring Party' means the Party who gives notice under paragraph 8.

AGREEMENT

This Agreement is made on the . . . . day of . . . . 20 . . . .

Between

1. . . . .

   of . . . .

   . . . . (the referring Party)

2. . . . .

   of . . . .

   . . . . (the other Party)

   (together called the Parties) and

2. . . . .

   of . . . .

   . . . . (the Adjudicator)

A dispute has arisen between the Parties under a contract between them dated . . .

in connection with . . . .

which has been referred to adjudication in accordance with the CIC Model Adjudication Procedure (the Procedure) and the Adjudicator has been requested to act.

The Parties and the Adjudicator agree that their rights and obligations shall be as set out in and subject to the terms of this Agreement:

1. The adjudication shall be conducted in accordance with the Procedure.

2. The Parties shall be jointly and severally liable to pay the Adjudicator's fees and expenses as set out in the schedule below and in accordance with the Procedure.

3. The Adjudicator and the Parties shall keep the adjudication confidential, except so far as is necessary to enable a Party to implement or enforce the Adjudicator's decision.

4. The Adjudicator may destroy all documents received during the course of the adjudication six months after delivering his decision, provided that he shall give the parties 14 days notice of his intention to do so and that he shall return the documents to the Parties if they so request.

5. The Adjudicator shall not be liable for anything done or omitted in the discharge or purported discharge of his functions as adjudicator (whether in negligence or otherwise) unless the act or omission is in bad faith, and any employee or agent of the Adjudicator shall be similarly protected from liability.

6. This Agreement shall be interpreted in accordance with the law of England and Wales.

Schedule

1. The Adjudicator shall be paid £........ per hour in respect of all time spent on the adjudication, including travelling time, with a maximum of £........ per day.

2. The Adjudicator shall be reimbursed his reasonable expenses and disbursements in respect of the cost of legal or technical advice obtained in accordance with the Procedure, travelling, hotel and similar expenses, room charges and other extraordinary expenses necessarily incurred.

3. The Adjudicator is / is not * currently registered for VAT (where the Adjudicator is registered for VAT, it shall be payable in accordance with the rates current at the date the work is done).

* delete as applicable.

Signed on behalf of the referring Party

. . . .

Signed on behalf of the other Party

. . . .

Signed on behalf of the Adjudicator

. . . .

Appendix 16
Association of Consulting Engineers
Conditions of Engagement

B9 DISPUTES AND DIFFERENCES

Mediation

9.1 The parties shall attempt in good faith to settle any dispute by mediation.

Adjudication

9.2 Where this Agreement is a construction contract within the meaning of the Housing Grants, Construction and Regeneration Act 1996 either party may refer any dispute arising under this Agreement to adjudication in accordance with the Construction Industry Council Model Adjudication Procedure.

Appendix 17
Institution of Chemical Engineers Red Book

45. Disputes

See Guide Note Q (Dispute resolution)

45.1 The **Purchaser** and the **Contractor** shall use all reasonable endeavours to avoid disputes both between themselves and with third parties including, but not limited to, **Subcontractors**.

45.2 In order to avoid the development of disputes and to facilitate their clear definition and early resolution, the procedures set out in Clauses 45 (Disputes), 46 (Adjudication), 47 (Reference to an Expert) and 48 (Arbitration) shall be applied as appropriate. The **Purchaser** and the **Contractor** undertake that they shall use reasonable endeavours to avoid the escalation of problems into 'disputes' as defined in Sub-clause 45.4. However the parties acknowledge and agree that this undertaking shall not prejudice either party's rights under Clause 46 to refer any dispute or difference to adjudication at any time, if that party so wishes, notwithstanding the definition of 'dispute' in Sub-clause 45.4.

45.3 If the **Contractor** is dissatisfied with any **Decision** or valuation of the **Project Manager**, or of any person to whom the **Project Manager** may have delegated any of his authority or responsibility, or if the **Purchaser** or the **Contractor** is dissatisfied with any other matter arising under or in connection with the **Contract**, either party may at any time refer such dissatisfaction to the **Project Manager** giving full details of the nature of the matter. The **Project Manager** shall give a written decision on the matter (giving the reasons for such decision) to the **Purchaser**, the **Contractor** and the **Contract Manager** within twenty-eight days of such reference to him.

45.4 The **Purchaser** and the **Contractor** agree that no matter shall constitute, nor be said to give rise to, a dispute, which shall include any difference, unless the same has been referred to the **Project Manager** under Sub-clause 45.3 and:

(a) the **Project Manager** has failed to give his decision on the said matter within the prescribed time; or

(b) a decision given within the prescribed time is either unacceptable to the **Purchaser** and/or the **Contractor** or has not been implemented within twenty-one days of the said decision;

and, as a consequence, either the **Purchaser** or the **Contractor** has served notice setting out the nature of the dispute (hereinafter called a 'Notice of Dispute') on the other (and on the **Project Manager**). For the purposes of the performance of the **Works** and all matters arising out of or in connection with the **Contract**, the word 'dispute' shall be construed in accordance with this Sub-clause 45.4.

45.5 Notwithstanding the existence of any dispute or any reference to the **Project Manager** under Sub-clause 45.3, the **Purchaser** and the **Contractor** shall continue to perform their obligations under the **Contract**.

45.6 Subject to any other provisions of the **Contract** the parties shall attempt to negotiate a settlement of any dispute in good faith.

45.7 If a dispute cannot be resolved by negotiation the parties may by agreement refer it to mediation in accordance with the procedures of the Centre for Dispute Resolution (CEDR) or some other body.

45.8 No **Decision**, opinion, direction or valuation given by the **Project Manager** shall disqualify him from being called as a witness and giving evidence before a third party, an **Expert**, adjudicator or arbitrator on any matter whatsoever relating to a dispute.

46. Adjudication

See Guide Note Q (Dispute resolution)

46.1 This Clause 46 shall only apply to disputes under a construction contract as defined in the Housing Grants, Construction and Regeneration Act 1996, or any amendment or re-enactment thereof.

46.2 Notwithstanding any provision in these General Conditions for a dispute to be referred to an **Expert** in accordance with Clause 47 (Reference to an Expert) or to Arbitration in accordance with Clause 48 (Arbitration), either party shall have the right to refer any dispute or difference (including any matter not referred to the **Project Manager** in accordance with Sub-clause 45.3) as to a matter under or in connection with the **Contract** to adjudication and either party may, at any time, give notice in writing to the other of his intention to do so (hereinafter called a 'Notice of Adjudication'). The ensuing adjudication shall be conducted in accordance with the edition of the 'Adjudication Rules' (the 'Rules') published by IChemE current at the time of service of the Notice of Adjudication.

46.3 Unless the adjudicator has already been appointed, he is to be appointed to a timetable with the object of securing his appointment within seven days of the service of the Notice of Adjudication. The appointment of the adjudicator shall be effected in accordance with the Rules.

46.4 The adjudicator shall reach his **Decision** within twenty-eight days of referral or such other longer period as may be agreed between the parties after the dispute has been referred.

46.5 The adjudicator may extend the period of twenty-eight days by up to fourteen days with the consent of the party by whom the dispute was referred.

46.6 The adjudicator shall act impartially.

46.7 The adjudicator may take the initiative in ascertaining the facts and the law.

46.8 The decision of the adjudicator shall be binding until the dispute is finally determined by legal proceedings, by arbitration or by agreement.

46.9 The adjudicator shall not be liable for anything done or omitted in the discharge or purported discharge of his functions as adjudicator unless the act or omission is in bad faith. Furthermore, any employee or agent of the adjudicator acting in connection with the carrying out of the adjudication shall be similarly protected from liability.

Guide Note Q (Dispute resolution)

Adjudication

If a review of the Works and the Plant leads to the conclusion that the Contractor will be required under the Contract to undertake activities which are defined as construction operations by the Housing Grants, Construction and Regeneration Act 1996 Part II, Clause 46 will apply to disputes arising from those activities, as will the IChemE Adjudication Rules. Whilst it is hoped that parties will generally follow Sub-clause 45.4, Sub-clause 46.2 makes it clear that for activities which do fall under the Act either party may call for adjudication at any time without first formally establishing a 'Dispute'.

Clause 46 has been modelled to correspond with Clause 108 of the Housing Grants, Construction and Regeneration Act 1996 under which an adjudicator's decision is binding until the dispute is finally determined by legal proceedings, by arbitration or by agreement. Sub-clause 46.8 of the Contract mirrors section 108(3) of the Act.

While there is, as yet, no judicial authority on the point, it is arguable that adjudication under the Act can proceed even if an Expert Determination is also in hand. If so, it would appear that an Expert's finding given before an adjudicator has given his decision will stay the adjudication, while a finding given after a decision will over-ride the latter. But it would be far better to avoid the problem by choosing one approach or the other at the earliest possible stage.

Appendix 18
Adjudication cases

The majority of the following cases appear on the www.adjudication.co.uk website, in most cases with a link to a copy of the full judgment.

*Cases that do not appear on adjudication.co.uk, as available on 15 December 2003.

No	Case	Date of judgment
1	Macob Civil Engineering Ltd v. Morrison Construction Ltd	12 February 1999
2	Rentokil Ailsa Environmental Ltd v. Eastend Civil Engineering Ltd	12 March 1999
3	Outwing Construction Ltd v. H. Randall & Son Ltd	15 March 1999
4	A. Straume (UK) Ltd v. Bradlor Developments Ltd	7 April 1999
5	A & D Maintenance & Construction Ltd v. Pagehurst Construction Services Ltd	23 June 1999
6	Allied London & Scottish Properties Plc v. Riverbrae Construction Ltd	12 July 1999
7	The Project Consultancy Group v. The Trustees of the Gray Trust	16 July 1999
8	John Cothliff Ltd v. Allen Build (North West) Ltd	29 July 1999
9	Palmers Ltd v. ABB Power Construction Ltd	6 August 1999
10	Lathom Construction Ltd v. (1) Brian Cross (2) Anne Cross	29 October 1999
11(1)	Homer Burgess Ltd v. Chirex (Annan) Ltd	10 November 1999
11(2)	Homer Burgess Ltd v. Chirex (Annan) Ltd	18 November 1999
12	Bouygues (UK) Ltd v. Dahl Jensen (UK) Ltd	17 November 1999
13	Sherwood & Casson Ltd v. Mackenzie	30 November 1999
14	Fastrack Construction Ltd v. Morrison Construction Ltd and Another	4 January 2000
15	VHE Construction Plc v. RBSTB Trust Co Ltd	13 January 2000
16	Northern Developments (Cumbria) Ltd v. J & J Nichol	24 January 2000
17	Samuel Thomas Construction Ltd v. Bick & Bick [aka J & B Developments]	28 January 2000
18	F.W. Cook Ltd v. Shimizu (UK) Ltd	4 February 2000
19	Workplace Technologies Plc v. E Squared Limited and Mr J.L. Riches	16 February 2000
20	The Atlas Ceiling & Partitions Co Ltd v. Crowngate Estates (Cheltenham) Ltd	18 February 2000
21	Nolan Davis Ltd v. Steven P. Catton	22 February 2000
22	Grovedeck Limited v. Capital Demolition Ltd	24 February 2000
23	Bloor Construction (UK) Ltd v. Bowmer & Kirkland (London) Ltd	5 April 2000
24	Bridgeway Construction Ltd v. Tolent Construction Ltd	11 April 2000
25	Tim Butler Contractors Ltd v. Merewood Homes Ltd	12 April 2000
26	Herschel Engineering Ltd v. Breen Property Ltd	14 April 2000
27	Nordot Engineering Services Ltd v. Siemens plc	14 April 2000
28	Edmund Nuttall Ltd v. Sevenoaks District Council	14 April 2000
29	Strathmore Building Services Ltd v. Colin Scott Greig t/a Hestia Fireside Design	18 May 2000
30	John Mowlem & Company Plc v. Hydra-Tight Ltd (t/a Hevilifts)	6 June 2000

31	Christiani & Nielsen Ltd v. The Lowry Centre Development Company Ltd	16 June 2000
32	R.G. Carter Ltd v. Edmund Nuttall Ltd	21 June 2000
33	Stiell Ltd v. Riema Control Systems Ltd	23 June 2000
34	Nottingham Community Housing Association Ltd v. Powerminster Ltd	30 June 2000
35	KNS Industrial Services (Birmingham) Ltd v. Sindall Ltd	14 July 2000
36	Absolute Rentals Ltd v. Gencor Enterprises Ltd	16 July 2000
37	Ken Griffin & John Tomlinson T/a K & D Contractors v. Midas Homes Ltd	21 July 2000
38	George Parke v. The Fenton Gretton Partnership	24 July 2000
39	Shepherd Construction Ltd v. Mecright Ltd	27 July 2000
40	Herschel Engineering Ltd v. Breen Properties Ltd (No 2)	28 July 2000
41	Bouygues (UK) Ltd v. Dahl Jensen Ltd (CA)	31 July 2000
42	ABB Power Construction Ltd v. Norwest Holst Engineering Ltd	1 August 2000
43	Whiteways Contractors (Sussex) Ltd v. Impresa Castelli Construction UK Ltd	9 August 2000
44	Discain Project Services Ltd v. Opecprime Developments Ltd	9 August 2000
45	Universal Music Operations Ltd v. Flairnote Ltd and Others	24 August 2000
46	Elanay Contracts Ltd v. The Vestry	30 August 2000
47	*Cygnet Health Care v. Higgins City Ltd (TCC – HT 00/285)	7 September 2000
48	Woods Hardwick Ltd v. Chiltern Air Conditioning Ltd	2 October 2000
49	Maymac Environmental Services Ltd v. Faraday Building Services Ltd	16 October 2000
50	Canary Riverside Development v. Timtec International	9 November 2000
51	Harwood Construction Ltd v. Lantrode	24 November 2000
52	ABB Zantingh Ltd v. Zedal Building Services Ltd	12 December 2000
53	Karl Construction (Scotland) Ltd v. Sweeney Civil Engineering (Scotland) Ltd	21 December 2000
54	A.J. Brenton t/a Manton Electrical Components v. Jack Palmer	19 January 2001
55	LPL Electrical Services Ltd v. Kershaw Mechanical Services Ltd	2 February 2001
56	Glencot Development & Design Co Ltd v. Ben Barrett & Son (Contractors) Ltd	13 February 2001
57	Rainford House Ltd (In Admin Rec) v. Cadogan Ltd	13 February 2001
58	Staveley Industries Plc (t/a EI.WHS) v. Odebrecht Oil & Gas Services Ltd	28 February 2001
59	Holt Insulation Ltd v. Colt International Ltd	
60	Joseph Finney Plc v. Gordon Vickers & Gary Vickers t/a Mill Hotel (A Firm)	7 March 2001
61	Watson Building Services Ltd (Judicial Review)	13 March 2001
62	Austin Hall Building Ltd v. Buckland Securities Ltd	11 April 2001
63	Discain Project Services Ltd v. Opecprime Developments Ltd No 2	11 April 2001
64	Farebrother Building Services Ltd v. Frogmore Investments Ltd	20 April 2001
65	RJT Consulting Engineers Ltd v. DM Engineering (NI) Ltd	9 May 2001
66	Re A Company (Number 1299 of 2001)	15 May 2001
67	Faithful & Gould v. Arcal Limited and Others	25 May 2001
68	Fence Gate Ltd v. James R. Knowles Ltd	31 May 2001
69	Mitsui Babcock Energy Services Ltd (Judicial Review)	13 June 2001
70	Barr Ltd v. Law Mining Ltd	15 June 2001
71	Sindall Ltd v. Solland and Others	15 June 2001
72	Ballast Plc v. The Burrell Company (Construction Management) Ltd	21 June 2001
73	Mecright Ltd v. T.A. Morris Developments Ltd	22 June 2001

74	C & B Scene Concept Design Ltd v. Isobars Ltd	June 2001
75	William Naylor, t/a Powerfloated Concrete Floors v. Greenacres Curling Ltd	26 June 2001
76	Bickerton Construction Ltd v. Temple Windows Ltd	26 June 2001
77	SL Timber Systems Ltd v. Carillion Construction Ltd	27 June 2001
78	British Waterways Board (Judicial Review)	5 July 2001
79	Barrie Green v. GW Integrated Building Services Ltd and G&M Floorlayers (Derby) Ltd	10 July 2001
80	City Inn Ltd v. Shepherd Construction Ltd	17 July 2001
81	Gibson Lea Retail Interiors Ltd v. Makro Self Service Wholesalers Ltd	24 July 2001
82	David McLean Housing Contractors Ltd v. Swansea Housing Association Ltd	27 July 2001
83	Yarm Road Ltd v. Costain Ltd	30 July 2001
84	Millers Specialist Joinery Co Ltd v. Nobles Construction Ltd	3 August 2001
85	Maxi Construction Management Limited v. Mortons Rolls Limited	7 August 2001
86	Stubbs Rich Architects v. W.H. Tolley & Son Ltd	8 August 2001
87	Parsons Plastics (Research & Development) Ltd v. Purac Ltd	13 August 2001
88	Britcon (Scunthorpe) Ltd v. Lincolnfields Ltd	29 August 2001
89	Durabella Ltd v. Jarvis & Sons Ltd	19 September 2001
90	Pro-Design Ltd v. New Millenium Experience Company Ltd	26 September 2001
91	Paul Jensen Ltd v. Stavely Industries Plc	27 September 2001
92	Sir Robert McAlpine v. Pring & St Hill Ltd	2 October 2001
93	Oakley and Another v. Airclear Environmental Ltd and Another	4 October 2001
94	Jerome Engineering Ltd v. Lloyd Morris Electrical Ltd	23 November 2001
95	Isovel Contracts Ltd (in administration) v. ABB Technologies Ltd	30 November 2001
96	Watkin Jones & Son Ltd v. Lidl UK GmbH	27 December 2001
97	Clark Contracts Ltd v. The Burrell Co (Construction Management) Ltd	January 2002
98	Watkin Jones & Son Limited v. Lidl UK GMBH (No 2)	11 January 2002
99	Shimizu Europe Ltd v. Automajor Ltd	17 January 2002
100	Karl Construction (Scotland) Ltd v. Sweeney Civil Engineering (Scotland) Ltd (Appeal)	22 January 2002
101	C&B Scene Concept Design Ltd v. Isobars Ltd (CA)	31 January 2002
102	Quality Street Properties (Trading) Ltd v. Elmwood (Glasgow) Ltd	8 February 2002
103	Earls Terrace Properties Limited v. Waterloo Investments Limited	14 February 2002
104	Solland International Ltd v. Daraydan Holdings Ltd	15 February 2002
105	Ashley House Plc v. Galliers Southern Ltd	15 February 2002
106	Total M & E Services Ltd v. ABB Technologies Ltd	26 February 2002
107	Gibson v. Imperial Homes	27 February 2002
108	Parsons Plastics (Research & Development) Ltd v. Purac Ltd (CA)	28 February 2002
109	RJT Consulting Engineers Ltd. v. DM Engineering (Northern Ireland) Ltd (CA)	8 March 2002
110	Edmund Nuttall Ltd v. R. G. Carter Ltd	21 March 2002
111	Chamberlain Carpentry & Joinery Ltd v. Alfred McAlpine Construction Ltd	25 March 2002
112	Fab-Tek Engineering Ltd v. Carillion Construction Ltd	March 2002
113	Balfour Beatty Construction Ltd v. The Mayor & Burgesses of the London Borough of Lambeth	12 April 2002
114	Martin Girt v. Page Bentley	12 April 2002
115	R.G. Carter Ltd v. Edmund Nuttall Ltd	18 April 2002
116	Impresa Castelli SpA v. Cola Holdings Ltd	2 May 2002
117	Barry D. Trentham Ltd v. Lawfield Investments Ltd	3 May 2002

118	Va Tech Wabag UK Limited v. Morgan Est (Scotland) Ltd	30 May 2002
119	Sim Group Limited v. Neil Jack and Others	5 June 2002
120	Levolux AT Ltd v. Ferson Contractors Ltd	26 June 2002
121	Gillies Ramsay Diamond v. PJW Enterprises Ltd	27 June 2002
122	J.T. Mackley & Company Ltd v. Gosport Marina Ltd	3 July 2002
123	Pring & St Hill Limited v. CJ Hafner t/a Southern Erectors	31 July 2002
124	Petition of Edinburgh Royal Joint Venture	2 August 2002
125	Hitec Power Protection BV v. MCI WorldCom Ltd	15 August 2002
126	Hart Builders (Edinburgh) Ltd v. St Andrew Ltd	20 August 2002
127	The Construction Centre Group Ltd v. The Highland Council	23 August 2002
128	Debeck Ductwork Installation Ltd v. T&E Engineering Ltd	14 October 2002
129	Guardi Shoes Ltd v. Datum Contracts	28 October 2002
130	Bovis Lend Lease Ltd v. Triangle Development Ltd	2 November 2002
131	Cowling Construction Ltd v. CFW Architects	15 November 2002
132	Carillion Construction Ltd v. Devonport Royal Dockyard Ltd	27 November 2002
133	Skanska Construction UK Ltd v. The ERDC Group Ltd and Another	28 November 2002
134	Baldwins Industrial Services plc v. Barr Ltd	6 December 2002
135	A v. B	17 December 2002
136	Ballast plc v. The Burrell Company (Construction Management) Ltd (Appeal)	17 December 2002
137	Picardi (t/a Picardi Architects) v. Mr & Mrs Cuniberti	19 December 2002
138	Hart Builders (Edinburgh) Ltd v. St Andrew Ltd	10 January 2003
139	Joinery Plus Ltd (in administration) v. Laing Limited	15 January 2003
140	Levolux AT Limited v. Ferson Contractors Limited (CA)	22 January 2003
141	Costain Ltd v. Wescol Steel Ltd	24 January 2003
142	Try Construction Ltd v. Eton Town House Group Ltd	28 January 2003
143	Dumarc Building Services Ltd v. Mr Salvador-Rico	31 January 2003
144	Pegram Shopfitters Ltd v. Tally Wiejl (UK) Ltd	14 February 2003
145	Vaughan Engineering Ltd v. Hinkins & Frewin Ltd	3 March 2003
146	Harvey Shopfitters Ltd v. ADI Ltd	6 March 2003
147	Jamil Mohammed v. Dr Michael Bowles	11 March 2003
148	R. Durtnell & Sons Ltd v. Kaduna Ltd	19 March 2003
149	Beck Peppiatt Ltd v. Norwest Holst Construction Ltd	20 March 2003
150	St Andrews Bay Development Ltd v. HBG Management Ltd	20 March 2003
151	Trustees of the Harbour of Peterhead v. Lilley Construction Ltd	1 April 2003
152	Hills Electrical & Mechanical Plc v. Dawn Construction Ltd	7 April 2003
153	Deko Scotland Ltd v. Edinburgh Royal Joint Venture *et al.*	11 April 2003
154	The Construction Centre Group Ltd v. The Highland Council	11 April 2003
155	Comsite Projects Ltd v. Andritz AG	30 April 2003
156	Galliford Northern Limited v. Markel Capital Limited	12 May 2003
157	City Inn Ltd v. Shepherd Construction Limited	20 May 2003
158	Orange EBS Ltd v. ABB Ltd	22 May 2003
159	Shimizu Europe Ltd v. LBJ Fabrications Ltd	29 May 2003
160	Brackee and Another v. Billinghurst	10 June 2003
161	RSL (South West) Ltd v. Stansell Ltd	16 June 2003
162	Hurst Stores & Interiors Ltd v. ML Europe Property Ltd	25 June 2003
163	Abbey Developments Ltd v. PP Brickwork Ltd	4 July 2003
164	Lovell Projects Ltd v. Legg and Carver	July 2003
165	The Highland Council v. The Construction Centre Group Ltd	5 August 2003
166	*Thomas-Fredric's (Construction) Ltd v. Wilson [2003] EWCA Civ 1367, CA	14 October 2003

167	*Dean & Dyball Construction Ltd v. Kenneth Grubb Associates Ltd [2003] EWHC 2465 (TCC)	28 October 2003
168	*Simons Construction Ltd v. Aardvark Developments Ltd [2003] EWHC 2474 (TCC)	29 October 2003
169	*Rupert Morgan Building Services (LLC) Ltd v. David Jervis and Harriet Jervis [2003] EWCA Civ 1563 CA	12 November 2003
170	*Tally Wiejl (UK) Ltd v. Pegram Shopfitters Ltd [2003] EWCA Civ 1750 CA	21 November 2003
171	*Galliford Try Construction Ltd v. Michael Heal Associates Ltd [2003] EWHC 2886 (TCC)	1 December 2003
172	*Costain Limited v. Strathclyde Builders Limited [2003] ScotCS 316	17 December 2003

Table of Cases

Table of Statutes

Table of Statutory Instruments

Index